Pharmaceutical Discovery, Development and Manufacturing Forum 2018

Core Programming Area at the 2018 AIChE Annual Meeting

Pittsburgh, Pennsylvania, USA
28 October - 2 November 2018

ISBN: 978-1-5108-7610-1

Printed from e-media with permission by:

Curran Associates, Inc.
57 Morehouse Lane
Red Hook, NY 12571

Some format issues inherent in the e-media version may also appear in this print version.

Printed by Curran Associates, Inc. (2019)

For permission requests, please contact AIChE
at the address below.

AIChE
120 Wall Street, FL 23
New York, NY 10005-4020

Phone: (800) 242-4363
Fax: (203) 775-5177

www.aiche.org

Additional copies of this publication are available from:

Curran Associates, Inc.
57 Morehouse Lane
Red Hook, NY 12571 USA
Phone: 845-758-0400
Fax: 845-758-2633
Email: curran@proceedings.com
Web: www.proceedings.com

TABLE OF CONTENTS

Some abstracts in this book refer to images that were not made available by the publisher

532296 A New High-Throughput Assay for Quantification of Antibiotic Penetration in Gram-Negative Bacterial Cells

Abstract:

A new high-throughput assay for quantification of antibiotic penetration in Gram-negative bacterial cells

Huan Gu[1,2], Zhaowei Jiang[1,2],and Dacheng Ren[1,2,3,4*]

[1]Department of Biomedical and Chemical Engineering, [2]Syracuse Biomaterials Institute, [3]Department of Civil and Environmental Engineering, [4]Department of Biology, Syracuse University, Syracuse, NY 13244, USA

Antibiotic resistance is a grand challenge with paramount societal impacts and requires immediate action to discover new antimicrobials. However, the discovery of new antibiotics has been at a standstill due to the lack of technologies to quantify antibiotic penetration in bacteria, especially the Gram-negative species. Here, we present a new assay that can rapidly quantify antimicrobial molecules in the periplasm and cytoplasm of Gram-negative bacteria. The premise of this method is separate lysis of outer and inner membranes, which allows the drug concentration to be determined using a reporter strain. Following the development of this assay, *Pseudomonas aeruginosa* was selected as the model microorganism for a proof-of-concept study because of its well-documented resistance to conventional antibiotics due to its double-layered membranes and narrow pores. By evaluating the penetration of ciprofloxacin (CIP, 200 µg/mL) into *P. aeruginosa* PAO1 using this assay, we found that its outer membrane is the major barrier to antibiotic penetration. In stationary phase *P. aeruginosa* PAO1 cells, CIP concentration in the periplasm is 300 times higher than that in the cytoplasm (107 vs. 0.3 µg/mL), where DNA gyrase, the target of CIP, locates. Penetration of CIP into the cytoplasm of stationary phase *P. aeruginosa* cells was significantly improved by mutating the gene *mexB* encoding an efflux pump protein in the inner membrane of *P. aeruginosa*. This new assay may have broad applications in antibiotic discovery and mechanistic study of antibiotic resistance.

525927 Deploying Automation to Execute Real-Time Feed Strategies in Bench-Scale Bioreactor Systems

Abstract:

Deploying Automation to Execute Real-Time Feed Strategies in Bench-Scale Bioreactor Systems

Kristin Oâ€™Neill, William Tran, Linda Hoshan, Sen Xu, and T. Craig Seamans

Biologics Process Development and Clinical Manufacturing, Merck Research Laboratories,

2000 Galloping Hill Road, Kenilworth, NJ 07033

Key to effective upstream process development is the presence of resource-efficient systems for examining complex process options. In this talk, we report on the progress of a completely automated sampling and feed control system, capable of simultaneously evaluating different cell density and metabolic-based feed strategies for bench scale cell culture processes. Though there are automated feed systems available for micro-scale vessels, to date there is not a complete 3-L scale sample and feed automation solution as described here, with capability to transfer small volume samples from multiple bioreactors (single-use or glass) to a suite of independent analyzers with concurrent feedback control of nutrient feed pumps.

The automated on-line sampling and feed system is capable of feedback control for a variety of complex feed strategies that incorporate data inputs from integrated third-party analyzers. Automated feeds can be delivered based on time as well as calculated with inputs from metabolite and cell density measurements. The completely automated system has supported experiments with two to eight 3-L bioreactors running in parallel, sampling each tank one to six times per day, and executing up to 4 independent feed pumps per vessel depending upon experimental needs.

In a recent typical experiment, eight bioreactors were sampled three times per day with feedback control for two separate nutrient feed solutions per vessel, based upon independent metabolite triggers. Over the course of the 14 day run, each bioreactor triggered feeds 35-40 times with feed volumes ranging from 2-47 mL, for a total of over 300 automated feeds across the eight vessels. In another case study, two bioreactors were sampled twice per day with feedback control for four separate nutrient feed solutions per vessel. In this study three different feed algorithms were employed with calculations utilizing inputs from metabolite and cell count analysis as well as real-time working volume data from integrated balances. Each vessel received ~80 automated feeds over the two week culture duration.

Increased use of automation for physical operations can increase efficiency in executing experiments but can shift the bottleneck to data capture. This issue was addressed with refined data extraction templates which streamlined data imports into an electronic notebook system and PI integration which enabled remote monitoring of sample analysis and feeding.

534083 Very Large Scale Microfluidic Droplet Integration for Continuous Industrial Scale Manufacturing of Monodisperse Biodegradable Micro and Nanoparticles

Abstract:

Microfluidics has been used to enable precise control of multiphasic flows to generate micrometer- and nanometer-scale materials with control and uniformity not possible using conventional technique. These micro- and nano-engineered materials have generated particular enthusiasm in the pharmaceutical industry where they have created new opportunities to generate novel drug formulations that offer unprecedented spatial and temporal control of drug delivery within the body. In comparison to conventional particle formation techniques, such as spray drying or ball milling, microfluidic-generated formulations have demonstrated increased particle monodispersity, more uniform composition of drug within the particles, increased drug yield, longer lasting formulations that are still injectable, and reduced burst release of drug.

Despite these advances, the inherently low production rate of droplet microfluidic devices < 100 milligrams per hour remains a key challenge to successfully translate the many promising laboratory-scale results of microfluidics to commercial-scale manufacturing. One promising direction to scale-up production has been the development of architectures that make it possible to operate many microfluidic droplet generators in parallel. While great progress has been made in these approaches, current chips with parallelized devices are limited to production rates max 1 L h^{-1}, have droplet homogeneities set by three-dimensional (3D) soft-lithography fabrication, are limited to low temperature and pressure operation, can only be used with the solvents compatible with the device's polymer construction, or are unable to be adapted to produce higher-order emulsions and particles that require multi-step processing.

In this talk, we present a novel microfluidic design of architectures for an individual droplet maker as well as at the device level to incorporate > 10,000 droplet makers in a single 4-inch silicon wafer. The design principles to integrate 2 0 droplet makers are developed analogous to the metal layers that are used in integrated circuits to connect transistors. The design architecture is implemented completely in silicon and glass. The device is fabricated using conventional photolithography and deep reactive ion etch (DRIE). Due to its silicon and glass construction, it can operate at high pressures (P>1000 psi), high temperatures (T> 200 C), and is compatible with a broad range of organic solvents, and thus greatly expands the library of micro- and nano-materials that can be generated. Moreover, it is robust for stable, continuous operation for industrial-scale production of micro and nanoparticles. Similarly, to design an individual droplet maker, the physics of multiphase flows is considered such that the device can generate monodisperse emulsions at maximum flow rate with a small footprint.

The massively parallelized microfluidic chip is used to generate oil in water emulsions. Hexadecane droplets were generated with varying sizes ranging from 22 um to 40 um with a coefficient of variation of 3% and a production rate of 1 trillion droplets per hour (7 Liters per hour dispersed phase, 9.2 Liters per hour continuous phase). To test the generation of United States Food and Drug Administration (US FDA)

approved biodegradable microparticles, polycaprolactone microparticles with sizes ranging from 8 um to 16 um with a CV < 5% and a production rate of 277 grams per hour were synthesized using this chip. In addition, we also show the industrial scale production of nanoparticles with precise control over their size and target density.

Acknowledgements:

National Science Foundation ((1554200), NIH (R01 EB022612) and Glaxo Smith Kline. D.L. acknowledge the support from NSF CBET 1604536.

References:

1. Silicon and Glass Very Larege Scale Microfluidic droplet integration for the terra scale generation of polymer microparticles, Sagar Yadavali, Heon Ho Jeong, Daeyeon Lee, David Issadore, Nature Communications, (2018) 9: 1222 (DOI: 10.1038/s41467-018-03515-2).
2. Sagar Yadavali, David Issadore &; Daeyeon Lee, "Large scale microdroplet generation apparatus and methods of manufacturing there of," PCT/US2016/066501, filed on December 14 th , 2016; US62/268,205

537520 Applying Automation and Data-Driven Modeling to Perform Rapid Reaction Optimization

Abstract:

In this presentation, we will discuss the application of automation technology and data-driven modeling to perform rapid reaction optimization for mid-phase pharmaceutical process development. The goal is to highlight what we have observed to be an effective approach, but also to share the concepts that have emerged from iterative application and refinement over the past year. We will present the strategy and ideas relying on data from real applications where we have tracked both the time-to-complete and also the accuracy of the model predictions and identified optimal conditions with follow-up experiments. Our data show, with automation and the right modeling algorithms, reliable reaction optimization can be completed on the order of days, compared to weeks.

536976 Piezoelectric-Based Spray Solvent Delivery System for Desorption Electrospray Ionization Mass Spectrometry: Design & Case Studies for High Throughput Reaction Screening

Abstract:

A Piezoelectric based Spray Solvent Delivery System (PieSDS) for high-throughput reaction screening using Desorption Electrospray Ionization Mass Spectrometry (DESI-MS) is described herein. Our goal is to evaluate the effects of DESI spray solvents and solvent mixtures on the ionization efficiency of DESI-MS. Using PieSDS, two solvent flows are simultaneously and independently controlled at a precisely low target flow rate and are then mixed prior to spraying onto the DESI substrate. This spray-solvent is delivered from a pressurized reservoir whose pressure is rapidly and dynamically adjusted using a piezoelectric pressure controller and a feedback algorithm based on the measured flowrate downstream. The piezoelectric technology ensures a fast reaction time (< 40 ms) and a short settling time (< 1 s). In addition, a piezoelectric valve system, which can handle up to eight solvent combinations per device, enables a quick solvent switch within both solvent streams. The integrated system is controlled in a LabVIEW environment, where we employ a decoupling algorithm between the two solvent flowrate control loops for a faster and more stable flow after each setpoint change. The PieSDS with its control algorithm enables near instantaneous change in each solvent flow, which then allows for a solvent mixture concentration gradient and, in turn, fast screening of spray-solvent. Specifically, PieSDS varies the polarity and the physico-chemical properties of the solvent mixture during DESI-MS that then impacts the reactivity of the screened reactions through solvent effects in the charged droplets and the desorption mechanism of the reaction mixture from the DESI substrate. This maximizes the information obtained from a single DESI-MS experiment. PieSDS also opens a door for a novel application of reactive DESI via *in-situ* manipulation of reagent composition - the volume of spray solvent and the reaction mixture concentration can be varied simultaneously by flowing and changing a solvent in one stream and the reagent in another. This feature illustrates that PieSDS is a versatile solvent delivery system that is applicable to other automated, high throughput chemical reaction screening system beyond DESI-MS.

528347 Closed-Loop Reaction Optimization in Microscale Oscillating Droplets: An MINLP Algorithm Applied to Suzuki-Miyaura Coupling Catalyst Selection

Abstract:

Among the many challenges of the notoriously-expensive drug development process is scaling up small molecule synthesis. Promising drug candidates must be produced in significantly larger quantities for preclinical toxicological and pharmacokinetic studies (10s of grams) compared to primary assays during discovery and lead optimization (sub-100 mg). While purity and speed is the primary concern during discovery, synthesis at the multigram scale warrants additional consideration of yield and cost.

The optimization of reaction conditions can help achieve this goal, but the inherent combinatorial nature of selecting reaction conditions (e.g., catalyst identity, concentration, reaction time, temperature) means that performing an exhaustive screen is counter to the goal of efficiency, both in terms of material usage and time. More focused experimentation is possible through the use of optimization algorithms that propose experiments based on accumulated data, so that only *useful* experiments need be performed.

Building on previous work in the group [1], here we report a mixed-integer nonlinear program (MINLP) algorithm for the simultaneous optimization of both discrete and continuous variables. The algorithm consists of two phases (1) an initial D-optimal design, which proposes a diverse range of experiments to regress a quadratic response surface model, and (2) an iterative branch-and-bound approach to discrete variable selection, where G-optimal experiments are proposed to minimize the variance at predicted optima. The algorithm is designed to optimize catalyst turnover number (TON) subject to a minimum yield constraint.

We first validate its performance using a suite of five simulated test cases representing variations on a catalytic bimolecular reaction. Next, we validate its performance experimentally using the exemplary Suzuki-Miyaura cross-coupling of 3-chloropyridine and 2-fluoropyridine-3-boronic acid pinacol ester. Experimental validation was conducted using an automated chemistry platform that performs reactions sequentially at the 20 microliter scale [2,3]. Yield and TON are calculated via online HPLC/MS and automatic peak integration. This closed-loop optimization proceeds without any human intervention. The new algorithm exhibits significantly faster convergence than the previous version and is found to be robust to measurement error.

[1] B. J. Reizman, K. F. Jensen, *Accounts of Chemical Research* **2016**, *49*, 1786-1796.

[2] Y.-J. Hwang*, C. W. Coley*, M. Abolhasani, A. L. Marzinzik, G. Koch, C. Spanka, H. Lehmann and K. F. Jensen, *Chemical Communications*, **2017**, *53*, 6649â€"6652.

[3] C. W. Coley, M. Abolhasani, H. Lin and K. F. Jensen, *Angewandte Chemie*, **2017**, *129*, 9979â€"9982.

524756 Reaction Screening in an Automated Droplet Screening System for Pharmaceutical Process Development

Abstract:

Pharmaceutical companies are increasingly incentivized to reduce the time to bring a drug to market. Automation offers one means to speeding up drug discovery and process development activities. For process development, it is important that reaction screening be performed in systems that allow for exploration of mass and heat transfer phenomena and for the extraction of intrinsic reaction kinetics. Droplet microfluidic screening systems, which allow for easy interchange of reagents and for precise control of temperature and mixing rates at micro- and millimeter length scales, have attracted recent attention in this space. Here, we describe the development and evaluation of an automated screening system focused on scalability and applicability towards a wide range of chemistries. Our system performs reactions at a 10-100 μl scale in droplets that are prepared by a retrofitted commercial autosampler controlled with LabVIEW. Droplets are transferred into a pressurized and heated reactor where reaction occurs. Oscillation of the droplet during the reaction provides thorough mixing and a tunable residence time. Crude reaction mixtures are diluted automatically and a second commercially modified HPLC performs online analysis and fraction collection with results being processed and exported automatically. The automated system is demonstrated by identifying optimal temperature, solvents, additives, and equivalencies for a deprotection reaction. The results are compared to a milliliter-scale batch screening methodology and to an automated reaction screening system using desorption electrospray ionization mass spectrometry.

526234 Methane (sI) Hydrate Crystallization and Dissociation in a Thermoelectrically-Cooled Microreactor

Abstract:

Clathrate hydrates are crystalline materials with hydrogen bonded water clusters entrapping small molecules. They are potential materials for energy, gas storage, and transportation. Even by the most conservative estimate, the amount of the energy in natural gas hydrate is equal to at least twice that of all other hydrocarbon fossil fuels combined. Furthermore, because of its excellent gas absorption capacity, (1 m^3 clathrate hydrate could store up to 180 m^3 methane under STP) and selectivity, clathrate hydrates could provide a safer and more efficient method of gas purification, storage, and transportation. [1,2]

In the present research, the crystallization and dissociation kinetics of methane (SI) hydrate were investigated in a microfluidic system by utilizing *in situ* Raman spectroscopy combined with optical microscopy. Rapid and precise control over the microreactor temperature allowed us to acquire reproducible data within minutes, while the nucleation of methane hydrates could take up to 24 hours in a more traditional setup.[3,4] The propagation rates of methane hydrate at the gas/liquid interface were investigated under different Reynolds numbers (0.7 to 68.9), pressures (30.0 bar to 80.9 bar) and sub-cooling temperatures (1.0 K to 4.0 K). While the Reynolds number was observed to have little influence on the propagation rate, higher pressure and sub-cooling temperature would contribute to increase the propagation rates. A theoretical model, derived from first principles, explained the non-linear relationship between the sub-cooling temperature and propagation rate. A transition from heat-transfer limited reaction to a mixed mass-transfer and intrinsic kinetic limited reaction was discovered. [5]

Microfluidics were also found to serve as an excellent platform to study the dissociation of methane hydrates. The dissociation kinetics were studied in order to better understand its mechanism. Remarkably, a switch from crystallization to dissociation was made possible by control of the isothermal reactor temperature of 0.1 K around the phase boundary. Better understanding of the mechanism of crystallization and dissociation are critical for technology that utilizes gas hydrates.

1 E. D. Sloan and A. C. Koh, *Clathrate hydrates of natural gases*, 3rd edn., 2007.

2 E. D. Sloan, *Nature*, 2003, **426**, 353–363.

3 P. Skovborg, H. J. Ng, P. Rasmussen and U. Mohn, *Chem. Eng. Sci.*, 1993, **48**, 445–453.

4 O. Fandiño and L. Ruffine, *Fuel*, 2014, **117**, 442–449.

5 W. Chen, B. Pinho and R. L. Hartman, *Lab Chip*, 2017, **17**, 2997–3192.

531078 C-H Activation By Ozone in Liquid CO_2

Abstract:

We demonstrate facile and safe ozonolysis of cyclohexane in liquid CO_2 with almost complete selectivity towards cyclohexanone. At 10 celsius, with an ozone: cyclohexane molar ratio of approximately 50%, cyclohexane conversion is between 18-25% with a KA oil yield between 10-12%. The dominant presence of CO_2 in vapor phase imparts safety by avoiding flammability at reaction condition. The experimentally observed product selectivity is confirmed by computations predicting that the reaction is initiated through hydrogen abstraction followed by cyclohexanone and hydrogen peroxide formation. The extension of this concept to the selective oxidations of linear alkanes will also be presented.

534992 pH Sensitive Colloidal Gold Nanoparticle Catalysts for Enhanced Recovery and Reuse

Abstract:

Gold nanoparticles (AuNPs) have attracted enormous attention due to their unique catalytic activities. Colloidal AuNPs provides the benefit of selectivity, greater surface area per mass of catalyst compared to supported catalysts, catalyzes reactions under mild conditions and are very effective for chiral catalysis. Recovery of colloidal catalysts involve tedious methods like pH or temperature induced aggregation, solvent extraction, etc which renders these catalysts uneconomical. For colloidal catalysts, surface functionalization with ligands is a prerequisite to prevent aggregation but surface passivation with ligands causes significant reduction in catalytic activity. On the other hand, functionalization of colloidal AuNPs with stimuli responsive ligands provides enhanced recovery and reuse of catalysts. In this work, colloidal AuNPs have been functionalized with pH responsive thiolated-polyacrylic acid (PAA-SH) to impart pH triggered recoverability in the gold nanoparticles (AuNP-SPAA). Two pH triggered recovery schemes were investigated: a) pH triggered aggregation-separation-redispersion and b) pH triggered phase transfer into chloroform by a phase transfer agent. In AuNP-SPAA with low thiol content, loss of activity was ~100% within 5 cycles of reuse when the reuse scheme was (a). On the other hand catalytic activity was maintained when a (b) reuse scheme was employed. In AuNP-SPAA with higher thiol content, loss in reaction rate was observed due to increased surface passivation but with much greater recoverability for scheme (a) and (b) recovery. Small angle neutron scattering was used to characterize the polymer conformation and ligand shell solvation to corroborate our observations.

530900 CO_2 Conversion Via RWGS-CL over La-Perovskite Oxide with Three Metals (Co, Fe, and Mn) in the B-Site

Abstract:

Carbon dioxide (CO_2) is one of the major greenhouse gases contributing to climate change. Countries around the world have already committed to reduce CO_2 emissions, for which it is imperative to develop CO_2 reduction technologies that are economically feasible. Reverse water gas shift â€“ chemical looping (RWGS-CL) uses perovskite oxides of the form ABO_3 to convert CO_2 to carbon monoxide (CO), which can be converted to liquid fuels. First the perovskite is reduced using hydrogen (H_2) forming an oxygen vacant perovskite $ABO_{3\text{-Î´}}$, subsequently CO_2 oxidizes the perovskite, regenerating the material and producing CO. Perovskite oxides are stable and can be used over several cycles. It has been shown in previous studies that cobalt (Co), iron (Fe), and manganese (Mn) have individual properties that facilitate CO_2 conversion. Here we show the CO_2 conversion performance of a La-perovskite with varying composition of the three metals in the B-site. Four compositions of $LaCo_xFe_yMn_{1-x-y}O_3$ were studied; Co-rich (x=0.5), Fe-rich (y=0.5), Mn-rich (x=y=0.25), and equal composition (x=y=0.33). Bulk oxygen vacancy formation energies (E_{vac}) were performed using density functional theory (DFT) via the Vienna ab-initio Simulation Package (VASP-5.3.3). The four compositions had E_{vac} within the range which has been reported leads to favorable CO_2 conversion. The Co and Fe-rich perovskites performed better, as predicted from E_{vac} calculations, producing 1786 µmol CO/g and 1431 µmol CO/g respectively. These results compare favorable to previous results in which two cations were used on both the A and B sites.

High Performance Non-Mercury Catalysts for VCM Production: From Theoretical Study to Industrialization

Hao Xu and Guohua Luo

Beijing Key Laboratory of Green Reaction Engineering and Technology,Department of Chemical Engineering, Tsinghua University, Beijing, China, 100084

Abstract: Acetylene hydrochlorination is the main reaction for the dominant coal based process for VCM (vinyl chloride monomer) production from coal resource in mainland China. The traditional activated carbon based mercury catalyst for this reaction causes severe human damage and environmental pollution. Gold is widely considered as the most active metal for acetylene hydrochlorination reaction and copper is the most common metal with obvious price advantage. These two metals show the best potential as the active component for environmentally friendly non-mercury catalysts. In our study, the performance of Au/Cu catalyst was dramatically enhanced by the introduction of negative ion ligands. KSCN (potassium thiocyanate) was confirmed to significantly enhance the initial activity of the catalyst and TCCA (trichloroisocyanuric acid) was proved to be an effective ligand for preventing Au^{3+} from reduction by acetylene, leading to longer catalyst lifetime. The findings were successfully verified by over 18000 hour's industrial plant-trial evaluation. Abundant characterization methods including TEM, XRD, TPR, XPS, ICP-AES and nitrogen adsorption-desorption were carried out to reveal the factors which affect the activity and stability of catalysts. Meanwhile, DFT calculations also gave atomic level understanding of the function of ligands. The experiment and DFT results provide reasonable methodology for the design of high performance non-mercury catalysts in acetylene hydrochlorination process, which demonstrate enormous potential for the industrialization in PVC (poly vinyl chloride) industry.

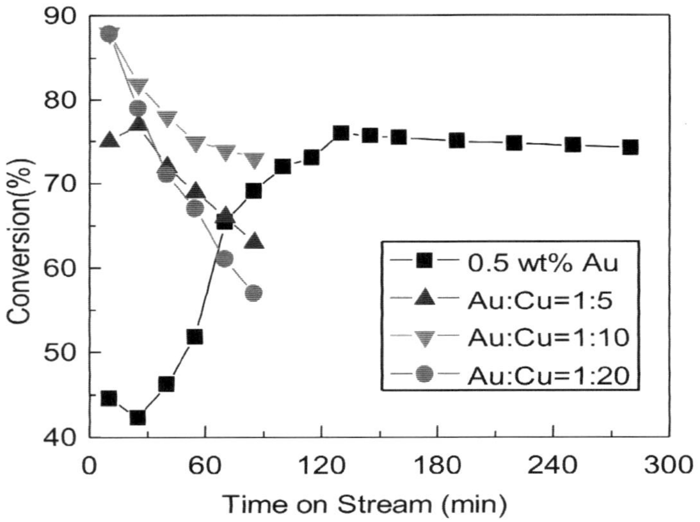

Fig. 1. Performance of 0.5 wt% Au catalysts with different amount of Cu addition, reaction conditions: T=180 °C, GHSV=600 h^{-1}, Au:Cu in molar ratio.

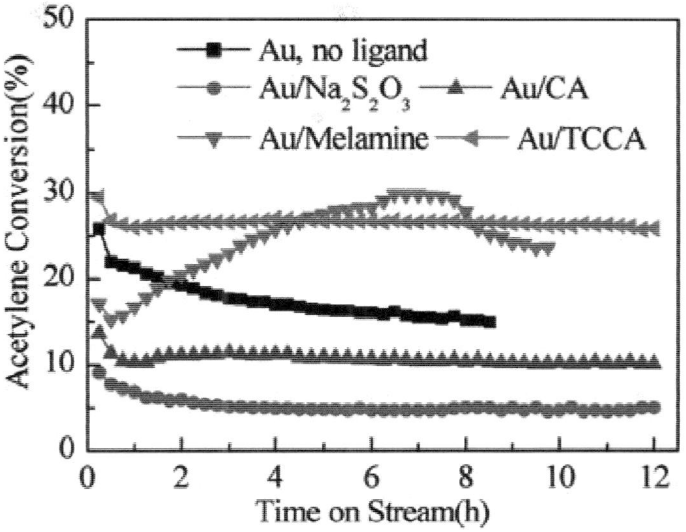

Fig. 2. Performance of 0.25 wt% Au catalysts with different ligands, reaction conditions: T=180 °C, GHSV=600 h^{-1}, Au:ligand=1:20 (molar ratio).

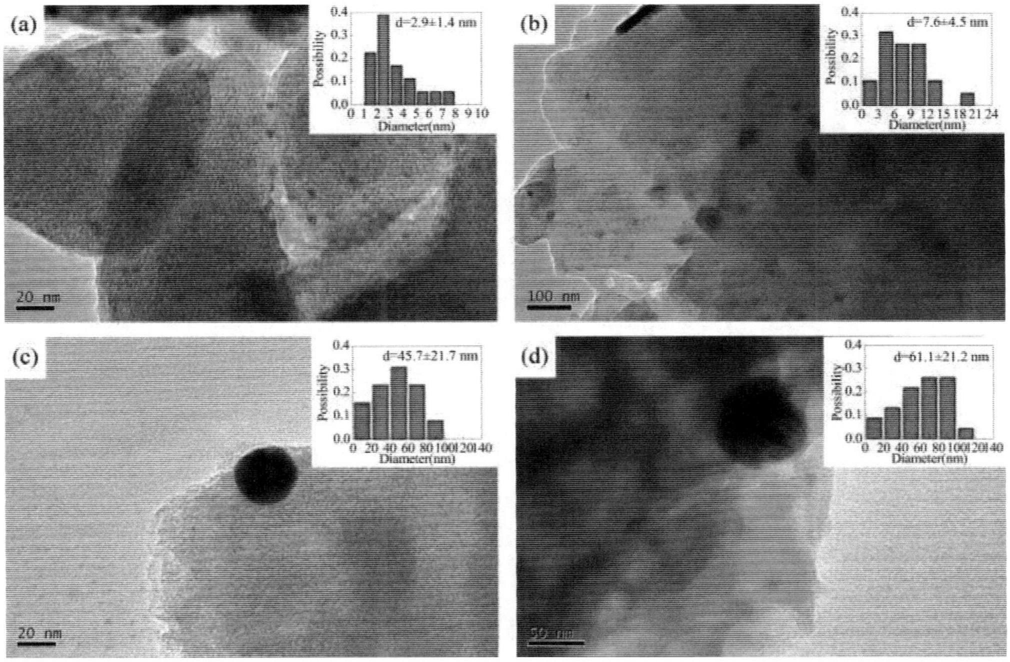

Fig. 3. TEM images and gold particle size statistics of (a) Au/TCCA, fresh, (b) Au/TCCA, spent, (c) Au, no ligand, fresh, (d) Au, no ligand, spent catalysts.

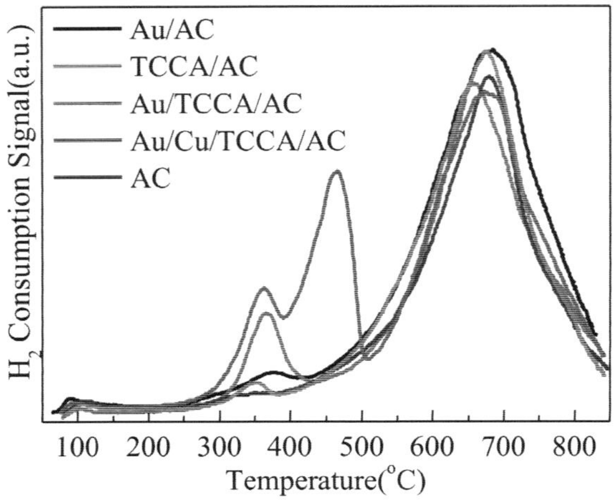

Fig. 4. TPR results of Au catalysts with different ligands.

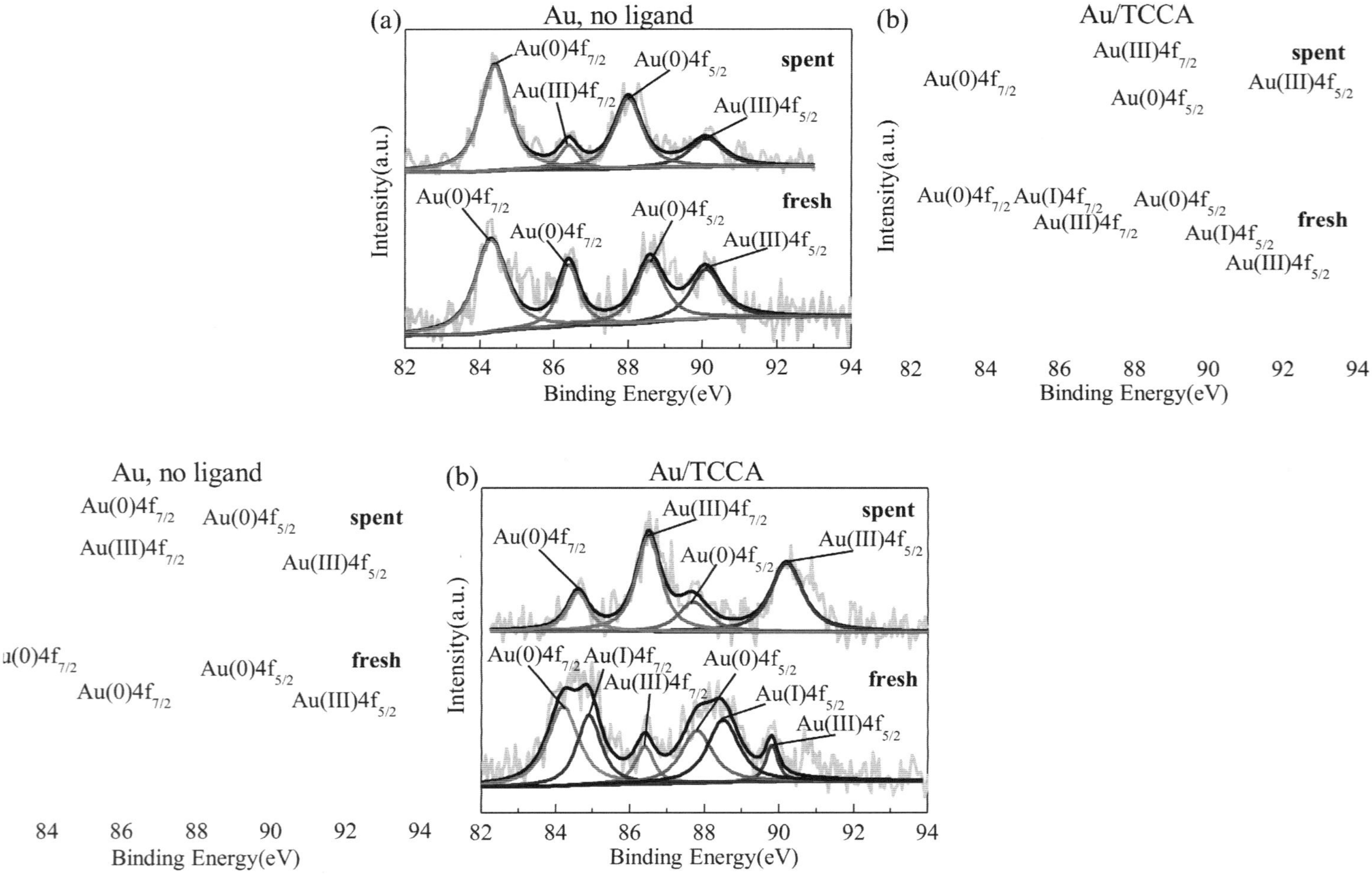

Fig. 5. XPS spectrum for Au and Au/TCCA system.

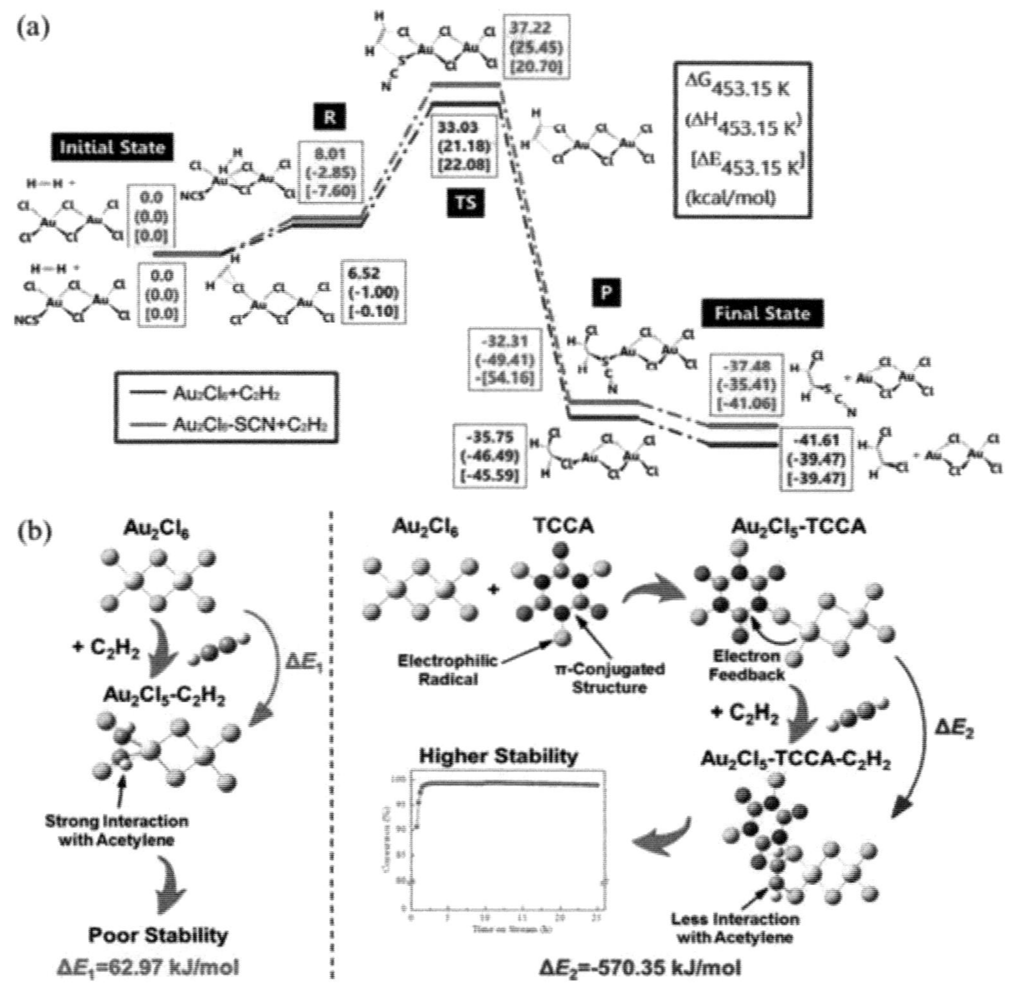

Fig. 6. DFT calculation for Au systems. (a)Au/KSCN system, (b)Au/TCCA system. Yellow balls for Au atoms, green for Cl, grey for C, blue for N, red for O and white for H.

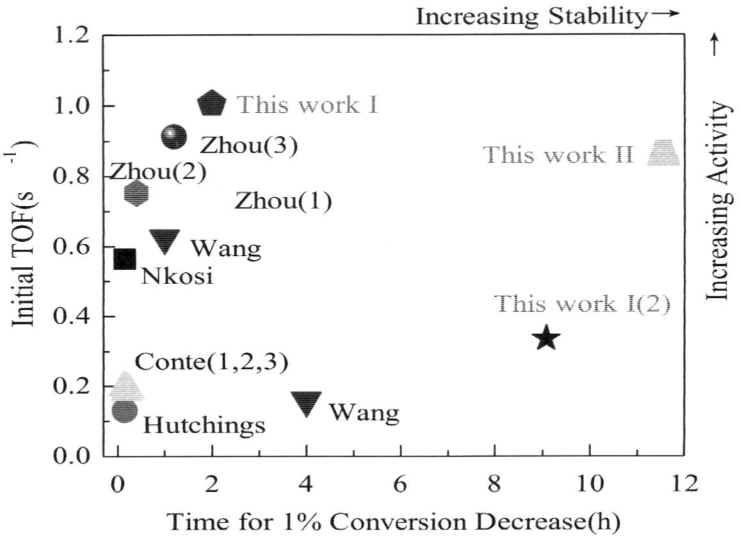

Fig. 7. Comparison with other literature results.

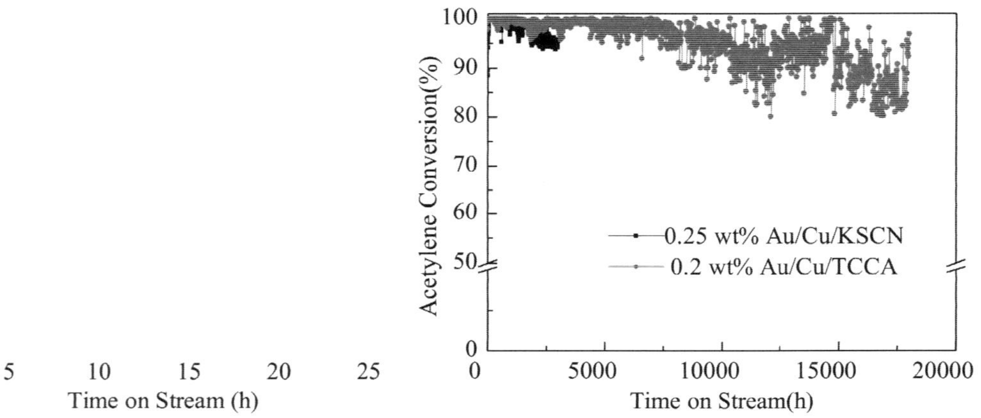

Fig. 8. Comparison of 18000 h pilot-trial evaluation between Au/Cu/TCCA and Au/Cu/KSCN systems, reaction conditions: T=180 °C, GHSV=30 h^{-1}.

References

[1]Xu H, Zhou K, Si J K, Luo G H. Ligand coordination approach for high reaction stability of an Au-Cu bimetallic carbon-based catalyst in acetylene hydrochlorination process. Catalysis Science & Technology, 2016, 6(5): 1357-1366.

[2] Xu H, Si J K, Luo G H. The Kinetics Model and Fixed Bed Reactor Simulation of Cu Catalyst for Acetylene Hydrochlorination. International Journal of Chemical Reactor Engineering, 2017, 15(4).

[3] Zhou K, Jia J C, Li C H, Xu H, Zhou J, Luo G H, Wei F. Low content Au-based catalyst: hydrochlorination of C_2H_2 for green industial PVC processing. Green Chemistry, 2015, 17(1): 356-364.

538985 Chloroplast-Inspired Artificial Photosynthetic Capsules for Efficient and Sustainable Enzymatic Hydrogenation

Abstract:

Inspired by the structure and function of chloroplast, an artificial photosynthetic alginate (ALG) capsule bearing carboxymethylcellulose (CMC) matric, red phosphorus quantum dots@g-C_3N_4 (P/CN) tubular photocatalysts and alcohol dehydrogenase (ADH) is prepared for efficient and sustainable synthesis of methanol from formaldehyde. CMC liquid core resembles grana in chloroplast to afford favorable micro-environment for stabilizing ADH, whereas P/CN photocatalysts mimic thylakoid in chloroplast to supply nicotinamide adenine dinucleotide (NADH, the hydride donor) for triggering enzymatic hydrogenation. Moreover, P/CN photocatalysts reinforce the mechanical stability and, thereby, inhibit the swelling of ALG capsules, thus suppressing the leaching of P/CN photocatalysts and enzymes. The kinetic processes of photocatalytic regeneration and enzyme hydrogenation are investigated, respectively. The optimal coupling of photo-enzyme reactions enabled by the capsules achieves the efficient and sustainable production of methanol. Our study offers some guidance for rational construction of integrated and recyclable artificial photosynthetic systems towards a broad range of chemical/biochemical reactions.

536946 Rational Design of Peptide Nucleic Acid Antibiotics Against Multidrug Resistant Bacteria

Abstract:

Multidrug-resistant (MDR) infections are an urgent global health concern which is exacerbated by a lack of new antibiotics in the pharmaceutical pipeline. Rational design of antibiotics can improve upon current antibiotic screening techniques and quickly create therapies that are specifically targeted to MDR bacteria. Here we rapidly design, synthesize, and test antibiotic peptide nucleic acids (PNAs), which can block translation of bacterial RNA via sequence-specific binding. In this study we sought to target PNAs to essential genes in MDR clinical isolates of carbapenem-resistant Escherichia coli, extended-spectrum beta-lactamase Klebsiella pneumoniae, New Delhi Metallo-beta-lactamase-1 carrying Klebsiella pneumoniae, and MDR Salmonella enterica. We created a library of 303 PNA sequencesâ€"originating from the sequences of E. coli essential genes of the Keio collectionâ€"and used genome alignment tools to predict which of these PNAs would both avoid incidental non-target inhibition within the E. coli genome and target an analogous gene within the other MDR strains. Additionally, we used a custom sequence analysis tool to filter for sequences with good solubility and low self-complementarity. The PNAs used in this study were selected from that filtered set based on their targeting of varied essential gene pathways including metabolism, cell signaling, and stress response. Though these MDR clinical isolates were found to be resistant to most classes of antibiotics, with genome sequencing revealing as high as sixteen resistance genes, the PNAs we designed were able to potentiate the activity of traditional antibiotics and inhibit growth as a monotherapy. Our full process from the design of the PNAs to in-house synthesis to testing can be achieved in 5 days or less, which represents an extremely rapid development in comparison to traditional antibiotics screening. These results demonstrate our ability to quickly and rationally design species-specific therapies for treatment of MDR infections via non-traditional pathways. Furthermore, this technologyâ€™s success has far-reaching implications for controlling protein translation in bacterial, viral, and mammalian systems in biopharmaceuticals, metabolic engineering, and other areas.

527903 Identifying CQAs of 3D Printed Extended-Release Tablets through the Optimization of Formulation and 3D Geometric Variables

Abstract:

Three-dimensional printing-3D printing (3DP) offers the potential to revolutionize the production of personalized drug delivery systems. In semisolid extrusion, 3D printing (SEP) a drug loaded paste is extruded through a small nozzle and allowed to dry to form a 3D object on build-plate SEP technology. This procedure has numerous advantages that include low cost, use of common excipients, and the capability to print personalized dose formulations with a wide variety of polymeric pastes. This study is aimed at identifying the critical quality attributes (CQA) of 3D printed extended release tablets prepared by this technology. Nineteen drug loaded pastes were prepared according to a fractional factorial design with three replicates of the center point: (1) to manufacture 3D printed round shaped tablets and (2) to evaluate the effect of formulation variability and (3) weaving patterns on the various tablet characteristics.

Diclofenac sodium, an NSAID, was used as a model drug. Drug-loaded carbopol based pastes were prepared by compounding Unguator using Avicel PH101, Avicel PH105 and spray dried lactose as diluents, glycerol as plasticizer and polyplasdone as the disintegrant. The pastes were extruded into 3D printed tablets using EnvisionTec 3D Bioplotter. The formulation variables included percentages incorporated of the drug and various soluble and insoluble excipients. The 3D geometric variables included nozzle diameter, weaving angle and distance between filaments. The 3D printed tablets were characterized in terms of drug assay and content uniformity, disintegration time, percent friability, tensile and axial breaking strengths, porosity, drug release characteristics and the uniformity of dosage units (thickness, weight and diameter).

Changing the formulation variables required variating in printing pressure from 0.5 bars to 4.5 bars to maintain a printing rate of 10 mm/secs for all pastes. Pre-flow and post-flow times of 0.1 seconds were required to avoid gaps in the internal structures and/or contours. The soluble components of the formulations affected the radial strength of the 3D tablets. However, the weaving parameters were more important to control the axial strength. The results indicate that 3D weaving pattern has a minimal influence on the porosity of the filament deposited while printing. Nevertheless, the distance between the printed filaments necessary to create the structure, determined the porosity of tablets. Dissolution results demonstrated that the weaving angle and lactose loading negatively affected the percent of drug release after 8 hours. Conversely, the drug release rate increased from 75% to 95% by increasing the distance between filaments from 0.5 mm to 1.5 mm, respectively.

In summary, this study demonstrated that formulation and weaving parameters should be controlled to produce robust 3D printed extended-release tablets by SEP. This or similar 3D printing (SEP) procedures have significant potential to contribute to personalized therapy and/or preparation-on-demand medications at health care settings.

535200 A Constrained Version of the Dynamic Response Surface Methodology for Challenging Time-Resolved Pharmaceutical Reaction Data

Abstract:

The newly proposed Design of Dynamic Experiments (DoDE)[1] and the Dynamic Response Surface Methodology (DRSM)[2] are two generalizations of Design of Experiments (DoE) and Response Surface Methodology (RSM). DoDE and DRSM methodologies have the advantages of allowing time-varying inputs, accommodating time-resolved output measurements, and modeling their interrelationship. Here, we apply the DRSM-2 algorithm[3] to some challenging example cases and further introduce knowledge driven constraints on the expected shape of the predictions. This constrained DRSM-2 algorithm consistently provides a better representation of the time evolution of species concentrations during reactions. We test the updated approach against a complex simulated pharmaceutical reaction network consisting of eight reactions and involving ten species. We also develop accurate DRSM models for two experimental data sets from the two collaborating companies.

The initial DRSM-1 approach[2] leads to oscillatory behavior in the modeling of some concentration profiles, especially for experiments with non-constant sampling intervals or for intermediate species whose final concentrations are comparable to the level of noise. To address these issues, the DRSM-2 approach uses an exponential transformation of time as the independent variable which results in more parsimonious models. This eliminates the occurrence of any oscillatory behavior on the prediction when none is expected, and it provides much more accurate models for the case of non-equidistant sampling strategies. The constrained improvements introduced here include: fixing the initial concentrations when they are known, prohibiting the model from temporarily predicting negative concentration for species that start at a zero value, and enforcing that all concentration predictions must be non-negative. We conclude with a demonstration of the effectiveness of the constrained DRSM-2 algorithm in cases that a substantial fraction of the concentration data are missing, but still leading to an accurate DRSM model.

References

1. Georgakis C. Design of Dynamic Experiments: A Data-Driven Methodology for the Optimization of Time-Varying Processes. *Industrial & Engineering Chemistry Research.* 2013;52(35):12369-12382.
2. Klebanov N, Georgakis C. Dynamic Response Surface Models: A Data-Driven Approach for the Analysis of Time-Varying Process Outputs. *Ind Eng Chem Res.* 2016;55(14):4022-4034.
3. Wang ZY, Georgakis C. New Dynamic Response Surface Methodology for Modeling Nonlinear Processes over Semi-infinite Time Horizons. *Industrial & Engineering Chemistry Research.* 2017;56(38):10770-10782.

535300 Characterizing Protein-Protein Interactions in Highly Concentrated Monoclonal Antibody Solutions Using Small Angle X-Ray Scattering and Molecular Dynamics Simulations

Abstract:

Monoclonal antibodies (mAbs) show great promise as effective therapeutics against diseases such as cancer. Highly concentrated solutions of mAbs are especially desirable in the biopharmaceutical industry for subcutaneous delivery of these proteins in patients. However, at these high concentrations, protein-protein interactions (PPI) including anisotropic short-range attractions due to hydrogen bonding, dipole-dipole, and hydrophobic interactions often cause prohibitively high viscosities and low storage stability. Recent studies have established Small Angle X-ray Scattering (SAXS) as an effective tool for probing mAb solution structure at varying length-scales and extracting information about the PPI involved. In this work, we present a set of generalized coarse-grained models capable of mimicking mAb geometry and interaction anisotropy. Through molecular dynamics simulations, we successfully fit our models to experimental structure factor ($S_{eff}(q)$) data obtained for mAbs across a series of concentrations and co-solute systems. Furthermore, our models highlight the importance of the slope of $S_{eff}(q)$ at low q for distinguishing between mAbs that exhibit weak diffuse attraction and mAbs that exhibit strong interactions between the Fab and Fc domains. Cluster size analysis shows that models with strong anisotropic attraction form large percolated networks in solution, and real systems fit by this model exhibit high viscosities, revealing an intriguing correlation between solution viscosity and cluster characteristics. Taken together, these findings provide a more fundamental understanding of the nature of PPI in highly concentrated mAb solutions.

527091 A Case Study and Design Considerations for the Robust Removal of Dissolved Hydrogen Chloride

Abstract:

A potential therapeutic asset utilized a chlorination reaction which generated hydrogen chloride (HCl) gas as the reaction progressed. The presence of low levels of HCl in the liquid phase was found to cause the crystallization of a hydrochloride salt when coupled with the next stepâ€™s reagent, thereby inhibiting the reaction and generating an impurity that would result in the product failing specifications. A simple nitrogen sweep of the reactor headspace, at laboratory scales, was initially deemed effective to reduce the dissolved HCl to acceptable levels within a few hours. However, severe reaction conditions (high temperature and a corrosive environment) inhibited the ability to quantify the degassing kinetics, which were required to predict the - degassing performance at scale. Additionally, mass-transfer effects at scale could provide a dominative resistance, perhaps slowing down the degassing efficiency. It was desirable to design an optimum protocol for the HCl degassing options to ensure successful degassing during scale-up. A mathematical model including the solubility of HCl in the liquid phase and mass-transfer rates was developed to assess and predict the performance of various options to remove dissolved HCl: 1) a nitrogen sweep of the reactor headspace, 2) nitrogen pressure/purge cycles with the reactor vent open to atmospheric pressure, and 3) nitrogen pressure/purge cycles with the reactor vent connected to vacuum. HCl degassing data was collected experimentally using a setup equipped with a pH meter and flow NMR; this data was used to verify the model predictions. This contribution will summarize the integrated approach of coupling experiments and model predictions to devise the most effective degassing methodology to be implemented at scale and ensure consistent product quality during scale-up.

525348 Real-Time Monitoring of Drug Concentration in a Dropwise Additive Manufacturing System

Abstract:

In recent years, a changing business environment, regulatory encouragement, and novel contributions from academic/industrial researchers have spurred the pharmaceutical industry toward realization of improvements in manufacturing efficiency and product quality through implementation of on-line process analytical technologies, predictive process models and advanced production methods. Presently, innovative technologies have been developed for both drug substance and drug product manufacturing, with research focused on potential uses in continuous pharmaceutical processing, personalized medicine, and early stages of drug development. Of the new technologies which have found applications in drug product manufacturing, the use of liquid-based systems for production of solid oral doses is especially prominent; the common feature of these processes is that they rely on the precise dispensing of small volumes of liquid to produce doses, and, as such, exhibit considerable flexibility in dose strength and formulation.

For liquid dispensing technologies that produce drug products by sequential deposition of individual drops onto a specified substrate, real time process monitoring has been achieved by combining an on-line photography system with automated image analysis to record the volume of each dispensed drop [1, 2]. Given the volume dispensed in each drop, number of drops, and the fluid composition, drug amount in each dose may be accurately estimated for processes operated with homogeneous fluids comprised of active ingredient, and potentially a polymeric excipient, dissolved in a liquid. The use of drug solutions as the dosing liquid is well-suited to production of low-dose drugs (active compound <5 mg); however, for drug products which require higher dose amounts, solutions may be limited by the available capacity of a substrate to accept liquid, or by prolonged evaporation times to remove the solvent. As an alternative to solutions, delivery of active compound as a suspension of drug particles in a carrier liquid enables production of high dose drugs by dropwise dispensing, though this presents unique challenges due to the effects of non-linear flow behavior on drop formation and system operation.

With specific relevance to pharmaceutical manufacturing, the inherent heterogeneity of particulate suspensions raises questions regarding the uniformity in composition of liquid drops dispensed by such a process due to the potential for flow- or gravity-driven segregation of particles and liquid. In the suspension rheology literature, segregation effects are related to suspension properties (e.g. particle fraction, carrier viscosity), system properties (e.g. conduit diameter) and operating conditions (e.g. flow velocity), which have been explored in our previous works to identify feasible conditions for dropwise printing of non-Brownian suspensions [3]. At these operating conditions suspension-based dosing liquids were dispensed into empty capsules to create high dose drug products, which exhibited low relative standard deviation (<2.5%) in mass. However, the potential for inhomogeneity in suspensions introduces uncertainty in the amount of drug dispensed to each dose. Since on-line monitoring of drop volume is limited to estimation of deposited volume, an additional monitoring system is required to provide information on particle concentration in the printing fluid.

Process analytical technologies capable on-line monitoring of particle concentration in multi-phase mixtures include, but are not limited to, measurement techniques based on turbidity, conductivity or density. Limitations exist, however, on application of these methods to a range of processing conditions; use of turbidity measurement for optically opaque suspensions is not practical, and conductivity measurement is restricted to use in fluids which are electrically conductive. Ultrasonic-based technologies, which relate sound velocity to density and compressibility, provide a robust alternative that is capable of measuring phase volume in concentrated particle-liquid mixtures, though not without some challenges. In two-phase systems the relation between sound velocity and density is complicated by multiple scattering effects, which cause the ultrasound propagation velocity to depend on particle size and particle volume fraction [4, 5]. The calibration procedure for measurement of phase volumes depends on the system properties â€“ when particle size distribution is fixed, correlation of ultrasound velocity with solid volume fraction has been successfully demonstrated [4, 5], but for processes (e.g. crystallizers) in which particle size distribution and particle volume fraction are expected to vary simultaneously, information obtained using an additional on-line measurement technique for particle size must be included in the model structure [6].

In this work the use of an in-line ultrasound probe is explored for monitoring of suspension density, and, by extension, particle concentration, in non-Brownian suspensions used with a dropwise additive manufacturing process for pharmaceutical solid oral dosage. For suspensions of particles with fixed size distribution, a modified form the Urick equation is used to fit experimental ultrasound velocity as a function of suspension density. Using these correlations, the in-line ultrasound probe can be used to monitor the concentration of drug particles in the suspension upstream of the printing nozzle. By conjoining real-time monitoring of particle concentration with on-line monitoring of drop volume, the drug amount in each dose can reliably be estimated. The accuracy of the on-line estimates compare very favorably to content analysis obtained using gravimetric measurements and chemical analysis performed using high performance liquid chromatography.

References

1. Hirshfield, L., Içten, E., Giridhar, A., Nagy, Z. K., & Reklaitis, G. V., 2015. Real-time process management strategy for dropwise additive manufacturing of pharmaceutical products. Journal of Pharmaceutical Innovation, 10(2), 140-155.
2. Clarke, A., Doughty, D., Khinast, J., & Rantanen, J., 2016. Development of liquid dispensing technology for the manufacture of low dose drug products. Continuous Manufacturing of Pharmaceuticals, 551-575.
3. Radcliffe, A. J., & Reklaitis, G. V., 2017. Dropwise Additive Manufacturing using Particulate Suspensions: Feasible Operating Space and Throughput Rates. In Computer Aided Chemical Engineering(Vol. 40, pp. 1207-1212). Elsevier.
4. Pinfield, V. J., Povey, M. J., & Dickinson, E., 1995. The application of modified forms of the Urick equation to the interpretation of ultrasound velocity in scattering systems. Ultrasonics, 33(3), 243-251.
5. Stolojanu, V., & Prakash, A., 2001. Characterization of slurry systems by ultrasonic techniques. Chemical Engineering Journal, 84(3), 215-222.
6. Pertig, D., Fardmostafavi, M., Stelzer, T., & Ulrich, J. 2015. Monitoring concept of single-frequency ultrasound and its application in dynamic crystallization processes. Advanced Powder Technology, 26(3), 874-881.

538138 Mechanistic Insights into a Process Intensification Strategy in Pharmaceutical Manufacturing

Abstract:

Spherical Agglomeration is a novel Process Intensification strategy that has been heralded as a revolution in pharmaceutical manufacturing. It combines the pharmaceutical unit operations of crystallization and granulation into a single unit operation thereby decreasing the cost of manufacturing Active Pharmaceutical Ingredients(APIs). Spherical Agglomeration is also very attractive from a particulate control viewpoint since the two rate processes â€“ crystallization and agglomeration are carried out sequentially, thereby giving greater freedom in the design space to control the process. Typically most drugs sold in the market as tablets also contain other inactive ingredients(excipients). A major bottleneck to industrial implementation of Spherical Agglomeration in Pharmaceutical Manufacturing is the fact that co-agglomerating drugs and excipients within a single stage using a binder is very challenging, albeit impossible in many cases.

Co-agglomeration of APIs and excipients has been achieved using Emulsion Solvent Diffusion(ESD). Despite its initial success, developing a robust ESD process still remains a challenging task. The success of ESD hinges on selecting the right combinations of solvents, antisolvents, API concentration and emulsifier concentration. This is further compounded by the fact that in ESD, the Primary Particle Size Distribution(PSD) and the Agglomerate Size Distribution(ASD) need to be controlled simultaneously during the diffusion process which gives less number of degrees of freedom from a process control standpoint.

Although much work has been done on finding the effect of process conditions(flow rate, concentration of API) on the ASD, there is still a huge knowledge gap in literature regarding the mechanisms of the fundamental rate processes behind ESD. This work delves into gaining a mechanistic understanding of the ESD by giving systemic perturbations to the process and studying the system response characteristics. Process Analytical Technologies(PAT) tools viz. FBRM have also been used in the process to monitor the evolution of the Chord Length Distribution. It has been shown that use of in-situ PAT tools gives novel insights into the mechanisms of ESD process that would have been otherwise unattainable by just studying the end of batch response characteristics to systemic perturbations.

533694 3D Printed Complex Dosage Forms Embedded with Engineered BCS Class II Drug Particles

Abstract:

Recent work suggests complex structured dosage forms can be manufactured where drug particles can be integrated via 3D printing technology. However, incorporating engineered drug particles into 3D printed structures so that various intermediate and final product properties can be enhanced, for example, good mechanical properties and drug content uniformity of feed materials, has yet to be explored. It is hypothesized that engineered particles enable uniform distribution of drug into molten polymeric matrix via twin screw extruder (TSE). This distinct feature of engineered particles leads to obtaining enhanced dissolution profile from 3D structured dosage forms. The printable filaments containing varying drug loading (DL) %, and different extent of particle engineering as well as filament their mechanical properties were considered in this investigation. Griseofulvin (GF) and Hydroxypropyl cellulose (HPC) SL grade was used as the model BCS Class II drug and the polymeric matrix, respectively. The feed materials having four different DL% (5, 10, 15, and 25 wt.% GF) incorporating engineered and non-engineered particles were extruded by HME and 3D structured tablets from the feed materials were successfully created via FDM 3D printer. No drug degradation was observed in TGA results for any cases. All the drug particles kept their crystalline structure from HME extrusion (140 °C) as well as FDM printing (170 °C). The feed materials containing engineered particles exhibited higher tensile strength and better modulus of elasticity, regardless of the drug loading. Moreover, 3D printed tablets containing engineered particles showed enhanced drug release for all DL%. Overall, the 3D printed tablets embedded with engineered drug particles resulted in tablets without any degradation/recrystallization issues and demonstrated enhanced dissolution profiles for four different dosage amounts.

529744 3D Printing for Rapid Prototyping of Innovative Process Equipment for Pharmaceutical Crystallization

Abstract:

The development of pharmaceuticals involves various stages with preclinical and clinical trials. A short time-to-market is essential to maximize the patent protection and to avoid any loss in potential revenues.[1] Furthermore, most pharmaceuticals in development do not reach commercial manufacturing. However, for those pharmaceuticals that do reach market, efficient and novel processing equipment can provide a competitive edge. Hence, rapid prototyping of processing equipment during development is important to minimize time-to-market, while maximizing early learning on processing. 3D printing is an attractive technology to fabricate customized and novel processing equipment almost instantaneously. However, despite this potential, the use of 3D printing during early pharmaceutical process development is still in its infancy.

3D printing has found applications for the fabrication of micro-reactors and to some extent for mesoscale reactors as well.[2] In contrast to reactors, the use of 3D printing to fabricate innovative designs of separators has received limited attention. For example, for crystallization, which is a key unit operation in many pharmaceutical processes, no studies on the use of 3D printing for rapid prototyping have been reported to the best of our knowledge. Pharmaceutical crystallization is complicated by the need to optimize intrinsic crystal quality attributes such as size distribution, which affects the performance of downstream processes and possibly the efficacy of the final product. Optimization of crystallization requires control over various mixing processes. 3D printing can be a flexible and rapid method to fabricate innovative structures of crystallizers for optimized mixing at all relevant scales during pharmaceutical development.

The objective of this work is to develop and characterize 3D printed prototypes of innovative crystallization equipment with different structures for mixing. Two cases will be investigated. A common feature of both cases is that the mixing mechanism does not require moving parts, which makes fabrication with 3D printing attractive.

The first case involves a 3D-printed airlift crystallizer for protein crystallization. Protein crystallization in an airlift crystallizer has not been reported yet, but is attractive as protein crystals are generally more fragile in comparison to crystals of small-molecule active ingredients.[3] Therefore, mild and homogeneous mixing conditions are required, which can be provided by airlift columns due to the low maximum shear forces.[4] An airlift crystallizer was designed and fabricated with a desktop stereolithography (SLA) printer. The hydrodynamics of the column have been characterized for lysozyme as model protein. A sufficiently high upward velocity in the column demonstrates practical feasibility for protein crystallization in an airlift crystallizer. Next, the crystallization performance of the airlift crystallizer is characterized in this work and compared to a conventional stirred tank crystallizer.

The second case involves antisolvent and reactive crystallization in continuous flow. Rapid mixing is achieved by using static mixers. The effectiveness of kenics-type of static mixers has previously been demonstrated for anti-solvent crystallization of ketoconazole [5]. In this work, Y-mixers and komax-type mixers have been designed and fabricated using a desktop stereolithography (SLA) 3D printer. Preliminary studies on reactive crystallization of barium sulfate have been conducted, which demonstrate practical feasible and reduced agglomeration when using the komax mixer in comparison to when using a Y-mixer. The stability of many stereolithography resins used for 3D-printed structures is good for aqueous-based system. However, pharmaceutical crystallization often involves organic compounds. Therefore, this case particularly focuses on the use of various 3D printing techniques to address stability issues for organic solvents. In the long term, this research aims to develop a platform around 3D printing, which allows users to select, design and fabricate innovative crystallization equipment rapidly to achieve shorter time-to-market and intensified pharmaceutical crystallization.

Acknowledgement: The work described in this abstract was supported by a grant from the Research Grants Council of the Hong Kong Special Administrative Region, China (Project No. 16242916).

References

[1] G. Reklaitis, J. Khinast, and F. J. Muzzio, "Pharmaceutical engineering science—New approaches to pharmaceutical development and manufacturing," Chem. Eng. Sci., vol. 65, no. 21, pp. iv–vii, Nov. 2010.

[2] A. K. Au, W. Huynh, L. F. Horowitz, and A. Folch, "3D-Printed Microfluidics," Angew. Chemie - Int. Ed., vol. 55, no. 12, pp. 3862–3881, 2016.

[3] S. Tait, E. T. White, and J. D. Litster, "Mechanical Characterization of Protein Crystals", Part. Part. Syst. Charact., vol. 25, no. 3, pp. 266-276, Sep 2008.

[4] A. Soare, R. Lakerveld, J. Van Royen, G. Zocchi, A. I. Stankiewicz, and H. J. M. Kramer, "Minimization of attrition and breakage in an airlift crystallizer," Ind. Eng. Chem. Res., vol. 51, no. 33, pp. 10895–10909, Aug 2012.

[5] A. J. Alvarez and A. S. Myerson, "Continuous plug flow crystallization of pharmaceutical compounds," Cryst. Growth Des., vol. 10, no. 5, pp. 2219–2228, May 2010.

529577 Direct Write of UV Curable Polymer Bonded Magnets

Abstract:

Today, almost all commercial magnet producers are using the conventional sintering-based procedure to manufacture magnets. Polymer-bonded hard magnets have recently gained popularity as a novel magnet manufacturing technique due to its design freedom regarding shapes and magnetizing structures. Comparing to the conventional sintering process, the maximum energy product $(BH)_{max}$ of the polymer-bonded magnets is lower. However, the fabrication of polymer-bonded magnets does not require high temperature treatment and thus can be more efficient. Furthermore, no current mass production technologies can produce magnets with complex geometries.

This work utilizes extrusion-based direct write (DW), combined with UV curing, to fabricate polymer-bonded magnets at room temperature. DW is an additive manufacturing technique enabling the deposition of electronic components and functional or structural patterns using different types of materials, directly without utilizing masks or subsequent etching processes. Compared to the molding process, DW provides intricate details and smooth surface finish. In this work, Nd-Fe-B powder are directly mixed with photo-curable binders to form a printable DW ink. Experiments were carried out to understand the relationship between magnetic powder solid loading, processing, and mechanical strength. Magnetic performance of printed parts was compared to other polymer-bonded and sintered magnets. The study opens a new pathway for preparing polymer-bonded magnets at room temperature for various magnetic applications.

523345 Before You Click "Print": Regulatory Considerations for 3D Printed Oral Drug Products

Abstract:

Additive manufacturing/3D printing of oral drug products and medical devices has become more attractive to shift the manufacturing strategies to support individualized medicines of the future. The US Food and Drug Administration (FDA) is charged with protecting and promoting the public health by ensuring these products are safe and effective. To that end, FDA's recent approval of a 3D-printed drug product in August 2015 (SPRITAM®) introduced a new chapter in pharmaceutical manufacturing for solid oral dosage form. Although the first approved 3D-printed drug product is a solid oral immediate release product, the majority of 3D printing research for oral delivery has been focused on controlled release, targeting and precise delivery for extremely low dose drugs. Various principles of 3D printing of oral drug products could support production of complex and personalized medicines on demand. However, standardization of some aspects of the 3D printed dosage form and associated manufacturing process may be required to achieve this objective. In this presentation, technical and regulatory considerations for development and control of 3D printed oral products will be discussed for various 3D printing technologies.

DISCLAIMER

This presentation reflects the views of the presenter and should not be construed to represent FDA's views or policies.

531710 Two-Color Photo-Inhibited Systems for Rapid Additive Manufacturing

Abstract:

Additive manufacturing (AM), also known as 3D-printing, is exciting disruptive technology that may upend contemporary manufacturing practices. Production in AM generally proceeds by layer-by-layer addition of materialâ €"typically thermoplastic or photopolymerizable resinâ€"to gradually build up the desired geometry, allowing for the production of highly individualized parts with near limitless complexity. The achievable print speed is still one of the major limitations of the current technologies with layer-by-layer devices only able to achieve vertical print speeds of several millimeters per hour. Recently, a layer-less (or continuous) method for resin based AM, termed continuous liquid interface production, was developed. The method relies on a thin diffusion-controlled oxygen-inhibited dead zone to eliminate adhesion to the projection window during printing, allowing for an order of magnitude increase in the achievable printing speeds when compared with traditional SLA devices.

Photo-inhibitor molecules that inhibit polymerization in the presence of light may be an attractive alternative inhibition mechanism to oxygen-inhibited systems. The advantages of this type of system are that diffusion of oxygen no longer controls the thickness of the dead zone, and the limit of this zone is theoretically much larger. Larger dead zones allow faster polymer reflow under the printing part and, therefore, faster build speeds. This talk will present a novel method using a two-color irradiation scheme to spatially control polymerization to generate dead zones for rapid AM and enable additional functionality over conventional continuous additive manufacturing methods.

531942 Glass-Forming Polymer Networks for Shape-Memory Contact Printing

Abstract:

Current methods for microscale pattern transfer fail to combine high feature resolution, ability to pattern large areas, high throughput and cost effectiveness. Over the past years, several researchers have demonstrated how films of shape-memory polymers can exhibit topographical memory, leading to switchable adhesion or optical properties. We are designing and developing new stimuli-responsive materials to ultimately enable high fidelity pattern transfer by thermal modulation of the elastomeric contact printing process.

Different glass forming polymers that can withstand high elongation and store large amounts of elastic strain energy in compression are under investigation. Two systems will be described: (i) small molecule thiol-ene networks that are softened by addition of dithiol reagents, and (ii) glass forming t-butyl acrylates that are lightly crosslinked. Their shape-memory behavior in tension and in compression will be reported. Furthermore, a custom-built materials testing / contact printing apparatus will be used to set compressive strain between two soft surfaces while measuring stress and contact area. Contact mechanics experiments are modeled and interpreted using finite element analysis to offer deeper insight to the nature of contact and detachment.

542079 Engineering Biocatalytic Cascades for the Production of Pharmaceuticals

Abstract:

Modern protein engineering has been accelerated, allowing the field of chemistry to realize the true value of biocatalytic synthesis. The rapid assembly of complex, high value molecules for pharmaceutical applications can be facilitated using biocatalytic cascade reactions. The exquisite selectivity of enzymes, combined with their overlapping and compatible reaction conditions allow for multistep organic synthesis in a single pot. This in turn enables us to engineer unit operations at the molecular level, avoiding protection/deprotection steps and significant labor and overhead in the manufacture of pharmaceutical products.

This talk will describe how protein engineering has been accelerated to enable modern biocatalytic synthesis. It will also highlight multiple examples of biocatalytic cascade reactions currently under development at Merck.

542081 Advanced Design of Experiment Methodologies for Enhanced Process Understanding

Abstract:

Motivated by recent robotic advances, a host of automated, high-throughput devises are enabling a data-rich experimental environment. This talk will address two recently introduced methodological advances that enable the design of information-reach experiments and model the time-resolved data that are collected. They both contribute significantly to the optimization of a multitude of processes and enhance our understanding of them. In such data-rich environment, the traditional Design of Experiments (DoE) and the Response Surface Methodologies are showing their limitations. We introduce the Design of Dynamic Experiments (DoDE) methodology, a generalization of DoE, and the Dynamic Response Surface Methodology (DRSM) to model time-resolved data. The first, DoDE, enables the design of experiments with time-varying inputs. The second one, DRSM, enables the estimation of a single model that incorporates all time-resolved data of a process output. Exploring the interrelationship among DRSM models of several measured reacting species, one can be led to the discovery of the stoichiometry of the reactions taking place. This enhances our understanding of complex kinetics and the discovery of optimal operating conditions that minimize the generation of unwanted byproducts. Examples from a series of industrial collaborations will be sited.

542080 Continuous Processing for the Manufacture of Drug Substance

Abstract:

Continuous processing has been used at Eli Lilly and Company for hybrid batch/flow processes at the 100 kg/day scale, and for fully continuous processes at the 3-10 kg/day scale. Safety, cost, and quality drivers for continuous versus batch, reactor designs, control strategies, automation and feedback control, on-line analytical, numerical modeling, and a new continuous processing facility will be discussed.

High pressure hydrogenation and Grignard formation reactions were run as hybrid batch/flow processes in batch manufacturing plants. The reactions were run continuously, and the workups and isolations were done in 2000 gallon batch tanks. Throughput was about 100 kg/day dissolved product in the continuous reactors. A 360 L vertical pipes-in-series continuous reactor was used for a reductive amination to make 2000 kg cGMP intermediate in a registration stability campaign. The manufacturing site had no high pressure hydrogenation capability prior to this campaign. The capital cost to install a continuous reaction system capable of 70 bar hydrogen pressure was about 1/10X the cost of batch. The continuous hydrogenation was deemed a low risk operation by the manufacturing plant because the volume of hydrogen headspace in the reactor was low and because the hydrogen source, reactor, and hydrogen stripping were all outside the building. A continuous Grignard formation reaction was run in a 50 L CSTR with sequestered Mg solids for reasons of safety and minimizing key impurities. Safety was improved because the Grignard reactor was about 100X smaller versus batch. Wurtz coupling impurity was minimized by maintaining higher stoichiometric ratio of Mg to aryl bromide starting material, proteo was minimized by minimizing water, and phenol was minimized by minimizing oxygen. All of these were easier to accomplish in a continuous reactor compared to batch.

A process was designed, developed and run in cGMP manufacturing with 8 continuous unit operations to produce a cytotoxic API at roughly 3 kg/day using small continuous reactors, extractors, evaporators, crystallizers, and filters in laboratory fume hoods. The results were published in Science. The continuous train was run in 3 separate segments one at a time, because there was not enough room in the 4-fume-hood facility to run it all simultaneously. A condensation reaction to form a pyrazole in step 1 used lower excess equivalents hydrazine in a superheated PFR compared to batch. The commercial scale PFR was only 1.5 L, and only about 20 g hydrazine was present in the reactor at any time. A counter-current multi-stage extraction removed residual hydrazine, acetic acid, and deprotected pyrazole product, while maximizing product yield. It also eliminated and isolation of step 1 product. Crystallization was needed after the step 2 S_NAr reaction to remove residual pyrazine, regioisomers, and low levels impurities. An â €œIsolation-Free Isolationâ€ was done via continuous crystallization and intermittent-flow filtration and dissolve-off. This gave the impurity rejection benefits of a crystallization but avoided the isolation and handling of cytotoxic solids. A formate salt isolation was eliminated in step 3 by using intermittent flow distillation with strip to dryness to remove formic acid. The distillation results could not be replicated in batch mode because batch distillation times were longer, which led to elevated impurity formation. On-line HPLC was used downstream from every reaction to ensure that controlled operations were maintained. On-line refractive index probes were used for monitoring startup transitions, quantifying *t* and axial dispersion. The business reason for 3 kg/day to 10 kg/day production by fully continuous processing is that most of Lillyâ€™s post-FHD portfolio is projected <1500 Kgs/year API. Therefore, a large portion of the portfolio may be delivered with the facility designed for fully continuous processing in laboratory fume hoods, which provide safety and containment advantages.

The successful lab hood continuous process led to the decision to invest in a new facility. In November 2017, Eli Lilly completed a 35 million euro facility for fully continuous processing at 10 kg/day throughput. The new facility was immediately used to run a 3-step fully continuous process for the cGMP production of 200 kg API. The process had an unstable Grignard reagent that was generated and used immediately in the next reaction in flow, which was a significant quality advantage over batch. Impacts of process holds were mitigated by on-line HPLC and divert valve, and the known kinetics of the impurity formations. Unreacted aryl bromide in a Negishi coupling reaction was to reject downstream, therefore accurate control of mass flow rates was necessary for maintaining stoichiometry. A combination of automated feedback control and manual feedback control was used throughout the continuous train. Strategically placed surge tanks were also important to the control strategy. Forward processing into the crystallization was approved via off-line analytical of drums that filled up about every 14 hours. Manual feed forward control based on the off-line analytical achieved the correct solvent ratio and API concentration in the downstream evaporator which fed continuous crystallization. Continuous crystallization proved superior for kinetic impurity rejection of a problematic dimer impurity. Numerical modeling with gPROMS was used for residence time distribution, surge strategy, lot genealogy, and determining allowable disturbances and pump stops. A silylation was run in a gas/liquid PFR designed to allow off-gassing without changing the liquid hold-up volume in the reactor, while maintaining low axial dispersion. Negishi coupling and acid deprotection reactions were accomplished in PFRs designed to handle solids, because of insolubilities at end-of-reaction conditions. The inexpensive PFA tube reactors were disposable, eliminating cross-contamination potential. On-line HPLC was utilized downstream from the silylation, Grignard formation, and Negishi coupling reactions. Based on the on-line LC, flow rates were adjusted to maintain high conversion, and return from divert decisions were made.

The presentation will cover manufacturing successes and failures. In GMP manufacturing, downtime and material loss events occurred because of improper pump selection, improper design of piping and manual valving systems, errors with positioning of manual valves, clogging because of restrictions in fittings and small bore valves, insufficient pressure relief downstream from positive displacement pumps, and air ingress due to in-line filters clogging and cavitating pumps.

Lab automation and on-line analytical were used to accelerate and enhance discovery and development. Continuous chemistry was used by the discovery scale up group to speed early phase material delivery using Newman Kwart and ortho-Claisen thermal rearrangements, imidazole cyclizations, and cryogenic lithiations. Automated intermittent flow stirred tank reactors were used for screening a wide range of conditions for hydrogenations and coupling reactions. Custom, in-house on-line HPLC systems have enabled tens of thousands of sample analyses on more than 50 different chemistries in development.

528751 In-Flow Production of Levoglucosenone from the Catalytic Dehydration or Cellulose Using Homogeneous Brønsted Acid Catalysts in γ-Valerolactone

Abstract:

Levoglucosenone (LGO) is an anhydro-sugar used in the organic synthesis of high value pharmaceutics (e.g., antibiotics and anticancer drugs); renewable solvents such as Cyrene®; 1-6-hexanediol used to produce polymers; and other specialty and high value chemicals [1]. Recently, our research group in collaboration with George Huber's at the University of Wisconsin – Madison studied the production of LGO from cellulose in a batch reactor at 483K using low concentrations of acid catalysts in THF. The study showed that similar LGO yields to those obtained with high concentrations of sulfuric acid [1] may be obtained using low concentrations of sulfuric acid and a solid acid catalyst. The study also showed that the water accumulation during the reaction limits the production of LGO. However, at those temperatures THF is flammable and high pressures are needed to keep it in the liquid phase. Here, we studied the catalytic conversion of cellulose to levoglucosenone in a flow reactor using homogenous Brønsted acid catalysts in a polar aprotic solvent. The flow reactor helps to minimize the concentration of water in the reaction. We used γ–valerolactone (GVL) as solvent because has similar properties to those of THF but it is safer. We demonstrated that using a flow-through reaction system and low concentrations of Brønsted acid catalysts such as sulfuric acid (H_2SO_4), phosphoric acid (H_3PO_4) or 1-propanesulfonic acid in GVL as solvent, high yields of LGO from cellulose are obtained. For GVL, the yield towards LGO passes through a maximum as a function of the reaction temperature at 445K. Thermal pretreatment of the cellulose at 455 K before the reaction decreases the yield to LGO. An increase in the catalyst concentration does not significantly increase the yield towards LGO, but decreases the yield towards LGA. We studied the effect on the LGO yield of the catalyst type, the cellulose loading and the crystallinity of the cellulose. Using a reaction temperature of 445K and 5mM of H_2SO_4 as catalyst, the total carbon and LGO yields were 81% and 65%, respectively. To our knowledge this LGO yield is the highest yield for cellulose conversion reported so far.

[1] F. Cao, T. J. Schwartz, D. J. McClelland, S. H. Krishna, J. A. Dumesic, and G. W. Huber, "Dehydration of cellulose to levoglucosenone using polar aprotic solvents," ***Energy Environ. Sci.***, vol. 8, no. 6, pp. 1808–1815, 2015.

526380 Cyclooctene Cooxidation-Facilitated Co-ZSM-5-Catalyzed Selective Oxidation of Ethylbenzene with Molecular O_2

Abstract:

Solventless activation of benzylic C-H bonds by oxidation is of great interest in the synthesis of various industrial chemicals[1]. However, common liquid-phase methods to oxidize benzylic C-H bonds either use stoichiometric amounts of organic oxidants or catalysts together with radical forming agents such as NDPI. For greener processes, it is much more desirable if neither of these needs to be used. We report here a new method for selective oxidation of ethylbenzene (EB) that is achieved by cooxidation of cyclooctene (COE) selectively to its epoxide, using O_2. In this process, no products from wasteful reagents are produced.

The solvent-free aerobic oxidation of EB was carried out in a batch reactor under atmospheric pressure at 100ËšC. The highest catalytic activity was achieved when Co-ZSM-5 and a Au-containing initiator solution generated from cyclooctene (COE) epoxidation reaction[2] were simultaneously present in the system. It was confirmed with a radical quenching experiment that radicals were essential in the reaction. However, EB oxidation did not involve a radical chain mechanism. Quantitative EPR experiment and spin-trap experiment were used to analysis the radical intermediates and to gain mechanistic insight of the reaction. The overall conversion of EB depended on the total amount of COE present in the system, which indicated that the two reactions were coupled together. The data suggested that the primary role of Co-ZSM-5 was to activate cyclooctene-3-hydroperoxide, an initiator in-situ generated by Au-cluster catalyzed-COE epoxidation, to form other radicals that accelerate the EB oxidation. This aspect is quite different from traditional initiators like tert-butyl hydroperoxide. Possible mechanistic explanations are presented.

1. A. M. Hossain et al., *Advanced Materials Research*. (2015) **1109**, p.248-252.
2. Qian, Z. Wang, E. V. Beletskiy, J. Liu, H. J. dos Santos, T. Li, M. do C. Rangel, M. C. Kung, and H. H. Kung. *Nat. Commun.* (2017) **8**, 14881.

521503 Dirhodium Immobilized Hollow Fiber Flow Reactor for Scalable and Sustainable C–H Functionalization in Continuous Flow

Abstract:

Dirhodium immobilized hollow fiber reactor for scalable and sustainable C-H functionalization in continuous flow

Chun-Jae Yoo,[a] Daniel Rackl,[b] Wenbin Liu,[b] Caroline B. Hoyt,[a] Brian Pimentel,[a] Ryan P. Lively,[a] Huw M. L. Davies[b] and Christopher W. Jones[a]

[a]*Georgia Institute of Technology, School of Chemical & Biomolecular Engineering, 311 Ferst Drive NW, Atlanta, GA 30332, USA*

[b]*Emory University, Department of Organic Chemistry, 1515 Dickey Drive, Atlanta, GA 30322, USA*

C–H functionalization has become an exciting new synthetic strategy that has brought significant innovation into total synthesis and new methodologies for drug development. [1] One particularly useful C –H functionalization procedure has been the C–H insertion chemistry of transition metal carbenes, typically generated from diazo compounds. [2] In recent years, rhodium-stabilized donor/acceptor carbenes have proven to be effective in a wide range of intermolecular C–H functionalization reactions, in which the site selectivity and stereoselectivity are dominated by catalyst control. [3-5] However, there are several critical drawbacks for this chemistry that present potential barriers to application in the pharmaceutical industry at scale. First, the metal carbenes are created by using relatively reactive and unstable diazo compounds, which are potentially explosive and highly toxic. Thus, such reactive compounds are ideally not stored in large quantities on site, and just-in-time syntheses are preferred. [6] The price of the soluble dirhodium catalysts (Rh_2L_4, L= ligand) is another barrier for industrial applications. Rhodium is a very expensive noble metal and the cost for the synthesis of the ligands for these catalysts also cannot be ignored. To recycle the homogenous dirhodium catalysts, additional separation steps are required and the catalytic activities are often not maintained after recovery and recycle. Thus, new methodologies for maximizing catalyst turnover number (TON) and recovering Rh for reprocessing are needed; one such approach involves immobilizing these catalysts on a support material. [7] In this research, we have developed tandem flow reactors where the reactive donor/acceptor-substituted diazo compounds are synthesized in flow from hydrazones, and immediately convert the diazo compounds in a second, downstream polymeric hollow fiber flow reactor containing supported asymmetric dirhodium catalysts. Enantioselective cyclopropanation and C-H functionalizations are described, giving similar performance as conventional batch reactions.

References

1. Davies, H. M. L. and Manning J. R., *Nature,* **2008**, 451, 417
2. Ford, A.; Miel, H.; Ring, A.; Slattery, C. N.; Maguire, R.; McKervey, M. A. *Chem. Rev.* **2015**, 115, 9981
3. Qin, C.; Davies, H. M. L. *Am. Chem. Soc.* **2014**, 136, 9792.
4. Guptill, D. M.; Davies, H. M. L. *Am. Chem. Soc.* **2014**, 136, 17718.
5. Liao, K.; Negretti, S.; Musaev, D. G.; Bacsa, J.; Davies, H. L. *Nature* **2016**, 533, 230.
6. Rackl, D.; Yoo, C.-J.; Jones, C. W.; Davies, H. M. L.; *Lett.* **2017**, 19 (12), 3055.
7. Moschetta, E. G., Negretti, S., Chepiga, K. M., Brunelli, N. A., Labreche, Y., Feng, Y., Rezaei, F., Lively, R. P., Koros, W. J., Davies, H. M. L., and Jones C. W. *Chem. Int. Ed.*, **2015**, 54, 6470

528972 Catalytic Dehydration of Levoglucosan and Cellulose to Levoglucosenone Using BrÃ¸nsted Solid Acid Catalysts in Tetrahydrofuran

Abstract:

The production of levoglucosenone (LGO) *via* levoglucosan (LGA) and cellulose dehydration was studied using Brønsted solid acid catalysts in tetrahydrofuran (THF). We evaluated the use of propylsulfonic acid functionalized SBA-15 (PS-SBA-15), commercial acid functionalized SiO_2 and zeolite catalysts for the dehydration of LGA. PS-SBA-15 showed the best catalytic performance to produce LGO from LGA. The highest LGO yield obtained from LGA using PS-SBA-15 was 58.6%. That yield is more than two times higher to that obtained using sulfuric acid, which was 24.5%. We showed that the conversion of cellulose to LGO using a solid acid catalyst must be assisted by a homogeneous acid catalyst such as sulfuric acid (H_2SO_4), to promote the depolymerization of cellulose into LGA. The main role of the solid Brønsted acid catalyst is the dehydration of LGA to LGO during the cellulose conversion to LGO. We identified that surface properties such as acid-surface interactions and hydrophobicity are important variables to explain the catalytic behavior of the solid acid catalysts evaluated. We also studied the stability of PS-SBA-15 in THF and 1.2 mM H_2SO_4 in THF. The physicochemical properties of the silica catalysts change for different treatment times in THF. However, these changes do not have a significant effect on the acidity and mesostructure of the silica catalysts. On the other hand, adding H_2SO_4 to the THF causes a decrease in the acidity of the PS-SBA-15 after treatment. We demonstrated that solid Brønsted acid catalysts produce higher yields to produce LGO from LGA and cellulose than sulfuric acid. Also, we identified that the catalysts surface properties are a crucial parameter for the dehydration of LGA to LGO.

536723 Composite Hollow Fiber Microfluidic Catalytic Reactors for Direct Conversion of Glucose to 5 HMF

Abstract:

The laboratory-scale continuous-flow processes have attracted a great deal of interest in view of the significance of economic and environmentally sustainable production of pharmaceuticals, fine chemicals, and agrochemicals, as well as upgrading of biomass feedstocks.[1,2] The immobilization of homogeneous catalysts such as palladium, ruthenium, and nickel for cooperative interactions on the surface of solid supports such as porous oxides and polymers offers a powerful catalytic system that exploits and enhances the advantages of both nanocatalysis and flow chemistry.[1-3] On the other hand, direct conversion of glucose to 5-Hydroxymethylfurfural (5-HMF) is an important step in upgrading biomass to value-added chemicals and fuels.[4] In this work, the highly hydrophilic and solvent-stable composite hollow fibers were created by crosslinking of porous polyamide-imide (PAI) fibers with 3-aminopropyl trimethoxy silane (APS). The obtained hollow fibers consist of bifunctional groups in the fiber wall that make them bifunctional catalysts for cooperative interactions (i.e., acid-base catalysis) were used as microfluidic reactors for continuous-flow process. Based on the optimized conditions for glucose conversion, we identified a robust process window to produce composite hollow fiber microfluidic reactors for sustainable carbohydrate dehydration reaction in flow.

References:

(1) He, Y.; Jawad, A.; Li, X.; Atanga, M.; Rezaei, F.; Rownaghi, A. A. Direct Aldol and Nitroaldol Condensation in an Aminosilane-Grafted Si/Zr/Ti Composite Hollow Fiber as a Heterogeneous Catalyst and Continuous-Flow Reactor. *J. Catal.* **2016**, *341*, 149–159.

(2) He, Y.; Rezaei, F.; Kapila, S.; Rownaghi, A. A. Engineering Porous Polymer Hollow Fiber Microfluidic Reactors for Sustainable C-H Functionalization. *ACS Appl. Mater. Interfaces* **2017**, *9* (19), 16288–16295.

(3) Jawad, A.; Rezaei, F.; Rownaghi, A. A. Porous Polymeric Hollow Fibers as Bifunctional Catalysts for CO 2 Conversion to Cyclic Carbonates. *J. CO2 Util.* **2017**, *21* (September), 589–596.

(4) Deshpande, N.; Pattanaik, L.; Whitaker, M. R.; Yang, C. T.; Lin, L. C.; Brunelli, N. A. Selectively Converting Glucose to Fructose Using Immobilized Tertiary Amines. *J. Catal.* **2017**, *353*, 205–210.

538257 Water Soluble Palladium-β-Cyclodextrin Complex and Its Catalytic Performance for a Suzuki-Miyaura Cross-Coupling in Flow

Abstract:

Palladium is routinely used in homogeneous catalytic carbon-carbon cross-couplings.[1,2] As an example, the Suzuki-Miyaura cross-coupling involves the reaction of an organic halide with an organoboron compound. Hydrophobic dialkyl phosphines have been demonstrated as effective ligands for the cross-coupling, yet inorganic salt by-products are problematic in continuous-flow processes.[3] The reactions are versatile in the preparation of fine chemicals, materials, natural products, and pharmaceuticals.[1]

The use of water soluble catalyst is attractive because it offers a way to ensure the solubility of inorganic by-products that clog reactors.[4,5] Palladium(II)-β-cyclodextrin is such a catalyst reported to be effective in Suzuki-Miyaura hetero-couplings in batch reactors.[5] Cyclodextrins (CDs) have a cyclic structure with a hydrophilic external rim and a hydrophobic internal cavity. The use of β-CD as a ligand allows competitive yields of cross-coupling reactions in water even at low catalyst loadings (~0.001 mol%). What's more, the catalyst maintains high catalytic activity after recycling.[5] The ligand has the potential for application in manufacturing processes, but the reaction kinetics are not well understood.

A capillary microreactor experiment was design to study the kinetics of the Suzuki-Miyaura using Pd(II)-β-cyclodextrin catalyst. Microreactors can expedite kinetic information compared to traditional batch reactors. Rapid mixing and heat transfer by conduction enable the design of ideal reactors for screening economical ligands. In the present study, a water-ethanol solvent system was explored as a potential reaction medium. The cross-coupling using natural product ligand was examined for the first time in continuous-flow. These results will be presented and discussed.

References:

1. Suzuki, A. Cross-coupling reactions of organoboranes: An easy way to construct C-C bonds (Nobel Lecture). *Angew. Chem. Int. Ed.* **50,** 6723–6733 (2011).
2. Negishi, E. I. Magical power of transition metals: Past, present, and future (Nobel Lecture). *Angew. Chemie - Int. Ed.* **50,** 6738–6764 (2011).
3. Domier, R. C., Moore, J. N., Shaughnessy, K. H. & Hartman, R. L. Kinetic analysis of aqueous-phase Pd-catalyzed, Cu-free direct arylation of terminal alkynes using a hydrophilic ligand. *Org. Process Res. Dev.* **17,** 1262–1271 (2013).
4. Hailes, H. C. Reaction solvent selection: The potential of water as a solvent for organic transformations. *Org. Process Res. Dev.* **11,** 114–120 (2007).
5. Kaboudin, B., Salemi, H., Mostafalu, R., Kazemi, F. & Yokomatsu, T. Pd(II)-β-cyclodextrin complex: Synthesis, characterization and efficient nanocatalyst for the selective Suzuki-Miyaura coupling reaction in water. *J. Organomet. Chem.* **818,** 195–199 (2016).

537896 Regioselective Epoxide Ring Opening with Alcohols Using Heterogeneous Lewis Acid Catalysts

Abstract:

Epoxides are versatile intermediates that can be ring opened using different nucleophiles to produce a wide variety of valuable chemicals. The key challenge for this reaction is selectivity. Indeed, the ring opened epoxide can form oligomeric byproducts. At the same time, the reaction can produce either a terminal ether or a terminal alcohol, resulting in low regioselectivity. Previous work with Bronsted acidic catalyst results in low regioselectivity. Higher regioselectivity has been observed for Lewis acidic catalysts such as Co-Salen and Co-Porphyrin. Unfortunately, these catalysts tend to deactivate through loss of the counterion used to maintain the Co in the +3 oxidation state.[1,2] Intriguingly, Lewis acidic zeolites have been demonstrated for the water ring opening of epoxides to produce a diol.[3,4] In this work, we will discuss the regioselective epoxide ring opening reaction with alcohols for Lewis acidic catalytic materials. Through comparing catalytic activities, it is found that Sn-Beta is more active than other similar Lewis acidic catalytic materials such as Sn-MFI and Sn-SBA-15. For all materials, high regioselectivity of >99% is obtained. With large substrates such as epoxy octane, diffusion limitations are encountered that can be overcome through creating zeolites with particle sizes less than 100 nm. These materials can be readily recycled while retaining high catalytic activity and selectivity. Hot filtration tests demonstrate that these catalysts are heterogeneous in nature and stable. Overall, these materials represent highly active and selective catalysts for the conversion epoxide ring opening reaction.

540773 2018 Outlook for Energy: A View to 2040

Abstract:

The Outlook for Energy is ExxonMobil's global view of energy demand and supply through 2040. It also highlights the dual challenge of providing affordable energy to support prosperity while reducing environmental impacts – including the risks of climate change. We use the Outlook to help inform our long-term business strategies and investment plans.
The Outlook anticipates significant changes through 2040 across the world to boost living standards, reshape the use of energy, broaden access to abundant energy supplies, and accelerate decarbonization of the world's energy system to address the risks of climate change.

- Global energy needs will rise about 25 percent over the period to 2040
- Electricity demand will nearly double in non-OECD countries
- Worldwide electricity from solar and wind will increase about 400 percent
- Natural gas will expand its role, led by growth for electricity generation
- Growth in oil demand will be driven by commercial transportation and chemical needs
- Efficiency gains and growing use of less-carbon intensive energy sources will contribute to a nearly 45 percent decline in the carbon intensity of global GDP

Energy is fundamental to modern life, and as the world's population approaches 9 billion people in 2040, we are challenged to improve living standards everywhere. Understanding the factors that drive the world's energy needs – and likely choices to meet those needs – is the mission of the Outlook. By sharing the Outlook with the public, we hope to broaden that understanding. Energy matters to everyone, and we all play a role in shaping its future.

541543 Energy Decarbonisation Scenarios

Abstract:

With the energy sector being associated with 2/3 of the annual greenhouse gas emissions, decarbonizing the energy is a key step to reduce their level. The National Determined Contributions (NDC's) announced so far by the participating countries do not meet the objectives of the Paris-climate agreement, as far as reaching a peak in emissions in the near future or decreasing to a level that is compatible with the stated long-term targets.

Using different scenarios for energy demand and supply, options for energy sector decarbonisation are evaluated. Meeting the 2-degree target requires a major transformation in the power sector, efficiency improvements – specially in the buildings and in the industry, and a revolution in transport towards fuel and modal shifts. With a significant increase in renewable energy, system integration and optimization will also be key in reducing the cost of the energy transition. Meeting more ambitious climate goals ("well below 2 degrees") will push further the technological innovation and implementation requirements throughout the entire energy value chain. A scenario aiming at zero net-emissions by 2060 is presented, along with a comparison to the 2-degree case.

542625 Fundamental Research Needs to Advance Energy Technologies

Abstract:

The U. S. Department of Energy (DOE), Office of Science, Office of Basic Energy Sciences (BES) supports fundamental research in chemical and materials sciences to provide the foundations for new energy technologies and to support DOE missions in energy, environment, and national security. This presentation will discuss opportunities for fundamental research to impact DOE's energy mission "to catalyze the timely, material, and efficient transformation of the nation's energy system and secure U.S. leadership in energy technologies" with a focus on the way we generate, store and use energy nationally. I will provide an overview of BES strategic planning over the past decade that identified priority research directions for advancing energy applications, highlight key scientific advances in these areas, and discuss some future opportunities for modern science, particularly chemical sciences, to accelerate the transformation of the U. S. energy portfolio.

540924 Nano- and Microfabricated Hydrogels for Regenerative Engineering

Abstract:

Engineered materials that integrate advances in polymer chemistry, nanotechnology, and biological sciences have the potential to create powerful medical therapies. Our group aims to engineer tissue regenerative therapies using water-containing polymer networks, called hydrogels, that can regulate cell behavior. Specifically, we have developed photocrosslinkable hybrid hydrogels that combine natural biomolecules with nanoparticles to regulate the chemical, biological, mechanical and electrical properties of gels. These functional scaffolds induce the differentiation of stem cells to desired cell types and direct the formation of vascularized heart or bone tissues. Since tissue function is highly dependent on architecture, we have also used microfabrication methods, such as microfluidics, photolithography, bioprinting, and molding, to regulate the architecture of these materials. We have employed these strategies to generate miniaturized tissues. To create tissue complexity, we have also developed directed assembly techniques to compile small tissue modules into larger constructs. It is anticipated that such approaches will lead to the development of next-generation regenerative therapeutics and biomedical devices.

Commercial Scale Manufacturing of a Pharmaceutical Product Using Powder-Liquid 3D Printing Technology

Timothy S. Tracy, PhD, Aprecia Pharmaceuticals, Blue Ash, OH

Abstract

Three dimensional printing (3DP) opens the possibility to manufacture pharmaceuticals differently from today and provides the opportunity to create novel dosage forms for patients. Despite significant advances in the field, an obstacle to commercialization of 3DP pharmaceuticals has been the difficulty in being able to manufacture at commercial scale. Recently, utilizing powder-liquid 3DP technology a pharmaceutical (levetiracetam, Spritam™) has been developed through the 505(b)(2) regulatory pathway at commercial quantities and has received FDA approval for manufacture and sale. In addition, the product is a novel dosage form that utilizes fast-melt technology (ZipDose™) that provides for rapid dispersion in the mouth (< 10 seconds) with only a sip of liquid, while containing doses of levetiracetam up to 1000mg. Products that rapidly disperse in the mouth are particularly useful for patients with dysphagia or difficulty swallowing, particularly of large tablets. This rapid dispersion at high doses was possible because of the selective porosity and hardness of the dosage form created by the 3DP layering process. To produce this unique dosage form, processes were developed to assure consistent and precise jetting of print fluid and careful control of powder deposition across up to 30 layers at commercial production speed to produce tablets that were highly accurate and reproducible with respect to dose of active drug. The regulatory approval and commercialization of this product demonstrated that unique pharmaceutical dosage forms could be created through 3DP technology that helped address unmet needs.

540925 Closing the Circle on Design, Hardware, and Materials for Additive Manufacturing

Abstract:

3D printing is a subset of additive manufacturing (AM) and represents an evolving set of tools and capabilities to prototype, and in some cases produce, new products in much faster iteration cycles than traditional subtractive processes. For 3D printing to reach its full potential as a viable large-scale manufacturing option, new advancements are still required across the three major segments of the industry â€“ design, hardware, and materials â€“ and to accelerate these advancements, highly collaborative efforts between the segments will provide 3D printing users with the quickest and best solutions for their production needs. The presentation will highlight the current state-of-the-art in materials for 3D printing, known existing property gaps, Dowâ€™s vision for the future development and performance of 3D printable materials, and how the broader industry can collaborate to shorten the path to wider scale adoption of these tools in 21st century manufacturing.

541541 Additive Manufacturing and Architected Materials

Abstract:

Material properties are governed by the chemical composition and spatial arrangement of constituent elements at multiple length-scales. This fundamentally limits material properties with respect to each other creating trade-offs when selecting materials for specific applications. For example, strength and density are inherently linked so that, in general, the more dense the material, the stronger it is in bulk form. We are combining advanced microstructural design, using flexure and screw theory as well as topology optimization, with advanced additive micro and nanomanufacturing techniques to create new material systems with previously unachievable property combinations â€“ mechanical metamaterials. The performance of these materials is fundamentally controlled by geometry at multiple length-scales rather than chemical composition. We have demonstrated designer properties of these mechanical metamaterials in polymers, metals, ceramics and combinations thereof. Properties include ultra-stiff lightweight materials, negative stiffness, and negative thermal expansion. Our additive micro-manufacturing techniques include Projection Microstereolithography (PmSL), Direct Ink Writing (DIW), and Electrophoretic Deposition (EPD). These tools are capable of generating the designed structures which are highly three-dimensional micro- and nano-scale architectures with multiple constituent materials in the same structure.

533315 Rapid Temperature Correction to Lattice Energy Landscape for Crystal Structure Prediction

Abstract:

The ability to solve solid-state structures through organic crystal structure prediction (CSP) would reduce risk and improve efficiency during drug development by guiding and complementing experimental screens for low free energy crystal forms. Recent advances in computational hardware and algorithms have enabled high accuracy crystal structure predictions to assist in pharmaceutical decision making for molecules with few low energy polymorphs or large energy gaps between polymorphs. These methods generally focus on crystal lattice energies computed through various quantum mechanics calculations, e.g. the widely used density functional theory, at 0 Kelvin, and thus the predicted stability ranking of polymorphs does not necessarily correspond to their ranking at physically relevant temperatures. This poses a challenge for compounds with many low energy polymorphs, as, for a given compound, tens to hundreds of polymorphs may be predicted by CSP, many of which cannot be crystallized experimentally, e.g. due to kinetic traps or low thermodynamic populations. The ability to determine the thermodynamic stability ranking of predicted polymorphs at physically relevant temperatures, as well as their kinetic accessibility, is required for CSP to have wide-spread applicability in drug development. We demonstrate the strength of our CSP method coupled with finite temperature analysis on compounds with multiple metastable polymorphs. Such predictions can guide and accelerate experimental polymorph screening for drug development and manufacture.

Corresponding authors' emails: shuhao@xtalpi.com, bruno.c.hancock@pfizer.com

* These authors contributed equally to this work.

538495 Understanding the EffectÂ of Changing Complexities of Potential Energy Functions on the Entropic Contribution to Free Energy Differences of Organic Polymorphs

Abstract:

Accurate and fast computational methods to predict the free energy differences of crystals can greatly impact the pharmaceutical industry. Experimental approaches to capture all polymorphs are costly and time consuming. Current computational methods are able to exhaustively search for the lattice minimum structures, but fail to include entropic effects. Our group has been focused on the testing and development of methods that include the entropic behavior of crystalline materials to determine the free energy ranking at temperatures of interest.

We compare the free energy differences of 13 polymorphic pairs of organic molecules using the OPLS-AA point charge potential and the AMOEBA polarizable potential, . using full molecular dynamics (MD) simulations as well as lattice dynamic (LD) approaches using isotropic expansion to compute the free energy differences. For the small and rigid crystalline systems, we find that there is little deviation (< 0.1 kcal/mol) between MD and LD approaches up to 300K, with somewhat larger differences for flexible molecules. Including anisotropy to the expansion in LD does not significantly improve the match to MD, suggesting that the deviations between the methods are because of fundamental anharmonicity in the free energy landscape.

We also calculate the free energy differences using isotropic lattice dynamics for the 13 polymorphic pairs at DFT-D3 levels of theory, and compare these free energy differences between the three different potential energy functions.

530278 Improved EfficiencyÂ in the *Ab Initio* Generation of Crystal Structures

Abstract:

The CrystalPredictor[1,2] and CrystalOptimizer[3] codes have been used to explore the space of crystal structures successfully in several crystal structure prediction (CSP) investigations in recent years, including in the series of blind tests organised by the Cambridge Crystallographic Data Centre[4] and in the prediction of the crystal structures of pharmaceutically-relevant molecules[5-7]. One of the key research challenges in developing CSP capabilities is to enable the investigation of increasingly flexible compounds within tractable computational times.

We present recent advances in CrystalPredictor that are focussed on addressing this challenge. Specifically, we discuss the smoothing of the intramolecular potential,[8] an innovation in CrystalPredictor II that allows the most efficient use of computational effort to cover a flexible moleculeâ€™s conformational space. We also report the recent implementation of torsional grouping, a feature that retains the near quantum mechanical accuracy of the Local Approximate Model technique whilst greatly reducing the computational effort for compounds with more than 5 flexible torsions. These improvements achieve greater accuracy in the initial ranking of potential crystal structures, while managing computational cost, so that a thorough exploration of the search space is possible. We present CSP results for the NSAID drug Acemetacin to demonstrate this capability.

[1] M. Habgood, I. J. Sugden, A. V. Kazantsev, C. S. Adjiman, and C. C. Pantelides, J Chem Theory Comput **11**, 1957 (2015).

[2] I. Sugden, C. S. Adjiman, and C. C. Pantelides, Acta Crystallogr B **72**, 864 (2016).

[3] A. V. Kazantsev, P. G. Karamertzanis, C. S. Adjiman, and C. C. Pantelides, J Chem Theory Comput **7**, 1998 (2011).

[4] A. M. Reilly, et al., Acta Crystallogr B **72**, 439 (2016).

[5] D. E. Braun, T. Gelbrich, K. Wurst, and U. J. Griesser, Cryst Growth Des **16**, 3480 (2016).

[6] D. E. Braun, V. Kahlenberg, and U. J. Griesser, Cryst Growth Des **17**, 4347 (2017).

[7] S. L. Price, D. E. Braun, and S. M. Reutzel-Edens, Chem Commun **52**, 7065 (2016).

[8] I. Sugden, C. S. Adjiman, and C. C. Pantelides, Acta Crystallogr B (In Preparation).

525352 Molecular Crystal Structure Prediction with Gator and Genarris

Abstract:

Molecular crystals are bound by dispersion interactions, whose weak nature produces potential energy landscapes with many local minima. Hence, molecular crystals often exhibit polymorphism, whereby the same molecule crystallizes in several structures. Polymorphs may exhibit markedly different physical and chemical properties. Crystal structure prediction is challenging due to the high accuracy required for the small energy differences between polymorphs and the high dimensionality of the configuration space. We present the genetic algorithm (GA) code, GAtor, and its associated structure generation package, Genarris. Both rely on dispersion-inclusive density functional theory (DFT) for geometry relaxations and energy evaluations.

Genarris generates random structures with physical constraints and uses a Harris approximation to construct the electron density of a molecular crystal by superposition of single molecule densities. The DFT energy is then evaluated for the Harris density without performing a self-consistent cycle, enabling fast screening of initial structures with an unbiased first-principles approach. Genarris creates a maximally diverse initial pool of structures by using machine learning for clustering based on structural similarity with respect to a relative coordinate descriptor (RCD) designed for molecular crystals.

GAs rely on the evolutionary principle of survival of the fittest to perform global optimization. GAtor offers a variety of crossover and mutation operators, designed for molecular crystals, to create offspring by combining/ modifying the structural genes of parent structures. GAtor achieves massive parallelization by spawning several GA replicas that run in parallel and read/write to a common pool of structures. GAtor performs evolutionary niching by using machine learning for dynamic clustering on the fly. A cluster-based fitness function is then used to steer the GA to under-sampled low-energy regions of the potential energy landscape. This helps overcome initial pool biases and selection biases (genetic drift).

528140 Combining Simulation and Experiment to Understand Polymorph Selection of Drug Molecules in Different Solvents

Abstract:

The importance of polymorph selection in drug discovery and development has been well recognized. Effects of polymorphism on multiple properties of drugs, such as physical and chemical stability, bioavailability and manufacturability. Nucleation is the first stage of crystallization that could determine the formation of a specific crystalline structure. Studying nucleation is challenging, both experimentally and using computer simulations. Nucleation is a stochastic process and only involves the order of only 100 to 1000 molecules, making it challenging to design experiments to study. On the other hand, nucleation is also a rare event, occurring on time scales far beyond those accessible to simple molecular simulation methods.

In this work, we combined simulations and experiments to study the nucleation of small molecules in different solvents. As the Ostwaldâ€™s rule stated that the polymorph with a lower nucleation energy barrier always crystallizes first in general, our simulations could provide information on polymorph selection that are consistent with the experimental data. Two polymorphs of sulfamerazine (Form I and II), and three different solvents, water, methanol and acetonitrile, were selected for our study. By implementing the String Method in Collective Variables (SMCV) in combination with the construction of order parameters, we sketched the minimum free energy paths of nucleation and calculated the energy barriers for both of the two forms. Simulation results showed that Form I has lower energy barriers in all the solvents than Form II, indicating that Form I would be easier to nucleate. PXRD data from our crystallization experiments showed consistent results that Form I was the favorable form in these solvents.

523089 Novel Computational Approaches to Support Pharmaceutical Solid State Chemistry Tasks

Abstract:

Given the complexity of the pharmaceutical solid state landscape and challenges facing the pharmaceutical industry, an accelerated Drug Development greatly benefits from guidance provided by computational methods. In particular, Computational Pharmaceutical Solid State Chemistry is used to support all steps related to the development of solid state pharmaceuticals.[1] This presentation will focus on novel computational approaches to support decrease of pharmaceutical crystals agglomeration,[2] and solvent selection for impurity purge *via* recrystallization.[3]

1. Abramov, Y.A., Ed. *Computational Pharmaceutical Solid State Chemistry*, John Wiley & Sons, **2016**.
2. Abramov Y.A. *Cryst. Growth Des.* **2017**, 17, 2873–2880.
3. Abramov Y.A. *Cryst. Growth Des.* **2018**, 18, 1208–1214.

525203 Prediction of Risk in Drug Substance Starting Material Selection

Abstract:

Regulatory approval for drug substance (DS) starting materials (SMs) represents a significant milestone in the progression towards approval of the proposed DS manufacturing process. Development of a robust data package in support of a proposed SM designation is a significant undertaking, and the requisite route optimization, control strategy development, method development, supplier qualification, and documentation for each proposed SM can translate to substantial resource utilization and financial investment.

To efficiently characterize the risk associated with a potential SM designation, Eli Lilly has developed a prototype tool that objectively predicts viability of a proposed SM, informed by an assessment of molecule complexity, genotoxic impurity control, and propinquity to the DS. To develop the tool, the team applied multivariate data analysis (MVDA) to identify molecular and route attributes that have previously correlated to acceptance or rejection of the SM by global regulatory agencies. As an outcome of the MVDA approach, two simplified metrics were empirically derived that applied broadly to the Lilly portfolio for which non-binding scientific advice or NME approval was available. These metrics classify high-risk SMs into at least one of two unacceptable classes: high molecular complexity (number of atoms and ring structures contributed to the DS) or a high number of genotoxic or stereospecific structures. The utility of the approach was tested and verified using both internal-to-Lilly and publicly-available external case studies.

525860 An Intelligent Machine for Document Preparation

Abstract:

This talk will describe a Proof-of-Concept project to develop an artificial intelligence-based solution to rapidly assemble an End-of-Phase II briefing document for molecule A. Both structured and unstructured data derived from electronic lab notebooks and technical report repositories was accessed via an data integration layer and analyzed using annotation-free natural language processing tools. A training corpus of 70,000 pdf files from five molecules, including molecule A was ingested and used as a the design corpus, and trained against. Prior EOPII documents from the four molecules (other than molecule A) was utilized to generate a set of training questions and answers. The design corpus was then utilized to help assemble the EOPII document given key questions that needed to be asked. The prototype tool featured accessing key textual phrases using Natural language search, a table builder, an image/plot finder and a free-text addition capability to collaboratively create an html or docx file as an output. This tool was compared against the current internal search tool and comparison metrics were generated.

529417 Data-Driven Real-Time Operation Support for Decontamination Processes in Biopharmaceutical Drug Product Manufacturing

Abstract:

Manufacturing and process control tools have been continuously evolving and becoming more extensive and sophisticated. An ever-increasing number of sensors are constantly monitoring production processes and providing real-time measurements of process variables such as pressure, humidity, and temperature. This large data inventory is mostly unused except for backtracking purposes in the case of an incident occurring or for good manufacturing practice (GMP) related revalidation efforts. This data could, however, be highly valuable for anticipating failures and for predictive maintenance applications. This work describes a framework for further utilizing this data. It presents an industrial case study for the application of the framework within the decontamination process of the sterile manufacturing of biopharmaceutical drug products at the Hoffmannâ€"La Roche facility, located in Kaiseraugst, Switzerland. Decontamination is performed after each product batch in the manufacturing environment such as clean rooms or isolators to kill any micro-organisms. The process includes dehumidification, conditioning, decontamination with hydrogen peroxide, and aeration.

Our suggested framework allows for probability-based performance evaluation and forecasting of processes in commercial operation. Application of the framework would enable operators to predict batch failures and to take pre-emptive action. The framework employs machine learning algorithms within a three-step procedure to classify the data based on batch quality, train a model to make predictions and finally a decision-making step based on the results of the previous two steps.

The first step of the procedure involves statistically defining quality boundaries using pre-recorded data, e.g., temperature, pressure, hydrogen peroxide content. Due to the high correlation of the recorded data, a principal component analysis is used to obtain a reduced set of independent variables (principal components). The results are used to define batch quality boundaries. The recorded data is then split into training and test data for a machine learning algorithm. Several algorithms such as k-nearest neighbors and random forest are compared to find the most suitable algorithm for the studied case data. The trained model should then predict process parameters for the remaining process duration based on the input from only a few seconds of measured operation parameters. The third step of the procedure provides a predictive monitoring tool which combines the previously defined operation boundaries with the predictions from the specified machine learning algorithm. The tool would allow for predicting failure probabilities along with the expected failure time. A reverse PCA is then performed to pinpoint critical process parameters. This allows operators to take precautionary preventive measures, e.g., adjustment of the pumps and valves or interruption of the process.

532255 Development of an End-to-End Data Management and Visualization System for Cell Culture Process Development

Abstract:

Cell culture processes generate extensive data sets originating from diverse systems. These data are collected on varying timescales, ranging from single metadata values to trends from on-line sensors generating thousands of data points per day. Additionally, data are stored in various formats and locations, making uniform retrieval and visualization challenging. Historically, we have used Microsoft Excel to manage data, perform simple analyses, and generate fixed visualizations. While Excel offers a familiar environment for handling data, it is limited in its processing capabilities, ability to directly compare differently sized or shaped data sets, and ease in flexibly creating and formatting graphics. Using Excel as a repository and analytical tool for comparison of data between experiments, bioreactor scales, laboratories, or manufacturing facilities requires an inefficient and labor intensive workflow of manual transfers and manipulations that is error prone, difficult to control, and time consuming.

To better maintain and use cell culture data, we developed an end to end data management and visualization system to meet user requirements while presenting familiar and user friendly interfaces that are accessible to both highly trained laboratory personnel as well as site leadership.

In the cell culture process development laboratory, the core of the system is a customized control system implementation for process control, coordination of run metadata, and automation of process execution. Users interact with simple templated Excel worksheets to deliver run parameters to the control system. A PI server archives data from the process control system and in-process analyzers; parallel data streams from pilot plant and clinical manufacturing campaigns are shared into the historian across enterprise level PI connections. PI data from active and historic runs across all scales can be queried and retrieved by PI client applications for real time visualization and comparison or ad hoc analysis. Archived data from all sources are processed through a routine to automate data interpolation, alignment, and calculations. Processed data is available in a relational SQL database for use by powerful visualization, business analytics, and analytical software.

This system has replaced Excel as the primary access point to cell culture data, reducing labor and risk of errors by automating the data management process. It has enabled flexible comparison and more valuable visualization and analysis of data across runs and scales which previously were not feasible. We envision this system to have applicability from process development studies to ongoing monitoring of manufacturing campaign data and to be expandable to be inclusive of process data from contract manufacturing facilities and product quality analytical data.

532436 Holmes: An Ontology-Based Knowledge Management System for Pharmaceutical Engineering

Abstract:

As we enter the "big data" era in knowledge management, old habits of modeling knowledge and decision-making are becoming inadequate. In the search for new modeling paradigms, we expect ontologies to play a big role. One of the critical challenges we face is the scarcity of semantically rich, properly populated, ontologies in most application domains in chemical and materials engineering. Developing such ontologies is a very challenging task requiring considerable investment in time, effort, and expert knowledge. One needs automation tools that can assist an ontology engineer to quickly develop and curate domain-specific ontologies. In this talk, we present a conceptual framework for populating scientific ontologies, and its implementation as the prototype HOLMES, as an early attempt towards such an automated knowledge management environment. Our approach integrates a variety of machine learning and natural language processing methods to extract information from journal articles and store them semantically in an ontology. In this work, identification of key terms (such as chemicals, drugs, processes, anatomical entities, etc.) from abstracts, and the classification of these terms into 25 classes are presented. Two methods, a multi-class classifier (SVM) and a multi-label classifier (HOMER), were tested on an annotated data set for the pharmaceutical industry. The test was done using two different versions of the same data set, one using the BIO notation and the other not. The F1 scores for HOMER, were better in the BIO notation while SVM performed better in the non-BIO version. However, the standard metrics did not consider the effect of the multiple answers that the multi-label classifier is allowed to obtain. As the results of our computational experiments show, while the performance of multi-label classifier is encouraging, much more remains to be done in order to develop a practically viable automated ontology-based knowledge management system.

524493 Sulfur Poisoning of SCR Catalysts

Abstract:

Over the last two decades, there have been significant advances in automotive emissions catalysis for lean-burn engines. Catalyst scientists and engineers have designed and developed new oxidation catalysts for lower temperature activity and longer term stability, new particulate matter filters and coatings for particulate oxidation, and NO reduction catalysts that can achieve high conversions in an oxidizing environment. One of the challenges that seems to perpetually trouble new catalyst design and application, especially in automotive catalysis, is sulfur poisoning. Catalysts are often oversized to handle longevity issues, and engine controls are adapted to ensure catalyst performance, sometimes through an active regeneration strategy that might include inducing sulfur release from the system. Here, I will focus on sulfur poisoning of metal-exchanged small pore zeolite selective catalytic reduction (SCR) catalysts. Specifically, Cu-SSZ-13 has become the candidate for NH_3-SCR in vehicle applications. Data clearly show that low temperature (<350 Â°C) activity of Cu-SSZ-13 can be significantly hindered by sulfur, but that activity can be readily regained. Via several groupsâ€™ efforts, this reversible poisoning mechanism has become understood, and will be the focus of this presentation. Data to be presented include those from experimental studies of S interactions with Cu^{2+} and $[Cu^{II}OH]^{+}$ and the associated SCR poisoning mechanisms - and from these experimental data, a kinetic model that describes the experimental behaviour of fresh, sulfated and regenerated catalysts.

534852 Overlayer Catalysts: Convincing Ourselves That the Observed Change Is Real

Abstract:

The binding strength of the catalyst on the products or reactants defines the success of a catalyst. The catalyst needs to bind the reactant and product strong enough to facilitate the desired reaction but not so strongly as to result in an undesired reaction or to poison the surface. Catalyst/species binding is controlled through the strain (distance) and ligand (electronic) properties of the metal. The periodic table provides a variety of element choices for catalytic application.

From an engineering perspective, the properties of certain metals for specific reactions are good, but improved performance is desired. A catalysts can facilitate a reaction of interest but a product or reactant might bind stronger than desired and block active sites. Overlayer bimetallic catalysts offer an opportunity to meet this binding strength challenge using a systematic approach. Overlayer (core@shell) can change the electronics of the overlayer and increase or decrease the binding strength depending on the choice of core metal. We have synthesized a series of bimetallic overlayer core@shell catalysts to investigate the ability of overlayer catalysts to improve reactivity when product and reactant binding issues are known to limit activity.

As we have worked on synthesizing, characterizing, and examining the reactivity of the bimetallic overlayer catalysts, we have also been repeatedly asking ourselves: how do we know this is real? What experiments can we perform that will demonstrate that the overlayer effect is real and not just something happening by chance? Repeatedly the answer has been to simply synthesize, characterize, and examine the reactivity of co-impregnated bimetallic catalysts. On multiple occasions, this simple approach has been very useful in providing experimental evidence that the bimetallic overlayer catalyst effect was real and not simply a random effect.

532717 Generating Novel Compounds through Bioprivileged Molecules

Abstract:

An important opportunity for biomass utilization is the generation of novel chemicals that cannot be readily accessed from traditional carbon sources. However, much of the effort in converting biomass to biobased chemicals has been driven by the retrosynthesis of target molecules, which limits the potential for identifying new chemicals. An alternative approach is the synthesis of â€œbioprivileged moleculesâ€ that are biology-derived chemical species that can be readily converted to a diversity of chemical products including known molecules ***and*** novel species, which can be used for next generation materials, specialty chemicals, nutraceuticals, antimicrobials, insecticides, herbicides, consumer goods, etc. This dual potentiality of a bioprivileged molecule is vital since innovative biobased chemicals represent a powerful driver for the development of biobased chemicals beyond just replacing fossil carbon with renewable carbon. Several examples of bioprivileged molecules being developed by the NSF Engineering Research Center for Biorenewable Chemicals (CBiRC) will be discussed.

525398 Interfacial Perimeter Sites in Au-TiO_2 Systems

Abstract:

The interfacial perimeter sites formed by small TiO_2 clusters on Au particles have been implicated to be active for reactions such as selective oxidation of propane to acetone and 2-propanol decomposition to acetone. Until now, there is practically no structural information on these sites. We found that it is possible to generate catalytic active interfacial perimeter sites on large Au particles (~20 nm) by decorating them with TiO_2 clusters derived from anionic, hydrolytically stable titanium (IV) complexes. Attractive interactions between charged Ti precursors and Au particle surface facilitated application of the Ti complex, which was subsequently converted to TiO_2 at 150â°C using O_3 to destroy the organic ligands. Au particles of this size exhibit Surface Enhanced Raman Scattering Resonance, resulting in substantial amplification of vibrational signals attributable to TiO_2 clusters. In combination with DFT calculations, this allowed estimating the upper limit of the size of TiO_2 clusters. Subsequently, catalytic activity of the resulting assemblies was correlated to availability of Au-TiO_2 interfacial sites. This methodology holds promise for application of small clusters of other metal oxides on metal nanoparticles in solution.

523115 Catalysis Researchers Caused Climate Change: What Can We Do to Reverse It?

Abstract:

In this lecture, I will discuss the root causes of climate change, suggest that climate change can be linked to new discoveries made by chemists and chemical engineers in the early twentieth century, and paint a vision for how chemical engineers can develop technologies that can mitigate or even reverse climate change.

532381 Insights into Catalytic Oxidation and Reduction Reactions at Metal/Solution Interfaces

Abstract:

Catalytic oxidation and reduction reactions are integral for the sustainable production of fuels, chemicals and clean water. The presence of protic solvents and the solution phase can significantly influence the catalytic activity and selectivity in metal catalyzed reduction and oxidation systems by stabilizing polar transition states. Such solvents can also directly participate in catalysis by facilitating proton and electron transfer reactions. Such reactions are pervasive throughout enzymatic, homogeneous and electrocatalysis. Much can be learned by comparing these different catalytic systems. Herein we explore the selective catalytic oxidation of alcohols to form acids and esters as well as the selective reduction of oxygen to form hydrogen peroxide and water and show the direct analogies to electrocatalytic reaction systems.

543062 Rapid Process Development - Bridging the Gap from Early Formulation Design to Integrated Continuous Drug Product Manufacture for a Dry Granulation Process

Abstract:

The adoption of integrated continuous manufacturing platforms has been pioneered in both academic and industrial environments in recent years.[1] The primary drivers for the adaption of integrated continuous drug product manufacture include: reduced processing time, reduced capital and operating costs, minimized scale-up risks, more consistent process control achieved via automation and opportunities for rapid process development.[2,3] Where the drug product is an oral solid dosage form, continuous tablet manufacture is made possible by a defined sequence unit operations that operate continuously â€“ e.g. continuous powder feeding, continuous powder blending, continuous granulation, milling, tablet compression and coating. Ensuring the integrated operation of such unit operations is critical to the successful implementation of a continuously manufactured drug product.

Early formulation design and development is often performed on a small scale where discrete unit operations that are either performed in batch mode or inherently continuous in their operation (e.g. tablet compression, roller compaction, etc.) are used to produce the product. Significant effort is required to translate these non-integrated unit operations to integrated continuous unit operations working in concert in order to produce the final product for commercial manufacture.

In this work, we present the necessary workflow highlighting considerations and efforts needed to transfer the knowledge from early formulation design through to a fully integrated continuous dry granulation process, specifically, excipient and API powder through to coated tablets. The continuous dry granulation process is described. The methods used for non-integrated testing of unit operations are outlined. The connection between non-integrated testing and measurement and the selection of appropriate process parameters is discussed. Product control strategy and residence time distribution measurements are highlighted. Early considerations for the generation of a desired manufacturing range in line with critical quality attributes are also discussed. A process workflow is generated that provides a means for rapid process development from early formulation design to Integrated Continuous Drug Product Manufacture for a Dry Granulation Process.

References

1-Moghtadernejad, S. *et al.* A Training on: Continuous Manufacturing (Direct Compaction) of Solid Dose Pharmaceutical Products. *Journal of Pharmaceutical Innovation* **13**, 155-187, doi:10.1007/s12247-018-9313-5 (2018).

2-TeÅ¼yk, M., Milanowski, B., Ernst, A. & Lulek, J. Recent progress in continuous and semi-continuous processing of solid oral dosage forms: a review. *Drug Development and Industrial Pharmacy* **42**, 1195-1214, doi:10.3109/03639045.2015.1122607 (2016).

3-Byrn, S. *et al.* Achieving Continuous Manufacturing for Final Dosage Formation: Challenges and How to Meet Them May 20â€“21 2014 Continuous Manufacturing Symposium. *J. Pharm. Sci.* **104**, 792-802, doi:http://dx.doi.org/10.1002/jps.24247 (2015).

525421 Developing a Loss-in-Weight Feeder Design Space Based on Performance and Material Properties

Abstract:

Abstract

Loss-in-weight feeders play a vital role in controlling content uniformity in continuous manufacturing. If, in a brief period of time, the feed rate of one ingredient changes with respect to the others, the change in concentration of the process stream can propagate throughout the system. Understanding how the feeder and the material behave is vital. Powder properties, such as density and cohesion, can cause large variability in the flow rate of ingredients fed from powder feeders. Also, one needs to understand the operational ranges of the feeder, i.e. percent motor power, which coincides directly with screw speed. Knowing an ideal range of operation and correlating powder properties to process performance can lead to determining the design space of the feeder accurately, and to faster optimization of feeder performance, which also saves money and material. In this work, a K-Tron KT-20 pharmaceutical loss-in-weight feeder has been evaluated with four common screw types (coarse auger, fine auger, coarse concave, fine concave), for various pharmaceutical (Avicel, Lactose Monohydrate, Magnesium Stearate, Acetaminophen, Prosolv, etc.) and catalytic powders, at multiple feed rates.

529956 Distillation in the Pharmaceutical Industry

Abstract:

Distillation in the pharmaceutical industry

Distillation is the process of separating the components or substances from a liquid mixture by selective boiling and condensation to increase the concentration of selected components of the mixture. The process exploits differences in the volatility of the mixture's components and this physical separation makes distillation a unit operation of practically universal importance.

Distillation plays an important role in the pharmaceutical industry where single stage batch distillations are frequently used and not the multistage distillation in columns studied at university. Distillation can be operated in batch or continuous modes, in single or multiple steps (put and take), and can be run for a variety of objectives: separation, solvent swap, drying from water, purification from volatile or non-volatile impurities and crystallization.

Distillation within the pharma industry is often used to solvent swap because the reaction solvent is usually different to the crystallization solvent or to concentrate a solution before crystallization. They offer many challenges, the environment of the Active Pharmaceutical Ingredient (API) and intermediates need to be controlled and the solubility limits known, impurity formation needs to be restricted by controlling the temperature and distillation time. Also, it is important to consider that dissolved species can affect phase equilibria.

Distillation can achieve different results depending on the conditions and equipment selected. There are a range of equipment and operating choices used at production and laboratory scale but often these are limited by the constraints from the chemistry involved. Phase equilibria is independent of scale, but other factors such as heat losses, vacuum or minimum and maximum stir volume are not. Distillation times can be increased considerably on plant, therefore it is difficult to scale up using experimental information. Phase diagram understanding is fundamental for distillation and simulations help in the decision-making process for optimal conditions and process specific requirements.

As an API moves through development, the scale and site of manufacture is subject to change. Understanding distillation and de-risking this operation by analysing the impact of different parameters such as pressure and equipment capability is fundamental in the technical transfer process. An overview on this fundamental unit operation focussing on different case studies showing continuous distillation, crystallization distillation and process optimisation through simulations will be given showing distillation lessons learn from the past and learnings for the future.

One case study will show a solvent swap simulated with Aspen Properties & Batch Modeller and demonstrate how modelling led to an improved process reducing both cost and environmental burden

Another case study will show how a put and take distillation which led to product oiling / gumming, was changed to a constant volume operation significantly improving robustness and product quality.

532501 Maintenance Management for the Sensor Network in Continuous Pharmaceutical Systems

Abstract:

The progress in the adoption of monitoring and control systems has accelerated the progress towards smart manufacturing in the pharmaceutical industry. Smart manufacturing necessitates reliable and accurate measurements from field devices combined with good quality estimates of unmeasured process variables. Thus, the sensor network which consist of equipment measurements, field devices, PAT tools, process models and data analytics is a fundamental requirement for robust process operations.

Data reconciliation exploits the redundancy in the sensor network together with process models to ensure the validity and accuracy of process measurements. The effective implementation of data reconciliation requires reliability in the raw measurements from PAT tools. However, since secondary pharmaceutical manufacturing primarily involves processing particulate materials, fouling of the inline and online PAT tools is inevitable. Corrective maintenance of fouled sensors can cause operational losses and uncertainties in manufacturing systems. Moreover, regulatory filings would require sensor network strategies for handling sensor faults where multiple sensors could fail simultaneously. For a smart pharmaceutical facility, it is imperative to anticipate and correct problems with the sensor network instead of only reacting to problems.

In this paper we investigate the impact of predictive maintenance of the sensor network in continuous tableting processes. The maintenance policy seeks to maintain measurement bias within acceptable limits such as to satisfy the conditions required for data reconciliation, using mechanistic models and material balances. The integrated framework schedules the anticipated downtime in maintaining every sensor â€“ disconnecting the sensor or a group of sensors, cleaning the sensors and connecting it back to the sensor network. It is important to note here that the maintenance policy depends on the physical setup, materials used, the sensor network of the process and must rely on and learn from process history.

The challenges of implementing a robust and reliable sensor network are discussed, followed by demonstration of the effectiveness of the maintenance framework for a direct compaction case study. By integrating predictive maintenance with data reconciliation, the framework enhances the reliability in the measured estimates of CQAs for practical realization of Real Time Release Testing.

References

1. Smart Process Plants, M. Bagajewicz (2010)

533431 Leveraging Mechanistic Models for Scale up and Optimization of Lyophilization

Abstract:

Leveraging mechanistic models for scale up and optimization of Lyophilization.

Abstract

Generic pharmaceutical industry has to operate in tight timelines of product development and scale up from lab to plant where time to market makes or breaks the viability of the business case. Getting it right the first time demands detailed product & process understanding and implementing appropriate mechanistic modelling and Quality-by-design (QbD) approaches in the product life cycle, to minimize downstream risks. This work describes example cases of such efforts involving the freeze/ vacuum drying unit operation with solvents.

Lyophilization or freeze drying with aqueous/ solvent systems is a routine, but expensive, unit operation carried out across a range of industries (pharmaceuticals, food, paper, etc). It's especially relevant for temperature sensitive materials (for ex., peptides, liposomes, and onco molecules) where traditional high temperature drying unit operations cannot be deployed. The process involves loading the solutions of the required products in vials or trays and drying to obtain typically a free flowing powder (without meltback or collapse or oiling out). Typical challenges in this process involve: defining the optimal drying conditions (temperature, pressure vs. time) for a given material on a certain equipment so that product of an acceptable quality is obtained; ensuring uniformity of this quality across the vials in a given batch; and translation of such conditions during scale-up from lab scale to plant scale. For both the aqueous and solvent based systems, this demands a quantitative understanding of the impact of the complex interactions (Figure 1) of the relevant material properties, equipment characteristics, process conditions on the quality of the output

Figure 1 – Lyophilization- CMAs, CQAs, and CPPs.

product. This is especially true as the cost of failure at plant scale is prohibitive as this is typically the last step in the manufacturing of these products. Substantial literature exists (for example, M.J. Pikal, et al 1984, Pisano et al 2013) in capturing the vial-level physics with quasi-steady state assumptions to describe aqueous systems and demonstrating the applicability. These are solved using one-dimensional coupled heat and mass transfer equations.

The present work focuses on leveraging vial and equipment level physics to help with the definition and scale-up of lyo cycles in aqueous and solvent based systems. The first part focuses via case studies on how thermal characterization of lyophilizers can be leveraged to: bring predictability in scale-up (from 100s of vials to 1000s of vials) and; optimize lyo cycles using a novel algorithm. Example scenarios show such optimization lowering the drying time by as much as 20% compared to the traditional cycle development strategy. The second part presents cases of how such efforts can be extended to solvent systems by incorporation of additional physics. The higher vapor pressure of the solvent based systems demands more attention to the downstream utilities such as condenser and vacuum pump during scale up.

522002 Incorporation of Process Intensification Options in a Fed-Batch Cell Culture Manufacturing Platform

Abstract:

Several cell culture process intensification approaches were explored at AbbVie for increasing the productivity and processing speed of our protein biologics, with minimal impact on product quality profiles. Included in this are: the use of 10 mL cryovials to shorten the duration of seed train expansion, the use of perfusion technology to facilitate high cell densities at the seed bioreactor stage, and the use of perfusion and alternating tangential flow technology to facilitate increased protein yields at the production bioreactor stage. These approaches were demonstrated individually across all cell culture unit operations at laboratory and pilot scales. The work detailed in this poster provides a snapshot of the future of our biologics manufacturing capability as we continue to incorporate â€œpieces of the puzzleâ€ of process intensification for our biologics pipeline.

521581 Round Granules of Dimethyl Fumarate By Three-in-One Intensified Process of Reaction, Crystallization, and Spherical Agglomeration in a Common Stirred Tank

Abstract:

Pure, isomorphic, round, and free-flowing dimethyl fumarate granules in a size range of 250−2000 μm were successfully produced directly from esterification through the three-in-one intensified process of three distinctive steps of reaction, crystallization, and spherical agglomeration (SA) in a 0.5 L sized jacketed glass stirred tank. Dimethyl fumarate was prepared by sulfuric acid catalyzed esterification of fumaric acid with methanol. The reaction temperature was below the maximum allowable limit of 65 °C as determined by reaction kinetics to avoid the runaway situation. The dissolution rate of primary crystals of dimethyl fumarate was inversely proportional to the particle size which was strongly affected by the antisolvent addition and temperature cooling schemes during crystallization. However, the dissolution rate of the round granules was mainly dependent on the exterior dimension of the granules and not so much on the primary crystal size inside the granules. The mechanical properties such as density, porosity Carr's index, friability, and fracture force of round dimethyl fumarate granules generated in (1) three-in-one processes with the final temperatures at either 5 or 25 °C (Three-in-one I and II) and (2) SA of dimethyl fumarate, which was done separately and disconnected from the train of reaction and crystallization process at either 5 or 25 °C (SA I and II), were thoroughly studied and compared. The concept of scale-up for Three-in-one I and II was also verified in a 10 L sized jacketed glass stirred tank. Powder manufacturability such as flowability, blend uniformity, and compressibility had been substantially enhanced by spherical agglomeration. The added values of free-flowing and easy-to-pack properties to dimethyl fumarate in addition to its original intrinsic slip planes in the crystal lattice would make direct compaction into tablets feasible.

534029 Systematic Approach to High Dosage Formulation Development for Continuous Direct Compression

Abstract:

Continuous direct compaction for high dosage formulations is a challenging topic. The understanding of the raw material properties and identification of the critical material and process parameters is key to predicting a continuous manufacturing process and implementing control parameters. A systematic strategy has been developed to enable formulation and process design for poorly flowing active pharmaceutical ingredients (API). Currently, high dosage formulations are manufactured mainly via wet granulations including various processing steps. The length of process makes it more time consuming, costly and challenging to minimise process deviations relative to direct compression A case study using a poor flowing model API at a high drug loading (~ 50 wt%) was used to evaluate the strategy. The approach adopted is material sparing, targeted and time efficient. The accumulated material database can be used to expedite future developments. A series of characterisation techniques have been used to analyse various pharmaceutical powders to identify their flow and compaction behaviour. This information was used to predict blend behaviour and inform formulation design. Independent lab scale predictors of manufacturing performance have been developed on a benchtop compaction simulator. The success of this strategy has been shown by scaling up to an R&D rotary tablet press maintaining all critical quality attributes (CQAs).

531789 Comparing High Drug Loaded Blend and Tablet Property Improvement Via Various Nano-Silica Dry Coating and Excipient Selection

Abstract:

It has been shown that dry coating cohesive active pharmaceutical ingredients (APIs) with nano-silica can improve packing and flow of their blends, facilitating high speed direct compression tableting. This paper examines the broader scope and generality of previous work by examining three fine APIs; micronized Acetaminophen (mAPAP), coarse Acetaminophen (cAPAP) and micronized Ibuprofen (mIBU), and considers dry coating with both hydrophobic or hydrophilic nano-silica to examine the effect not only on packing density and flow of their blends, but also dissolution and tensile strength of their tablets. The impact of the excipient size on blend and tablet properties are also investigated, indicating blend flow is most improved when matching API particle size with excipient particle size. In all cases where the API is dry coated, the blend packing and flow improve, so as to suggest such high drug loaded blends could enable direct compression. Using dry coated API along with finer excipients in blends lead to improved hardness of the corresponding tablets. Interestingly, dissolution profiles show dry coated API tablets generally have faster dissolution rates, regardless of silica hydrophilicity, suggesting API powder deagglomeration via nano-silica coating plays a crucial role. The most significant conclusion is that, although there are differences in properties of blends that depend on the API, hydrophobic or hydrophilic nano-silica coating, as well as large or fine excipients, in all cases, dry coating of APIs significantly improves the possibility of using the specific blend at high drug loading in direct compression tableting.

530191 Evaluation of Segregation Intensity of Pharmaceutical Blends Using Near Infrared Spectroscopy

Abstract:

Segregation of components, especially of an active pharmaceutical ingredient (API), in a pharmaceutical blend during secondary manufacturing processes is considered to be one of the major bottlenecks in achieving tablet content uniformity, ultimately leading to batch rejections. The main reasons for segregation such as differences in particle size, density, shapes, solid-solid interactions and segregation induced due to external interferences such as gravitational, shear or vibrational forces are well established, but there are limited tools available to quantify the API segregation intensity at a manufacturing scale. Determining segregation intensity using spectroscopic technique can help in identifying segregation risks early on in product development. This research work focusses on quantifying segregation intensity of pharmaceutical blends using SPECtester (an instrument with an in built near-infrared (NIR) capability).

The instrument has two different modules capable to induce segregation to the pharmaceutical blends. The first module uses vibrational force and the second module uses compressed air to fluidize the particles in the system. In the first module, a simulating condition to that of large scale manufacturing process was created by dropping the blend from the highest point (high drop chute -16.69â€)of the equipment into the NIR testing section of the instrument. Around 0.8 to 1.25 liters of the blend is required to fill the hopper. The material was dropped from the hopper through the feed chute to form pile using vibratory feeder. The speed of material flow in the chute was controlled by regulating the frequency of vibration. In the second method, the blends to be analyzed were placed in the Fluidization Analysis Module (FAM) test hopper. Appropriate fluidization settings (the flow required to fluidize the material) were selected and it was confirmed that materials reached fluidized bubbling regime. The pile height of material after fluidization were recorded and fed in the SPECTester software. The NIR analysis methodology in both these segregation modules were followed in the same way. The system allows the measurement of blend containing up to 6 components. Trays filled with individual components were placed adjacent to the testing hopper section. These component trays were used for collecting NIR spectra of pure samples of the components in the mixture. The instrumentâ€™s operational parameters such as pile height, the size of the view box (window) to scan, and number of measurements in a square box were optimized to suitable values to acquire segregation intensity data. Spectra of the blend was taken about 6 mm below the pile surface horizontal to the top. SPECTester primarily measures the concentration of components at these discrete number of square spaces in the poured pile and collectively computes the segregation intensity. The instrument superimposes the spectra of pure components on the spectra of components of the segregated blend and the concentration of components in the blend in each of these data points or measurements were then computed by tracking the changes in the absorption intensities in each of these spectrum.

Formulation blends using different types of APIs with different particle size ranges were explored. 200gm of blend weight was found to be adequate to ensure reliable data collection. Blends of these APIs with the excipients were prepared by mixing the constituents for 15min in a Turbula blender (This formulation composition was also used to make tablets in a manufacturing scale by direct compression process and to further study its content uniformity).The segregation induced due to the above mentioned modules were then evaluated for individual APIs and excipients at different API concentrations (w/w) ranging from 2 to 15%. The concentration profile of formulation components (along the pile length) were also determined. API segregation intensity (obtained using SPECTester) for corresponding formulation blends were compared with tablets content uniformity produced in the manufacturing scale.

The measured segregation potential corroborated with the content uniformity results. In other words, formulation blends with high segregation potentials generated tablets having higher deviation from the content specifications. Hence, this small scale prediction tool assessed the segregation potentials of formulation blends in an early formulation development phase and also proved its ability to predict the segregation issues experienced at large scale manufacturing.

521895 Powder Flow Analysis Using an Automated Milli-Scale Powder Flow Test

Abstract:

A novel milli-scale powder flow testing method has been developed that allows an understanding of the active pharmaceutical ingredient (API) powder flowability earlier in particle engineering or API development and with only 1 gram or less of material. The milli-scale flow test method requires as little as 50 mg of material per replicate, leverages automation technology, and provides a new tool that pragmatically evaluates powder flow characteristics. Results demonstrate discrimination between a broad range of materials. Least squares regression modeling calibrates the milli-scale mass flow rates to USP standardized powder property tests. The level of flow risk is determined by a comparison of the flow rates to the Hausner ratio. A case study will be presented showing the ability of the test to identify flow risks of numerous lots of API manufactured at different crystallization and milling conditions executed through development and optimization experiments.

525142 Understanding the Compression Behavior of Blends: The Application of Percolation Threshold Theory and Multivariate Analysis

Abstract:

Introduction

Quality assurance of drug product manufacture and requires strategies to manage raw material variability, thereby ensuring consistent processing and product quality. Problems related to raw material variability are often encountered following development, during routine production. At this point, formulation modification is undesirable due to regulatory complexity and related costs. Hence it is important to design formulations that are robust with respect to raw material variability encountered during production.

Aim

The aim of this project was to apply the percolation threshold theory to determine a critical drug concentration. Above this critical concentration, given as percolation threshold, the level of drug substance could undermine the compactability of the blend, and therefore it should be avoided for a robust formulation of a direct compression blend. Multivariate analysis of compression data for formulations above and below the calculated threshold value was employed to demonstrate its validity.

Experimental

Blends of Ibuprofen and microcrystalline cellulose (MCC) were compacted under fourteen different pressures between 20 MPa and 400 MPa in a Piccola rotary tablet press. Two MCC grades with differing physical properties were employed: Vivapur® 102 (air stream dried quality), and Emcocel® 90 (spray dried quality) in order to evaluate raw material variability effects. Ibuprofen and MCC were analysed for particle size distribution (Malvern 3000), particle morphology (Morphologi G3-ID), bulk and tapped density and true density (Accupyc). Tablet blends containing 0%, 2.5%, 5%, 10%, 15%, 20%, 30% and 40% w/w ibuprofen were prepared. The blend with 40% ibuprofen did not flow from the hopper (rat-holing), therefore could not be compacted.

A range of mathematical models were implemented. Firstly, compaction triangle and out-die Heckle model were applied for screening direct compression behavior. Secondly, the percolation threshold theory was employed. This is a statistical model that has been identified to analyse and control raw material variability and its effect on tableting process during development. The model outputs a solid fraction range where properties of the blend undergo changes, and therefore should be avoided for robust tablet production and performance. Percolation threshold ($\ddot{I}_c$) is mathematically given by $X = S\hat{A}\cdot(\ddot{I}-\ddot{I}_c)^q$ (Leuenberger, 1999), which describes the behavior of a property *X* when the relative density ($\ddot{I}$) varies. Tensile strength was the reference property chosen due to its importance for tablet quality control. Coefficient q is dependent on the blend. At last, multivariate analysis was carried out in order to link compression behavior and the value of percolation threshold modeled by using compaction data to compare the formulations below and above the threshold.

Results

As ibuprofen concentration was increased in the direct compression blend, tabletability and compactibility decreased and compressibility increased. Comparing MCC grades, both grades revealed similar performance. The yield pressures determined using the Heckel model were 167 MPa for Vivapur and 182 MPa for Emcocel. Particle size and morphology were the main differences observed between these MCC grades. Morphologi G3-ID analysis showed that Vivapur had a greater number of needle shaped particles compared to Emcocel. Laser diffraction confirmed the particle size differences between grades.

For percolation threshold modelling, the coefficient q had not been previously determined for blends of Ibuprofen/MCC. Therefore, one of the goals of this study was to model the coefficient q, based on a modified Heckel equation. The value calculated was $q = 3.5 \pm 0.2$ and there was no significant difference between the values found for the different MCC grades. Percolation threshold was modeled, not only with the empirical value from the previous step, but also using a theoretical coefficient (Guyon et al., 1987) for comparison. The best fits were obtained with the empirical coefficient resulting in the percolation threshold as solid fraction of 0.2056 for Vivapur and 0.1720 for Emcocel. An additional calculation transforms this value in drug mass concentration according to dilution capacity theory, which gave the critical mass fractions of 15.6% w/w and 17.5 %w/w ibuprofen in Emcocel and in Vivapur, respectively.

A principal component analysis (PCA) using the algorithm Singular Value Decomposition was performed for all the formulations, including the blanks of excipients using the software The Unscramble X 10.4.1, CAMO. The model performed applying full cross validation with 42 segments was proven efficient to distinguish between samples below and above the percolation threshold. Above the percolation threshold calculated, the tensile strength was lower, flowability poorer and tablet weight variability and compressibility higher, which shows that the overall properties of the blend were overwhelmed by ibuprofen and proves the percolation threshold determined to be correct. The model could also distinguish between blends containing Vivapur or Emcocel. These blends mainly differed in bulk and tapped densities, with Emcocel being denser than Vivapur.

Conclusions

The percolation coefficient for ibuprofen/MCC binary blend was calculated and the percolation threshold successfully modeled. Excipient variability had minimal impact on the percolation threshold determined. PCA analysis of compaction data demonstrated that the threshold value determined was correct. Percolation threshold modelling was shown to be a useful approach for developing robust tablet formulations.

534828 Multi-Scale Modeling in Immuno-Oncology to Support Immunotherapy Drug Development

Abstract:

The innate and adaptive immune responses are complex and encompass a large range of spatial (from molecular interactions to cellular processes and organ functions) and temporal (from milliseconds to minutes and years) scales. Multi-scale modeling of the immune system provides a wholistic view of emergent properties from the dynamic interplay of components across levels of organization. Such models can be used to qualitatively and quantitatively address various immunological questions in support of the development of novel antibody-based constructs and adoptive cellular therapies (e.g., dendritic cell vaccine, chimeric antigen receptor T-cell, and checkpoint therapies). In addition, studying the implications of dynamical interactions between immune cells and cancer cells may provide a basis for discerning individual patient responses to immunotherapy. Multi-scale mathematical models can be utilized to guide hypothesis generation and testing in drug discovery and clinical development in order to discover novel immunotherapeutic targets and rationally design combinatorial treatment strategies. However, there are only a limited number of multi-scale models of immune responses that reach this important goal of providing new insights into the conception, design, and implementation of these promising immunotherapeutic strategies. In this survey, we discuss the current development of multi-scale modeling in immuno-oncology, and in particular, we provide an overview of the current multi-scale models that may be used to identify novel drug targets and combinatorial approaches for the treatment of cancer. A strategic use of multiple model types will be presented.

534369 Evaluation of Microfluidic Device Designs for a Potassium Release Toxicity Assay

Abstract:

With the integration of mixing systems, microfluidic devices offer potential applications in the fields of reactor design and assay development [1]. This project investigates key design parameters of commonly-used mixing systems to optimize a microfluidic device for use with the potassium release toxicity assay.

The potassium release toxicity assay offers a measure of drug toxicity by exposing red blood cells (RBCs) to various concentrations of drug solution; as RBCs are exposed to more toxic environments, levels of potassium released from the cells increase towards a maximum value [2]. Drug toxicity is quantified by the drug concentration corresponding to half-maximum potassium release.

Insufficient mixing of RBCs and drug solution can artificially reduce the drug toxicity determined by this assay. Additionally, inconsistent mixing between trials can cause random fluctuations in assessed drug toxicity. Incorporating the mixing step of the potassium release toxicity assay into a microfluidic device can ensure that assessment of drug toxicity is conducted consistently, even with low fluid volumes.

During the study, CFD simulations in COMSOL Multiphysics are being used to simulate flow of RBCs and drug solutions through various device designs. Among the design parameters being tested are the lengths, widths, and shapes of channels and baffles integrated into the device. Following simulation, fluid velocities, compositions, and pressure drops will be collected and assessed to quantify the mixing efficiency accompanying each design. The combination of design parameters corresponding to the greatest mixing efficiency will be recommended for microfluidic device manufacturing and use in the potassium release toxicity assay.

[1] Losey, M.; Jackman, R.; Firebaugh, S.; Schmidt, M.; Jensen, K. *Journal of Microelectromechanical Systems* **2002**, *11* (6), 709â€"717.

[2] Jensen, G. M.; Skenes, C. R.; Bunch, T. H.; Weissman, C. A.; Amirghahari, N.; Satorius, A.; Moynihan, K. L.; Eley, C. G. S. *Drug Delivery* **1999**, *6* (2), 81â€"88.

538292 Effects of Unsaturated Phospholipid Dilinoleoylphosphatidylcholine on Degradation of Phospholipid Vesicles Catalyzed By a Model Phospholipase A2

Abstract:

Secreted phospholipase A2 ($sPLA_2$) is a lipolytic enzyme that catalyzes the degradation of phospholipids and is widespread in the venoms of insects and reptiles and extracellular spaces of mammalian tissues. Research has shown that dysregulation of $sPLA_2$ is related to a variety of inflammatory diseases. The intrinsic relationship of in vivo $sPLA_2$ levels to certain diseases can be exploited for medical diagnosis and treatment. For example, overexpression of $sPLA_2$ in cancerous regions is a potential trigger for increasing local release of drugs from lipid-based drug carriers. Understanding the interactions of phospholipid membranes with $sPLA_2$ is necessary to reveal the roles it plays in related physiological processes and to design lipid vesicles for drug delivery. To date, the research on $sPLA_2$ activity on membranes has mainly focused on the saturated phospholipids. The interactions of unsaturated phospholipids with $sPLA_2$ are also essential in the human body, yet there is a lack of systematic in situ investigations of these interactions.

In this study, dilinoleoylphosphatidylcholin (DLPC) was used as a model unsaturated phospholipid. Its acidic degradation product, linoleic acid (LA), is one of the two essential fatty acids and is important to our inflammatory defenses. Through inclusion of DLPC in the dipalmitoylphosphatidylcholin (DPPC) saturated membranes, the effects of unsaturation on the degradation of giant unilamellar vesicles (GUV) were investigated by recording the vesiclesâ€™ deformation using fluorescence microscopy. The mechanisms of the observed differences in the degradation of the DPPC and DPPC-DLPC GUVs were analyzed by comparing the phase behaviors of a series of DPPC-DLPC monolayers (x_{DLPC} =0, 0.2, 0.4, 0.63, and 1) using a Langmuir trough-coupled fluorescence microscope. The Angstrom-length scale analysis of the interfacial packing structures of these films before and after degradation was conducted using X-ray surface scattering techniques.

We found that inclusion of unsaturated phospholipids in the membranes not only enhanced the $sPLA_2$-induced degradation of vesicles, but also changed the vesicle deformation processes and the productsâ€™ configurations. The higher enzyme activity on DPPC-DLPC GUVs compared to DPPC GUVs resulted from the differing phase behaviors and molecular packing structures of DPPC and DLPC moleculse at the interface. On the other hand, the differing deformation processes and final configurations of DPPC and DPPC-DLPC GUVs were related to the distinct interfacial organization of the acid degradation products of DPPC and DLPC. These study findings advance the understanding of the influence of unsaturated phospholipids on the interaction of phospholipid membranes with $sPLA_2$ at the molecular level. Thus the study sheds light on the roles of $sPLA_2$ in related physiological processes and supports development of a systematic mechanism-based approach for designing lipid-based drug delivery systems.

536754 Peptoid JPT1A Reduces RAGE Expression and Attenuates Inflammatory Response: A Potential AD Therapeutic

Abstract:

Introduction: For several decades, amyloid b (Ab) has been a key therapeutic target for Alzheimer's disease (AD) related therapies, but more recent studies have recast Ab as one of several participants in the disease rather than its sole etiology. AD therapies that aim to inhibit Ab production and/or aggregate formation alone have yielded little success in clinical trials. At the same time, Phase II clinical trials that target AD-associated inflammation using an antagonist for the receptor for advanced glycation end products (RAGE) demonstrate improved outcomes in patients with mild to moderate AD. We envision that a therapeutic capable of affecting both of these targets would exhibit enhanced efficacy.

A peptoid mimic of Ab's KLVFF hydrophobic core, JPT1a, was designed as a potential peptoid therapeutic for AD, with prospective capabilities to both antagonize RAGE via ligand mimicking sequences and inhibit aggregation of Abvia self-recognition. Peptoids are peptidomimetics that attain invulnerability to proteolytic degradation through repositioning of the side chain from the a carbon to the amide nitrogen. This invulnerability to proteases, as well as other qualities such as diminished immunogenicity, enhanced cellular permeability, and capacity for intranasal administration, make peptoids immensely attractive as neurotherapeutic agents. We previously demonstrated that JPT1a has the capacity to modulate Ab aggregation as well as alter the morphology of Ab aggregates. In the current study, we examine JPT1a for protection against inflammatory responses associated with RAGE signaling.

Materials and Methods: THP-1 monocytes were differentiated into macrophages through exposure to phorbol 12-myristate 13-acetate (PMA) for three days prior to experimentation; THP-1 monocytes differentiated in this manner closely resemble the behavior and phenotype of primary human monocyte-derived macrophages. To assess the capacity of peptoid JPT1a to modulate RAGE expression, differentiated cells were incubated (3 days) with 0 -50 µM JPT1a in the presence of low-dose LPS (2 ng/mL) or Ab_{1-42} oligomers (0.01 mM). The capacity of peptoid JPT1a to reverse RAGE expression was assessed in THP 1 macrophages following chronic (48-h) exposure to low dose LPS (2 ng/mL) or Ab_{1-42} oligomers (0.01 mM) , at which time media was removed and replaced with media containing 25, 5, 1, or 0 mM JPT1a (24-h incubation). Following each treatment, supernatant was harvested for cytokine analysis via ELISA and cells were stained for quantitative image analysis of RAGE expression.

Results and Discussion: Co incubation of the peptoid JPT1a with a chronic pro inflammatory stimulus (LPS or Ab oligomers) significantly reduces RAGE upregulation in a dose dependent manner within this model that utilizes THP 1 macrophages (Figure 1A). Upon ligand-binding, RAGE can activate multiple signaling pathways to influence the magnitude and character of the immune response, as well as initiate a positive feedback cycle that upregulates its own transmembrane expression. Membrane-bound RAGE expression is altered as part of a feed-forward mechanism for environmental stress or exposure; therefore RAGE is a means through which cells anticipate and interpret their local environment, and RAGE may also be a source of dysfunction in chronic inflammation. In parallel, JPT1a also reduces the production of pro-inflammatory cytokines IL 1b, IL 6, and IL 8 within this model (Figure 1B – D). Addition of JPT1a to THP 1 macrophages pre treated with chronic pro-inflammatory stimulus (LPS or Ab oligomer) also significantly reduces RAGE expression but does not significantly impact the generation of pro inflammatory cytokines.

Conclusions: When paralleled by the ability of JPT1a to reverse RAGE expression, these results demonstrate the potential of JPT1a as a dual-target therapy in AD, modulating both inflammation and Ab aggregation. Moreover, as RAGE has also shown promise as a therapeutic target in several pathologies, a RAGE inhibitor such as JPT1a may yield a new therapeutic option to a wide array of illnesses.

533476 Rapid and Efficient Development of Downstream Bio-Pharmaceutical Processing Alternatives

Abstract:

Downstream processing in pharmaceutical and bio-pharmaceutical production is becoming increasingly important due to the higher product yields desired as the minimum operating cost. As such, from a process development point of view, there is a need to rapidly and efficiently synthesize optimal downstream processing routes while taking into account a very narrow time line between approval of a product from a regulatory body (such as FDA) to the beginning of full scale production. This work in particular focusses on the design and synthesis for the recovery and purification of bio-pharmaceutical molecules produced through fermentation processes. To this end, this work adapts and extends the domain of application of the process synthesis methodology based on thermodynamic insights[1] for traditional chemical processes.

The first task in the methodology is to deal with a mixture analysis which allows to immediately screen out some separation techniques that are not suitable with the type of mixture considered. A binary ratio matrix is then computed to store the properties of the binary pairs of molecules: this method allows in determining the extent to which a separation technique is feasible for a binary separation task. As such this step allows the identification of all physically feasible separation techniques and the first separation tasks. Further screening on separation techniques can be done considering the conditions of operation (temperature and pressure) of single unit operations. This means that a particular separation technique, even if considered as physically feasible in previous steps, can be rejected whether extreme condition of temperature or pressure or both of them need to be achieved to carry out the operation. In the second and final step in the methodology, further insight on the mixture properties are employed to find a set of physically feasible process pathways with an estimation of the conditions of operations for each flowsheet. The outcome of the work is thus to design a sequence of feasible processing routes. After the process synthesis, models of the different unit operations are used to simulate the process and to predict the behavior of the system. The methodology was then been applied for the synthesizing the downstream processing pathway of Lovastatin a molecule that is mainly exploited as an anti-hyperclorestemia drug.

(1) Jaksland, C. a.; Gani, R.; Lien, K. M. Separation Process Design and Synthesis Based on Thermodynamic Insights. *Chem. Eng. Sci.* **1995**, *50* (3), 511â€“530.

523821 Pump-Down the Mycobacterium Tuberculosis: A DNA Gyrase/P-Glycoprotein Combined Inhibition Approach

Abstract:

Tuberculosis (TB) is one of the most common infectious diseases worldwide, with about one third of world population infected with Mycobacterium tuberculosis (M.Tb.). More frightening is the recent emergence of multidrug and extensively drug resistant M.Tb. strains (MDRM.Tb. and XDRM. Tb., respectively). In the setting of drug resistance or intolerance to firstline agents (e.g. isoniazid, rifampicinâ€¦), secondline agents may be used. Indeed, fluoroquinolones have been classified as secondline antituberculous drugs since they are active on isolated M. Tb. expressing resistance to both isoniazid and rifampicin. Recently we synthesized a new series of quinolones derivatives, endowed with a good antimycobacterial activity (against H37Rv and H37Ra and further clinical isolates of MDRTB/ XDRTB strains) coupled with low cytotoxicity. In this work we present a combined in vitro/in silico approach with the aim (i) to improve knowledge on the structure-activity relationship (SAR) of the new compounds, (ii) to identify a pharmacophoric map and (iii) enhance the biological activity against the biological target. According to the computational results, new derivatives will be designed and synthesized to obtain compounds able to inhibit replication in H37Rv and clinically isolated M.Tb. strains. The identification of a new action mechanism by our derivatives compared to classical quinolones could be used to generate a new class of drugs, with the aim to create an innovative treatment plan able to reduce possible pharmacological resistances. In parallel a series of new quinoxaline derivatives, previously identified as Pglycoprotein (Pgp) inhibitors, will be investigate on M.Tb. strains, in combination with new compounds and classical drugs. In fact, efflux pump inhibitors could enhance a latent biological activity and implement the DNAgyrase inhibitors activity against resistant M. Tb. strains. Even in this topic the computational and the experimental tecniques will be applied in order to improve the level of knowledge on the mechanism of action of these molecules againt their biological target. All obtained data will be analyzed and promising compounds will be optimized to obtain selected gold compounds.

535990 Continuous Processing of Doxorubicin-Loaded Liposomes

Abstract:

PURPOSE

Complex dosage forms such as liposomes have tremendous potential to revolutionize the pharmaceutical industry and provide significant health benefits. One challenge is to process large quantities of these nanoparticles while also maintaining physicochemical properties to ensure a high-quality formulation. One approach to overcome challenges was to turn to continuous processing. Continuous processing can achieve improved process control (*via* process analytical technology), while avoiding inherent issues associated with batch processing to ultimately improve product quality and reduce manufacturing cost. Accordingly, we developed a continuous process for liposomes. In this work, we outline important steps in the particle formation process, in-line/ at-line measurements, in-line concentrating and drug-loading of doxorubicin-HCl.

METHODS

Risk Assessment

Ishikawa diagrams were formed to outline process parameters and material attributes that had the likelihood to effect quality attributes. Some common critical quality attributes (CQAs) for nanoparticle formulations include particle size distribution (z-average particle size and polydispersity index), zeta-potential, nanoparticle concentration and encapsulation efficiency. A custom-built system was designed to incorporate process control strategies to ensure CQAs were within user specifications.

Nanoparticle Formation Process

The nanoparticles were formed using an ethanol injection method with a custom-built continuous processing system that is controlled using a program written using National Instruments LabVIEW. Briefly, the continuous processing system consists of multiple tanks that contain lipid dissolved in ethanol and another set of tanks for the aqueous components (*e.g.* DI water or buffer salts). The system further consists of an injection port such that a turbulent jet is formed under controlled flow conditions with downstream conditioning modules and concentrating modules.

Active-loading Doxorubicin-HCl into Liposomes

Hydrogenated soy phosphatidyl-choline (HSPC), cholesterol and DSPE-mPEG2000 or DSPG were prepared at a molar ratio of 5.6:3.8:0.52 and were dissolved in ethanol. The lipid solution was subsequently injected into the aqueous phase (250 mM Ammonium Sulfate) *via* the co-flow injection port. Liposomes were then concentrated downstream and doxorubicin-HCl was injected into the pre-formed liposomes under controlled conditions. Offline UV/HPLC measurements were performed to confirm loading percentage.

RESULTS

HSPC:Chol:DSPE-mPEG2000 and HSPC:Chol:DSPG liposomes were successfully formed under different processing conditions. The z-average particle size could be controlled from approximately 40.0 to >150 d.nm with a PDI of less than 0.10 under most conditions. For the doxorubicin loading studies, high encapsulation efficiency was achieved using the liposomes prepared on our continuous processing system.

CONCLUSION

Continuous processing of liposomes that incorporates a turbulent jet in co-flow enables a fine-control of nanoparticles. On-line/ at-line measurements techniques are important to characterize and provide feedback during the continuous process and can easily be implemented into a process control strategy. Lastly, we have provided groundwork on continuous processing of liposomes that can be used to make doxorubicin-loaded liposomes with similar quality attributes of existing drug products.

ACKNOWLEDGEMENTS

This work was supported by the FDA (HHSF223201610121C and 1U01FD005773-01).

DISCLAIMER

This article reflects the views of the authors and should not be construed to represent FDAâ€™s views or policies.

536980 Experiments and Multi-Scale Models to Understand Liposome Processing Using a Turbulent Jet in Co-Flow

Abstract:

PURPOSE

The continuous processing administers considerable advantages over batch manufacturing to improve scalability, automated control and incorporation of process analytical technology (PAT). Conventional methods to form liposomes do not provide for precise control of the particle size and particle size distribution both of which can directly impact encapsulation efficiency, cellular uptake and biodistribution. The continuous manufacturing platform provides precise control over liposome formation including monitoring of particle size and size distribution. The present study focuses on coaxial turbulent jet flow design to control liposome properties and quantify them as a function of the flow conditions (velocity, temperature, Reynoldâ€™s number). Computational Fluid Dynamics (CFD) based models have been developed to verify the dynamics of the micro-environment at site of liposome formation. Molecular Dynamics (MD) simulation being further performed to analyze the lipid-lipid interactions and lipid-salt interactions at the liposomal formation site.

METHODS

Liposomes were formed using an ethanol injection method with a custom-built continuous processing system that is controlled using a program developed using National Instruments LabVIEW. Briefly, the continuous processing system consists of multiple tanks that contain lipid dissolved in ethanol and another set of tanks for the aqueous components (*e.g.* DI water or buffer salts) along with at-line particle size measurements using a Malvern Zetasizer Nano. Fluid flow profiles (i.e. Flow Velocity ratio (FVR) vs Reynolds number (Re) were related to the liposomal size distribution (*i.e.* Z-Average particle size and PDI). CFD geometry similar to the continuous liposome manufacturing unit was developed using ANSYS 19.0. Species Transport Equations were implemented to predict the fluidic microenvironment at the site of liposome formation and was validated with experimental data by image analysis and fluid flow dimensions (FVR and Re) of the co-axial jet formed from ethanol water mixture.

RESULTS

The continuous processing system at UConn can form monodispersed liposomes. Process parameters can be controlled to regulate the velocity, density, and viscosity at the co-axial jet. The CFD studies have successfully be used to predict the experimental flow properties observed at the co-axial jet. Based on the FVR and Re, the model provides insight on the alteration of the micro-environment at the liposome formation site. The liposome mean particle size can also be controlled and has dependence on the Reynolds number of the mixed fluid. In addition, MD modeling can be used to understand the differences in liposome formation based on factors such as type of lipid and aqueous phase additives.

CONCLUSION

The liposome size and PDI can be systematically controlled using a co-axial turbulent jet in co-flow. The non-linear increment of the ethanol-water mixture viscosity at the site of liposome formation (predicted from the CFD studies), plays a significant role to alter the fluid properties and can be directly related to the liposome dimensions and distribution. MD simulations are being performed to predict the mechanism of liposome formation due to variability of fluid flow properties and physical properties of the formulation.

ACKNOWLEDGEMENTS

This work was supported by the FDA (1U01FD005773-01).

DISCLAIMER

This article reflects the views of the authors and should not be construed to represent FDAâ€™s views or policies.

533257 Continuous Protein Crystallization of Lysozyme

Abstract:

The current work investigates the technical feasibility of performing continuous protein crystallisation in an oscillatory flow platform with lysozyme as the model protein. Different crystallisation conditions were generated by varying the protein concentration (30 to 100 mg/mL) as well as the oscillation amplitude (10 to 25 mm) and frequency (0.05 to 0.25 Hz). Based on the Reynolds number (Re) under different flow conditions, the shaking batch crystallisation was successfully transferred to the batch oscillatory flow crystallisation and subsequently the continuous oscillatory flow crystallisation. Experimental results show that the continuous crystallisation reached equilibrium state at a resident time of approximately 200 min and consistent crystal shape and size were achieved at the outlet of the crystalliser. This work presents a robust method for the development of continuous protein crystallisation in an oscillatory flow system, which is useful for the industrial downstream processing of proteins.[1,2]

Acknowledgement

We acknowledge the EPSRC (EP/N015916/1) for funding.

Reference

1. Roberts, M. M., Heng, J. Y. Y. & Williams, D. R. Protein crystallization by forced flow through glass capillaries: Enhanced lysozyme crystal growth. *Cryst. Growth Des.* **10,** 1074â€"1083 (2010).
2. Yang, H., Yu, X., Raval, V., Makkawi, Y. & Florence, A. Effect of Oscillatory Flow on Nucleation Kinetics of Butyl Paraben. *Cryst. Growth Des.* **16,** 875â€"886 (2016).

533679 Development of a Cation Exchange Chromatography Step for Robust Impurity Clearance and Improved Polysorbate 80 Stability in the Drug Product of a Monoclonal Antibody

Abstract:

It has been reported that polysorbate 80 (PS80) degradation not only occurs due to auto-oxidation, but also due to lipases generated by the host organism. Specific lipases linked to PS80 degradation, such as phospholipase B-like 2 (PLBL2) and lipoprotein lipase (LPL), are difficult to remove in the downstream purification process, causing concerns for PS80 stability in the drug product (DP) as well as potential immunogenicity in humans.

A case study is presented on the development of a cation exchange chromatography (CEX) step for mAb A. After adding CEX as the second polishing step in the purification process, additional lipase clearance was achieved and PS80 stability in the DP improved. To further characterize the CEX step, a design of experiment (DOE) was conducted which elucidated critical step parameters and impurity clearance robustness. The DOE was a split plot central composite design involving 40 column runs operating in bind-and-elute mode. Five CEX parameters were studied: pH and conductivity of the elution buffer and load as well as resin loading. All 40 runs were evaluated for step yield and product quality. Results from HCP ELISA, UP-SEC, and HP-IEX analysis indicated high statistical dependence on elution buffer pH and conductivity ($p<0.001$). Overall, as elution buffer pH and conductivity increased, impurity clearance decreased and yield increased. LC-MS results indicated non-detectable lipases levels in the CEX pools for all runs in the elution buffer design space studied. These results suggest CEX is a robust process step to remove lipases, and hence, improve PS80 stability in the DP.

532212 Production of Anti-CD20 Monoclonal Antibody Biosimilar

Abstract:

The monoclonal antibodies (mAbs) targeting CD20, such as ofatumumab and obinutuzumab, have been successfully applied to treat chronic lymphocytic leukemia. The development of biosimilar anti-CD20 mAb with improved production efficiency and reduced production cost is very important. In this study, CHO DG44/anti-CD20 mAb production cell line was developed using *dhfr* deficient expression system after serial selection and single cell cloning. The top three clones produced mAb with titer 200 mg/L in batch culture, while a high mAb titer of 2,000 mg/L was achieved in fed-batch culture using stirred-tank bioreactor. The purified mAb showed strong and specific binding and efficient toxicity to cancer cells with CD20 expression.

533230 Multi-Stage and Multi-Objective Design Tool for Process Design in Sterile Filling of Biopharmaceuticals

Abstract:

Sterile filling is a process within biopharmaceutical manufacturing, where compounded drug solutions are filled into vials or syringes. The strict consideration of product quality necessitates that manufacturing equipment be kept clean and sterile. The conventional use of fixed stainless steel equipment, termed multi-use technology (MUT), conducts multiple cleaning procedures following every batch of pharmaceutical products. Recently, a set of resin-made pre-sterilized equipment have been introduced as disposable alternatives, known as single-use technology (SUT). Introduction of SUT offers a large number of alternatives regarding possible combinations of SUT or MUT per unit operation and the relative sizes in a production line. A few tools have been already developed to support the decision-making process concerning SUT alternatives. Such tools however, are based on the economic evaluation of the alternatives and offer limited functionalities for generating process alternatives or understanding sensitivity to process parameters.

In this work, we present a multi-objective decision-making tool for sterile filling of biopharmaceuticals that can be used across multiple design stages. The basis of the work is the evaluation models developed by the authors to compare SUT and MUT in terms of the economic performance, environmental impacts, product quality, and supply robustness. The weighting factors of design objectives were defined for aggregating multi-objective evaluation results, where sensitivity analysis of the factors were conducted within the possible combinations. We will conduct a questionnaire survey to industrial practitioners, and the results can be used to set the weighting factors and provide a solution that best fits the users. Also, the tool will provide rigorous and tailored solutions across design stages by varying underlying model complexities in accordance to user requirements and available data at each design stage. The tool will be supported by data collected through industrial databases and expert knowledge. We aim to implement the tool in the form of a web-based application to enhance the practicability of the design tool for the users.

531341 Comparison of Batch and Continuous Biopharmaceutical Antibody Production Based on Techno-Economic Analysis

Abstract:

Continuous processing has advantages compared to batch mode of operation including reduced equipment sizes, increased productivity, improved product quality and high facility flexibility. These are the main reasons that continuous manufacturing has become an interesting alternative for biologic production where high productivity and process robustness is of great significance [1, 2]. In the recent years, a few studies have been published that compared batch and continuous biopharmaceutical processing including upstream, downstream and integrated process analysis that provide solutions to achieve continuous biopharmaceutical processing. Excel-based process and cost modeling software Bio-Solve (Biopharm Services, UK) [3] and a discrete event simulation tool ExtendSim (ExtendSim v8, Imagine That! Inc. San Jose, CA) [4, 5] have been applied to simulate different modes of operation and provide guidance using economic analysis to evaluate cost-effectiveness. However, many biopharmaceutical companies are still skeptical and need more analytical help to de-risk their decision-making.

This study uses data from a realistic case study as base scenario and builds a continuous biopharmaceutical manufacturing simulation flowsheet with detailed mass balance for all the streams and operation units using SuperPro Designer (Intelligen, Scotch Plains, NJ). Based on detailed process flow diagram with energy and material balance, the overall cost is calculated. This includes capital cost consisting of working capital, start-up and validation cost, and operating cost comprising of maintenance costs, equipment depreciation, labor costs, medium costs, consumables costs, waste and utilities costs [6]. Detailed economic analysis is performed using preliminary cost information that results in the evaluation of the cost-effectiveness of the alternative processes including quantifiable metrics with different cost categories. Investment profitability considering time value of money is analyzed to show the future cash flow and determine the discount rate with expected return of investment under different level of risks. Global Sensitivity analysis is applied to evaluate the importance of different model parameters and help with the risk assessment [7]. In sensitivity analysis, operating conditions and process parameters including upstream bioreactor perfusion rate, downstream process breakthrough capacities, integrated process running time and failure rate caused by contaminations are considered. This framework provides the platform for comparing cost sensitivity of different operating conditions in batch and continuous manufacturing processing to assist in decision-making.

Reference

1. Konstantinov, K.B. and C.L. Cooney, *White Paper on Continuous Bioprocessing. May 20â€"21, 2014 Continuous Manufacturing Symposium.* Journal of Pharmaceutical Sciences, 2015. **104**(3): p. 813-820.
2. Jungbauer, A., *Continuous downstream processing of biopharmaceuticals.* Trends in Biotechnology, 2013. **31**(8): p. 479-492.
3. Xenopoulos, A., *A new, integrated, continuous purification process template for monoclonal antibodies: Process modeling and cost of goods studies.* J Biotechnol, 2015. **213**: p. 42-53.
4. Pollock, J., et al., *Integrated continuous bioprocessing: Economic, operational, and environmental feasibility for clinical and commercial antibody manufacture.* Biotechnol Prog, 2017.
5. Pollock, J., S.V. Ho, and S.S. Farid, *Fed-batch and perfusion culture processes: Economic, environmental, and operational feasibility under uncertainty.* Biotechnology and Bioengineering, 2013. **110**(1): p. 206-219.
6. Petrides, D., *Bioprocess Design and Economics*, in *Bioseparations Science and Engineering* P.W.T. Roger G. Harrison, Scott R. Rudge and Demetri P. Petrides, Editor. 2015.
7. Pannell, D.J., *Sensitivity analysis: strategies, methods, concepts, examples.* Agric Econ, 1997. **16**: p. 139-152.

526015 Increasing Capacity and Lifetime of Reverse Phase Resin

Abstract:

Development time and cost are two major impact factors for the biopharmaceutical industry in bringing a new molecule into the market. Chromatography resin is one of the more cost-intensive items used in drug substance manufacturing. One important consideration for chromatography resin in particular is its lifetime, since operating costs can be significantly reduced if the resin can maintain its resolution and capacity across multiple cycles of use. Column lifetime can vary significantly between different resins and especially between resins of different chromatographic modalities. This research focuses on reverse phase chromatography, which is a critical step in the purification of a wide variety of therapeutic proteins. A single, packed reverse phase chromatography column can be used for approximately 50 cycles without suffering from diminished performance. However, column lifetime can be adversely affected if feeds contain compounds that have the potential to foul the resin, either by precipitation or by strongly binding to the resin such that it will not elute during the method. This fouling was observed in one such program, which utilized an enzyme to improve specific productivity of an upstream reaction. This enzyme, in combination with residual sodium citrate in the feed, was found to cause column fouling of a Kromasil C8 column used for purification. The fouling resulted in dramatically reduced resolution in subsequent cycles, and therefore the effective lifetime of the column would end once this fouling occurred. The reverse phase columns could not be successfully regenerated using typically employed cleaning techniques, such as the use of sodium hydroxide. This research screens multiple potential regeneration solutions using both high throughput slurry plate screening as well as small scale column runs. A cleaning method with ethanol and acetonitrile in series was developed that successfully removed 98% fouling caused by citrate and enzyme. At this level of reduction, resolution on the column in subsequent cycles was similar to column resolution prior to fouling.

539792 Liquid Phase Synthesis of Monodisperse PEGs By Nanostar Sieving

Abstract:

Polyethyleneglycol (PEG) is the go-to, FDA-approved polymer for modifying proteins, bio-molecules and nano-carriers. Monodisperse, or discrete, PEG is especially desirable because the resultant PEG-conjugates are unique molecular entities with narrowly defined physical, biological and pharmacokinetic properties. Monodisperse PEGs are produced by stepwise, or iterative, synthetic methods, and there is an active market in shorter monodisperse PEG derivatives. However, many applications still resort to polydisperse PEG because higher molecular weights (> ~1.5 kDa) of monodisperse PEG are difficult to synthesis and are not commercially available.

We have developed a scalable synthetic platform for the manufacture of sequence-defined heteropolymers and monodisperse homopolymers â€“ *Nanostar Sieving* technology â€“ that combines liquid phase coupling reactions with purification through a size-selective polymer membrane. Three strands of polymer are grown unidirectionally from a central hub giving a polymer nanostar. A rigid central hub was developed that both maximises membrane rejection of the growing polymer-star and provides a distinct chromatographic marker (lambda max = 290 nm).

This approach has been applied to oligonucleotides [Gaffney *et al. Chem. Eur. J.* **2015**, *21*, 9535-43] and peptides, but is even more appropriate to polyethers. Solid phase methods and supports are poorly compatible with Williamson etherification. By contrast, solution phase chain extension of PEG-stars is unaffected by mass transfer kinetics, the hub is chemically inert, and slow reactions (hours) can be continuously monitored to completion. PEG-star chain elongation was explored using Thp or Dmtr protected dodecagol (Eg12) toluene sulfonate building blocks, up to mono-methyl Eg60 (mPEG-2675) and Eg112 (PEG-4951), respectively. Synthesis of and purification of monodisperse PEG by nanostar sieving is ideal for efficient heterobifunctionalisation of the chain termini [Livingston *et al. Chem. Eur. J.* **2014**, *20*, 10038-51], and new building blocks are being developed to permit intra-chain functionalization at precise positions along the sequence.

This technology is being commercialised by EXACTMER, a new start-up from Imperial College London.

521874 Virtual Screening of Process Parameters for Pharmaceutical Drying Operation: A Combined DoE-CFD Approach

Abstract:

In recent years, the mechanistic modeling of pharmaceutical unit operations has made significant progress. Modern process modeling has been recognized as an important tool for developing fundamental understanding of processes, optimization, scale-up, sensitivity analysis, transient dynamics and development of predictive control strategies. There has been growing interest in quality-by-design (QbD) based thinking for various purposes including cost reduction and ensuring consistent product quality. Process modeling gives a perfect platform to achieve these objectives by reduction in batch cycle time, improvement in energy efficiency and miniaturization of equipment sizes.

For example, spray and fluidized bed drying operations are carried out with short residence times due to excellent gas-particle contact, very high heat and mass transfer rates. These help in uniform drying of all particles and also significantly reduce thermal degradation of product. Although the operation is simple, the design engineer has to make choices related to large number of operating parameters including inlet gas humidity, temperature and flow rate, incoming slurry temperature, slurry concentration, position and type of atomizer, size of the equipment and permissible exhaust air temperature. The effect of these variables on product quality and throughput needs systematic investigation. However, the number of experiments/simulations can be extremely high. Design of experiments (DoE) plays a key role in addressing this issue.

In this work, operating conditions of existing spray and fluid bed drying were optimized to improve drying of drug products and drug substances with known drying kinetics. The optimization was carried out using virtual screening of operating parameters using a combined design of experiments (DoE) and computational fluid dynamics (CFD) approach. The inlet gas humidity, temperature and flow rate are considered as manipulated variables. The set of CFD simulations were shortlisted using central composite design (CCD). For each parameter, three levels were considered and their impact on the final moisture content has been assessed. The CFD simulations were carried out in a Eulerian-Lagrangian framework. In the constant and falling rate periods, the drying kinetics were modelled by modified vapor pressure approach. The simulation results were validated with the actual data in terms of product moisture and outgoing air temperature. In addition, the residence time distribution (RTD) of gas and solid phase has been studied using CFD. A simplified process model for spray and fluid bed drying supplemented with RTD information has been proposed and incorporated in Dynochem. The Dynochem predictions were found to correlate well with the rigorous CFD approach.

The methodology has been demonstrated using following case studies:

The first case study is the use of spray drying to obtain crystal morphologies that could not be obtained using conventional solution phase crystallization. We show how we can predict, from the droplet size and residence time distribution in the spray dryer, the distribution of the crystal morphologies finally obtained.
The second case study involves drying in a fluidized bed. We have used the aforementioned approach to predict characteristics of the dried material as a function of temperature, flow-rate and moisture content of the drying air and the drying time. Also, this approach has been combined with the models of nonnewtonian fluid processing for extending these calculations to wet granulation.

535656 Stoichiometry Identification in Pharmaceutical Reactions Using Dynamic Response Surface Methodology and Target Factor Analysis

Abstract:

In many pharmaceutical reactions under development, one knows the main reactionâ€™s stoichiometry, but when the reaction experiments are performed, the presence of several other nonidentified compounds are measured. Then there is an urgent need to identify the possible additional reactions that generate the unwanted secondary compounds and to find ways to minimize their presence. These secondary compounds are either intermediates or reaction by-products. To identify the possible reaction stoichiometries that will agree with the data and those that can be refuted, we perform a target factor analysis (TFA)[1,2] in combination with the development of the dynamic response surface methodology (DRSM) model[3,4].

The DRSM model is a newly proposed data-driven modeling methods which generalize the Response Surface Methodology (RSM) used to quantify the results of Design of Experiments (DoE). We use the constrained version of the DRSM-2 model and the TFA to identify the stoichiometry of the reactions under study. We update the TFA approach by utilizing a statistical test for the determination of the number of significant singular values, and we propose a new test to statistically evaluate whether the projection of candidate stoichiometries is successful.

The use of the DRSM model enables the generation a more informative set of data on the rate of concentration changes than directly relying on the initial concentration data. This is especially crucial if there are infrequent measurements and missing data. We use an F test[5] to decide the number of significant eigenvalues and we propose a new statistical test for deciding whether the difference vector, subtracting the response vector from the target vector in TFA, is close to the zero vector. We also compare it against a statistical test previously proposed [5]. These two statistical tests are:

1. A *t*-test that each individual component of the difference vector is inside the interval, with being a value close to zero.
2. An F test on the hypothesis that the difference between the target and the response vectors is of the same magnitude as the normal variability of the process.

We apply the projection algorithm both in a parallel or in a sequential mode, in a blind test of a complex pharmaceutical reaction network, involving ten species in eight reactions. In the parallel test we are able to identify six of the eight reactions when the measurement error is 0.005 mol/L. Under the same conditions we refute all six of the proposed false reaction stoichiometries. One more reaction is identified in the sequential TFA. The two reactions that were not initially identified proceed in slow rates and the corresponding concentrations of the participating species have concentrations of the same order of magnitude as the measurement error.

References

1. Bonvin D, Rippin DWT. Target Factor Analysis for the Identification of Stoichiometric Models. *Chem Eng Sci.* 1990;45(12):3417-3426.
2. Malinowski ER. *Factor analysis in chemistry*. 3rd ed. New York: Wiley; 2002.
3. Klebanov N, Georgakis C. Dynamic Response Surface Models: A Data-Driven Approach for the Analysis of Time-Varying Process Outputs. *Ind Eng Chem Res.* 2016;55(14):4022-4034.
4. Wang ZY, Georgakis C. New Dynamic Response Surface Methodology for Modeling Nonlinear Processes over Semi-infinite Time Horizons. *Industrial & Engineering Chemistry Research.* 2017;56(38):10770-10782.
5. Malinowski ER. Statistical F-tests for abstract factor analysis and target testing. *J Chemom.* 1989;3(1):49-60.

521863 Heat Transfer Transients in Semi-Batch Systems: A Computational Approach to Process Intensification and Mitigating Process Hazards

Abstract:

Localized thermal pockets can have a significant impact on yield and quality if thermally labile compounds are present in the reaction mixture. They can also present a formidable process hazard in the form of runaway scenarios. Hence, semi-batch operations with controlled addition protocols are popular in the chemical and pharmaceutical sectors for highly reactive and/or exothermic systems.

For example, in the semi-batch operation the liquid level in the vessel keeps changing with time resulting in a continuous change of the mixing profiles and therefore changes in effective heat and mass transfer rates resulting into heat removal from the reactor.

The resulting change in relative positions of agitators with respect to the liquid surface affects the flow pattern significantly. For example, agitators rotating near the liquid surface or in the vapor phase do not contribute to mixing and heat transfer which can cause in-homogenous mixing resulting into localized thermal pockets. These effects are neither accounted by simple empirical relationships nor by standard commercial software tools.

Computational fluid dynamics (CFD) is an effective tool to design and/or troubleshoot systems as stated above. Our work using CFD is a simple and effective approach using modified CFD simulators data combining with the process design tools (Dynochem). The approach is not only computationally faster but can also provide excellent results in terms of predicting liquid phase mixing and heat/mass transfer transients. This tool has been used effectively to plan batch operation schedules and thermal profiling of reactions posing a runaway process hazard. It has also been used to intensify batch distillation/evaporation operations. Furthermore, the CFD based approach can be integrated into quality-by-design (QbD) frameworks in pharmaceutical manufacturing and scale up activities.

531818 Plasmonic Nanocatalysts for Continuous Synthesis of Drug Substances: An Example of Visible-Light Mediated Cross Coupling Reactions

Abstract:

Approximately 25% of all the reactions performed in the pharmaceutical industry consist of cross-coupling reactions. The cross-couplings have been traditionally carried out by homogeneous palladium (Pd) catalyzed batch processes. The National Science and Technology Council (NSTC) has recently identified the continuous manufacturing of pharmaceuticals as one of five manufacturing areas of emerging priority in the USA, revealing that continuous manufacturing of pharmaceuticals has the potential to reduce manufacturing cost by up to 40-50%. Given these advantages, in recent years, metal nanoparticle catalyzed cross-couplings have been of great interest since they can be used as a platform to switch cross-couplings from traditionally used batch mode to continuous flow mode. Additionally, heterogeneous metal nanocatalysts can be immobilized on a rigid support in the form of a packed-bed reactor, and therefore, can avoid the need for subsequent difficult steps to separate the catalyst and product.

In this contribution, we show that heterogeneous plasmonic metal nanocatalysts build on silver (Ag), copper (Cu) and gold (Au) can be used as alternate catalysts to drive cross couplings. Specifically, we demonstrate that visible light can be used as a powerful tool to make plasmonic metal nanocatalysts as efficient photo-catalysts with activity and stability superior to that of traditionally used homogeneous Pd based catalysts. These plasmonic metal nanocatalysts can be used to design efficient cross-coupling flow processes to produce drug substances.

538571 Development and Scale-up of a Robust Impinging Jet Process Under GMP Conditions

Abstract:

To address the formation of a problematic dimer impurity in a mixing sensitive reaction network, an impinging jet flow process was developed to be used under GMP conditions. Efficient process development of the robustness of the impinging jet required an intricate understanding of the kinetics, transport phenomena, and analysis of flow patterns within these jets. Rational exploration of these factors enabled us to scale-up this process to process multi-kilogram amount of API.

The work presented here was sponsored by AbbVie. AbbVie contributed to the design, research, and interpretation of data, writing, reviewing, and approving the presentation. All authors are employees of AbbVie.

524115 Single-Step Continuous Purification from Liquid-Liquid-Solid Mixture: Design and Experimental Implementation for an HIV Drug Intermediate

Abstract:

An objective in pharmaceutical manufacturing is to increase process efficiency and yield, while reducing capital and operating costs (e.g., am Ende 2010; Simon et al., 2015). As a typical pharmaceutical manufacturing process consists of multiple chemical reactors that are separated by multiple other types of separations such as solids filtration, liquid-liquid extraction, evaporation, crystallization, and precipitation, a strategy for reducing both capital and operating costs is to redesign the process to reduce the number of unit operations. Moving from batch to continuous unit operations creates opportunities for reducing the number of unit operations, such as demonstrated for the end-to-end manufacturing of aliskiren hemifumarate which reduced the number from 21 to 14 (e.g., Mascia et al., 2013). As such, the capital and operating costs can be reduced substantially from the use of continuous-flow technologies.

While much progress has been made in the development of effective continuous-flow options for some unit operations, such as for crystallization (e.g., Nagy and Braatz, 2012; Simon et al., 2015), other continuous unit operations, such as filtration, are less well developed (e.g., Gursch et al., 2015). One of the guiding principles during the design of continuous-flow processes is to avoid the creation of conditions in which particles are present unless suspended within a liquid (e.g., Mascia et al., 2013). The reasoning behind this principle is that particle-particle interactions are much less predictable and more likely to cause processing problems when liquid is not present. Particles suspended within a liquid, however, can be transported through a continuous-flow process in the same manner that liquids are transported. As such, there is an especial interest when reducing the total number of unit operations to also avoid the use of unit operations such as filtration that involves particles that are not suspended in a liquid.

This presentation describes a unit operation for controlled continuous-flow separation that improves manufacturing efficiency while reducing the number of unit operations. The unit operation replaces the typical process sequence of solids filtration followed by liquid-liquid extraction, while having particles interact with particles only while being suspended within liquid. As such, the unit operation is highly reliable and robust. The unit operation does not require high pressure for its operation and does not employ any form of membrane or filtration media, and so cannot clog. The controlled continuous-flow separation does not involve crystallization nor precipitation on a surface (e.g., Zhou et al., 2015), and has high efficiency by retaining a high surface area to volume ratio for the particles throughout the process. The process also operates at low temperature, which minimizes potential oxidation or generation of impurities during processing.

The unit operation is experimentally demonstrated for the purification of an intermediate (Compound #9 in Wang et al., 2015) in the manufacture of a potent HIV integrase inhibitor. The process is controlled by a combination of a physical feedback control system designed into the process and a computer-controlled feedback control system implemented using open-source hardware commonly used by the Maker Community (Lahart, 2009). The cost of the computer-controlled system is less than $200. The actuator and sensor are also inexpensive off-the-shelf equipment. The use of commodity equipment common to the Maker Community results in very low cost for the controlled continuous-flow separation. The presentation will describe the unit operation in some detail, including the design strategies that led to its development and the specific equipment and software used in its experimental implementation.

References

J. am Ende, Editor. Chemical Engineering in the Pharmaceutical Industry: R & D to Manufacturing. Wiley, Hoboken, NJ, 2010.

Gursch, R. Hohl, G. Toschkoff, D. Dujmovic, J. Brozio, M. Krumme, N. Rasenack, and J. Khinast. Continuous processing of active pharmaceutical ingredients suspensions via dynamic cross-flow filtration. J. Pharm. Sci., 104(10):3481–3489, 2015.

Lahart. Tinkering makes comeback amid crisis. The Wall Street Journal, November 13, 2009.

Mascia, P. L. Heider, H. Zhang, R. Lakerveld, B. Benyahia, P. I. Barton, R. D. Braatz, C. L. Cooney, J. M. B. Evans, T. F. Jamison, K. F. Jensen, A. S. Myerson, and B. L. Trout. End-to-end continuous manufacturing of pharmaceuticals: Integrated synthesis, purification, and final dosage formation. Angewandte Chemie, 52(47):12359–12363, 2013.

K. Nagy and R. D. Braatz. Advances and new directions in crystallization control. Annu. Rev. Chem. Biomol. Eng., 3:55–75, 2012.

L. Simon, H. Pataki, G. Marosi, F. Meemken, K. Hungerbuhler, A. Baiker, S. Tummala, B. Glennon, M. Kuentz, G. Steele, H. J. M. Kramer, J. W. Rydzak, Z. Chen, J. Morris, F. Kjell, R. Singh, R. Gani, K. V. Gernaey, M. Louhi-Kultanen, J. O'Reilly, N. Sandler, O. Antikainen, J. Yliruusi, P. Frohberg, J. Ulrich, R. D. Braatz, T. Leyssens, M. von Stosch, R. Oliveira, R. B. H. Tan, H. Wu, M. Khan, D. O'Grady, A. Pandey, R. Westra, E. Delle-Case, D. Pape, D. Angelosante, Y. Maret, O. Steiger, M. Lenner, K. Abbou-Oucherif, Z. K. Nagy, J. D. Litster, V. K. Kamaraju, and M.-S. Chiu. Assessment of recent process analytical technology (PAT) trends: A multiauthor review. Org. Process Res. Dev., 19:3–62, 2015.

Wang, M. D. Kowalski, A. S. Lakdawala, F. G. Vogt, and L. Wu. An efficient and highly diastereoselective synthesis of GSK1265744, a potent HIV integrase inhibitor. Org. Lett., 17(3), 564–567, 2015.

L. Zhou, M. Su, B. Benyahia, A. Singh, P. I. Barton, B. L. Trout, A. S. Myerson, and R. D. Braatz. Mathematical modeling and design of layer crystallization in a concentric annulus with and without recirculation. AIChE J., 59:1308–1321, 2013.

531937 Crystallization Kinetic Measurement and Parameter Estimation Utilizing Population Balance Model in a Dynamic/Oscillatory Baffle Crystallizer

Abstract:

Crystallization has become a predominant technique in particle design technology and it is widely applied in the pharmaceutical industry. Oscillatory systems have been shown to enhance mass and heat transfer capabilities while imposing less shear on crystals.[1-2] A novel commercial dynamic baffle crystallizer(DBC), also known as an oscillatory baffle reactor(OBR), has been studied. Crystallization kinetics of paracetamol(PCM) in ethanol are measured in the DBC to develop the population balance model(PBM) of batch and continuous crystallization in the DBC. PBM is a first-principle model often used to describe cell dynamics, polymerization processes and particulate systems such as crystallization.[3] Coupling with mass balance, PBM model describes the size distribution as well as concentration profile over time during crystallization processes which aid the understanding and optimization of crystallization. With the help of in-situ process analytical technology(PAT) tools such as Infrared(IR), and focused beam reflectance measurement(FBRM), nucleation and growth kinetic parameters are estimated separately by solving the PBM multiple times using high resolution finite volume(HRFV) method to fit experimental data.[4-5] Experiments are repeated for different oscillation conditions to account for the effect of mixing intensity on crystallization kinetics into the PBM by incorporating oscillation-dependent terms into the kinetic expressions. The PBM model is then validated with batch and continuous experiments and showed agreement. The kinetic parameters are then compared to those of a traditional stirred tank crystallizer reported in the literature to further understand the impact of the oscillatory mixing dynamics on crystallization kinetics.

1. Ni, X.; Mackley, M. R.; Harvey, A. P.; Stonestreet, P.; Baird, M. H. I.; Rama Rao, N. V., Mixing Through Oscillations and Pulsations - a guide to achieving process enhancements in the chemical and process industries. *Trans IChemE* **2003,** *81* (A), 373-383.
2. Hewgill, M. R.; Mackley, M. R.; Pandit, A. B.; Pannu, S. S., Enhancement of gas-liquid mass transfer using oscillatory flow in a baffled tube. *Chemical Engineering Science* **1993,** *48* (4), 799-809.
3. Ramkrishna, D., *Population Balances: Theory and Applications to Particulate Systems in Engineering*. Academic Press: San Diego, CA, 2000.
4. Simone, E.; Szilagyi, B.; Nagy, Z. K., Systematic model identification and optimization-based active polymorphic control of crystallization processes. *Chemical Engineering Science* **2017,** *174*, 374-386.
5. Gunawan, R.; Fusman, I.; Braatz, R. D., High resolution algorithms for multidimensional population balance equations. *AIChE Journal* **2004,** *50* (11), 2738-2749.

533286 Nucleation Kinetics of Pharmaceutical Co-Crystals

Abstract:

Nucleation Kinetics of Pharmaceutical Co-crystals

Hannah McTague

Abstract

Approximately 40% of all drug molecules on the market are suffering from physio-chemical property (PCP) problems such as low solubility and dissolution rates, poor thermal stability and bioavailability. These properties can be largely attributed to the molecular arrangement and intermolecular forces within the solid material. The formation of co-crystals consisting of an API with a co-former or two APIs can offer a solution by allowing one to modify the PCPs without changing the chemical properties of the drug molecule.

Although there have been many studies on co-crystals, little work has been done regarding the nucleation and growth kinetics of pharmaceutical co-crystals.

The theophylline:salicylic acid (THP:SA) co-crystal has been synthesized and exists as a 1:1 co-crystal. THP:SA co-crystal exhibits favourable properties in the solvent chloroform wherein the system dissolves congruently in the same 1:1 stoichiometric ratio. The solid co-crystal was characterised according to PXRD, Single Crystal XRD, TGA and DSC and compared with the properties of each individual coformer. The solubility of the co-crystal and each individual component has also been identified.

As a method of studying the nucleation kinetics of THP:SA induction time measurements have been undertaken. These experiments have initially been performed at 6 different concentrations allowing for a gradient of supersaturation ratios to exist. In order to measure the induction times the following protocol has been followed: 500ml of solution at the desired concentration is created. Upon 24 hour equilibration at a chosen dissolution temperature, which is 5°C above the saturation temperature, the solution is filtered into 20ml aliquots in pre-heated vials and allowed to equilibrate at dissolution temperature. The vials can then be simultaneously submerged at a fixed nucleation temperature of 10°C and recorded by a HD video recorder. Following nucleation the video is analysed by eye and the moment of first detectable crystals is recorded for each vial. 40-80 vials are recorded for each concentration. These results provide a median induction time, where the induction time is defined as the time period between the moment the vials are submerged at nucleation temperature and crystals are detected.

A probability distribution has been created displaying these results in which the median induction times can be seen to be representative of the driving forces present. On the distribution it is obvious that at the same concentrations there exists large variation in induction times. This is attributed to the stochastic nature of nucleation at small volumes which holds valuable information regarding the crystal nucleation rate. A set of induction times under equal supersaturation and temperature results in the crystal nucleation rate (J).

In order to analyse according to classical nucleation theory the median induction times have been used to identify the kinetic and thermodynamic parameters A and B allowing for the subsequent calculation of interfacial energy. B is seen to have a value of 3202766.31 s/K^{-3}, which represents the dimensionless energy barrier for nucleation. A is found to have a value of 32.51 m^{-3}s^{-1} and this describes the molecular kinetics of the nucleation process i.e. it reflects how building units attach to the nucleus.

Using the obtained values of the thermodynamic parameter B the value of the interfacial energy Î³ can be calculated by assuming spherical nuclei and using the corresponding shape factor. The molecular volume for the THP:SA co-crystal was obtained from the CSD file. The interfacial energy for the co-crystal has been found to be 1.62 mJ/m^2.

The same experiments have been performed on the individual components namely theophylline and salicylic acid revealing interfacial energies of 0.85 mJ/m^2 and 0.71 mJ/m^2 respectively.

The calculated interfacial energies can be compared with that of the THP:SA co-crystal and analysed such as to examine how the nucleation kinetics of the API may be altered due to the formation of a co-crystal. Based on the current results there is an indication that due to the higher interfacial energy value of the co-crystal, versus both of the individual components, the co-crystal is harder to nucleate at these small volumes.

527873 Extrudability Analysis of Drug Loaded Pastes for 3D Printing of Modified Release Tablets

Abstract:

The rheological characteristics of 3D printing pastes cannot be fully described by the traditional rheological tests suitable for other pastes. In the present study, extrudability testing of carbopol based 3D printing pastes was performed to establish a constitutive rheological model for micro-extrusion. This model was developed for pastes that follow a plasto-viscoelastic constitutive relationship and the associated generalized Herschelâ€"Bulkley flow rule. An analytical model was then applied to the extrudability data obtained by extrusion through nozzle of 400 and 600 Âµm diameters. For this purpose, nineteen paste formulations were prepared per a fractional factorial design using various concentrations of the active ingredient and soluble and insoluble excipients. Critical constitutive material parameters including uniaxial yield flow stress (*Ïf*$_{\text{â,€}}$), uniaxial flow consistency index (*k)*; and uniaxial flow index (*n)* related to viscosity of pastes were then calculated. This was performed by interpreting extrudability data into mathematical equation relating flow rate, nozzle and cartridge diameters, printing pressure and slip-flow angle. The accuracy of the constitutive model to predict yield stress, consistency and flow indices were evident by low RMSE values of 0.0691 bar, 0.034 and 6.3 bar/s^n, respectively. The percent loading of soluble and swellable excipients was found to significantly affect the rheology of various pastes. However, nozzle diameter was found to affect significantly the flow index rather than consistency index. Hence, this study helped develop a mechanistic model to interpret data for characterizing the complex intrinsic rheological behavior of various pastes, for 3D printing, by employing the micro-extrusion process.

533956 One Step Purification of Curcumin from Its Lower Grades Via Particle Mediated Crystallization

Abstract:

One step purification of curcumin from its lower grades via particle mediated crystallization

K. Vasanth Kumar, KiranA.Ramisetty, Rama Krishna Gamidi,C. Heffernan, K. Renuka Devi, B.K. Hodnett

and Åke C. Rasmuson

Synthesis and Solid State Pharmaceutical Centre, University of Limerick, Limerick, Ireland.

E-mail: vasanth.kannuchamy@ul.ie; ake.rasmuson@ul.ie

Purification of organic molecules via crystallization is a challenging process. Conventional methods rely on several recrystallization steps which are time consuming and laborious. [1] Alternatively, crystallization can be performed with highly super-cooled solution and at very low temperature that can force the crystallizing molecule to crystallize via a unique particle aggregation pathway (Figure 1) mediated by perfectly and imperfectly aligned crystallites. In the presentation we will propose the general mechanisms involved in this process (see Figure 1) and will show experimental evidence to show particle mediated crystallization can produce >99% and new form of pure curcumin in one-step from its lower grades that contains at least 20% of structurally similar impurities. We will also show the characteristic properties of the final crystals obtained using microscopic and spectroscopic techniques (PXRD, SCXRD, SEM/TEM and HPLC).

In the presentation we will also show, how this unique pathway self-regulates the particle size and offers a totally new strategy to produce spherical shaped curcumin particle with remarkably high external surface area with faster dissolution kinetics versus the Form I curcumin crystals.

Figure 1: Mechanism showing the formation of pure curcumin spherulites from its impure solutions: (1) curcumin molecules in a highly supersaturated solution, (2) formation of pure and defect-free nanocrystals, (3) imperfectly oriented attachment of the nanocrystals through mechanical relaxation of solid/solid interface and (4) near perfectly oriented attachment of crystallites via crystallite docking followed by elimination of a specific solid/solid interface and growth of new fibre like structures via low angle-branching; also shown is the SEM image of spherulite shaped curcumin crystals.

Literature cited

De Yoreo J J, Gilbert PUPA, Sommerdijk NAJM, Penn RL, Whitelam S, Joester D, Zhang H, Rimer JD, Navrotsky A, Banfield JF, Wallace AF, Michel FM, Meldrum FC, Cölfen H, Dove PM. Crystallization by particle attachment in synthetic, biogenic, and geologic environments. Science. 2015; 349(6247): 498-509.

Acknowledgement

This work was supported by Science Foundation Ireland and is co-funded under the European Regional Development Fund under Grant Number 12/RC/2275.

520589 Ternary Phase Diagram and Population Balace Model for Solvent-Mediated Phase Transformation of Lansoprazole

Abstract:

The combination of *in situ* techniques and mathematical simulations would be helpful for better understanding the crystallization process of pharmaceuticals which exhibit polymorphic phenomenon. In this work, two new polymorphs (monohydrate and ethanol solvate) of lansoprazole, with better flowability and higher solubility, were first discovered and fully characterized. The selected area electron diffraction (SAED) was also used to analysis the structure of different polymorphs. The ternary phase diagram of lansoprazole in binary solvent mixtures of ethanol and water at 15 , 25 and 35 was established in order to determine the relative stability of the two polymorphs. Raman spectroscopy was used to *in situ* monitor the unseeded solvent-mediated transformation from monohydrate to ethanol solvate of lansoprazole. During transformation, the solution concentration was measured by UV spectroscopy. Furthermore, the population balance model PBE of the transformation process was built to describe the dissolution, nucleation and crystal growth process during transformation. This model was used to simulate the transformation process. By combining the experimental and simulation results, the rate-controlling step, the changing profiles of the size distribution and the solid compositions of both polymorphs were investigated and discussed. The valuable information of the process parameters which are difficult to determine by experiments are also obtained.

532549 Advancing Smart Manufacturing in Pharmaceutical Systems

Abstract:

The advancement in continuous manufacturing, process monitoring, and control systems has accelerated the progress towards smart manufacturing in the pharmaceutical industry. Smart manufacturing necessitates reliable and accurate measurements from field devices combined with good quality estimates of unmeasured process variables. Further, effective real-time communication between equipment and sensors is critical. Thus, the sensor network which comprises of equipment measurements, field devices, PAT tools, process models, DCS and SCADA systems, and data analytics is a fundamental requirement for robust process operations.

In this poster, we discuss the components of an effective sensor network for such a process. The challenges arising from an integrated system comprising of multiple unit operations, the measurement technologies and tools for systems integration are presented. A case study on a continuous roller compaction system is utilized to highlight the integrated system, along with demonstrating the application of data reconciliation for ensuring data accuracy. Finally, the concept of quality-by-control is demonstrated as the integrated platform to advance smart manufacturing in pharmaceutical systems.

538511 Dropwise Manufacturing of Oral Solid Dosage Forms Using Powder Slurries

Abstract:

As part of the shift toward continuous processing in the pharmaceutical industry, there has been considerable interest for adaptation/development of advanced manufacturing methods due to potential advantages in product quality, flexibility of scale-up and drug delivery. Inkjet printing technology, or adaptation thereof, has been the focus of much academic/industrial research efforts for use in oral solid dose manufacturing due in part to the extent of control over product properties enabled by precision deposition of small volumes of active ingredient or excipient [1]. Such applications of inkjet printing to pharmaceuticals have examined the use of several methods for liquid dispensing that may be distinguished based on factors such as actuation method (e.g. thermal, piezoelectric), the mode of fluid deposition – as individual drops, liquid jet, or continuous filament – and by physical dimensions of the ejected liquid volumes, which range in diameter from a few micrometers (e.g. inkjet printing, 3D printing) to a few millimeters (e.g. this system) [1]. The development of multiple liquid dispensing methods has been motivated by the variety of applications related to production of low dose drugs, creation of novel drug formulations and integration with continuous or semi-continuous manufacturing strategies. Fluid formulations for these processes may be adjusted to the performance requirements of the equipment, or may be tuned to achieve specific properties in a drug product; flexibility in this aspect is documented in the literature, as formulations based on drug solutions, eutectic melts and suspensions of drug nanoparticles have been used to produce solid oral doses. [1, 2, 3]

In this work, a dropwise manufacturing platform is used to produce high dose drug products using concentrated suspensions of non-Brownian particles. Pharmaceutical powders are selected based on achievable particle sizes (1-500μm) from controlled crystallization unit operations, and suspensions are formed with volatile/non-volatile carrier liquids for delivery onto suitable substrates by dropwise addition. The resultant solid oral dosages are analyzed for their mass/composition content uniformity and for the extent of preservation of particle size distribution before/after printing. Process performance metrics for dropwise manufacture of solid oral dosage forms using particle suspensions are presented in terms of drop volume/mass statistics and process throughput (number of doses per unit time); the effects of suspension properties (e.g. particle loading, particle size) on drop volume/mass and process throughput are clarified using the rheological behavior observed through high speed photography of the drop formation process [4].

References

1. Daly, R., Harrington, T. S., Martin, G. D., & Hutchings, I. M., 2015. Inkjet printing for pharmaceutics – A review of research and manufacturing, International Journal of Pharmaceutics, 494, 554-567.
2. Içten, E., Purohit, H. S., Wallace, C., Giridhar, A., Taylor, L. S., Nagy, Z. K., & Reklaitis, G. V., 2017. Dropwise additive manufacturing of pharmaceutical products for amorphous and self emulsifying drug delivery systems. International journal of pharmaceutics, *524*(1), 424-432.
3. Bonhoeffer, B., Kwade, A., & Juhnke, M., 2017. Impact of formulation properties and process parameters on the dispensing and depositioning of drug nanosuspensions using micro-valve technology. Journal of pharmaceutical sciences, *106*(4), 1102-1110.
4. Radcliffe, A. J., & Reklaitis, G. V., 2017. Dropwise Additive Manufacturing using Particulate Suspensions: Feasible Operating Space and Throughput Rates. In Computer Aided Chemical Engineering(Vol. 40, pp. 1207-1212). Elsevier.

537226 Effect of Process Parameters on Stability of Lactate Dehydrogenase duringÂ Bulk Freeze-Thaw

Abstract:

Nearly half of the commercial biotherapeutics are stored in frozen state. Storage in the frozen state provides operational flexibility during the drug product fill-finish, and extends the protein solution shelf life. However, the freezing and thawing (F/T) processes themselves can destabilize the complex structure of proteins. Cold-denaturation, osmotic stress, and ice-interfacial stress are inevitable consequences of freezing. Formulations and processes, therefore, must be designed to be protective against all these freezing-induced stresses. Hence, the overall goal of this research project was to investigate the impact of F/T process parameters on the stability of a model protein under pharmaceutically relevant conditions, at a large-scale (1 L polycarbonate bottle), through a DoE approach. To properly uncover how process parameters affect the response, i.e., the protein stability after F/T, a response surface model was designed and results evaluated using JMP® software version 12.1.0 (SAS Institute, Cary, NC). 1 L square polycarbonate bottles containing the protein solution of 10 μg/mL lactate dehydrogenase (LDH) in 20 mM histidine buffer at pH 7.0 were frozen at -50°C. All F/T runs applied a full loading configuration, i.e., total of nine bottles in a 3x3 array configuration. The following parameters were varied during the study: solution fill volume (50 %, 70 %, and 90 %), thawing rates (thawing was performed at room temperature with or without forced air flow, inducing either a fast or slow thawing rate, respectively), and F/T loading distance (containers were placed at 1, 4, or 10 cm apart during both F/T). Moreover, the F/T rate was monitored using type T thermocouples inserted at different axial and radial locations inside the solutions. The stability of the model protein formulated solutions was evaluated after each F/T cycle. Samples were analyzed by size-exclusion chromatography (SE-UPLC), and FlowCam® to study the effect of F/T process variables on the tetramer purity and sub visible particle formation, respectively. Moreover, protein concentration by UV-spectroscopy and LDH enzymatic activity bioassay analysis was also performed. Analysis from 22 experimental runs indicated the correlation between the required time to complete the F/T process, the distance among the containers, and the fill volume. The array with the smallest separation distance and higher fill volume required the longest time to both F/T processes to complete. In addition, a direct correlation was observed between the F/T rate and protein stability: process parameters yielding faster F/T rates showed the highest LDH stability; increased stability was observed as an increased enzyme activity and percentage of protein native structure. This study confirms the feasibility of a faster freezing rate and forced-air procedure during thawing to enhance the stability of proteins exposed to F/T unit operations. Data obtained may lead to a more thorough understanding of how F/T critical process parameters impact on protein stability. Also, the obtained results can be applied to establish optimal manufacturing conditions, to generate F/T process guidelines, and to guide the evaluation of novel bulk storage technologies.

538589 Effect of Solvent in Strip Film Manufacturing Containing BCS CLASS II Drugs VIA Solution Casting

Abstract:

Oral strip films are an emerging dosage form for drug delivery due to their many advantages over other oral dosage forms, especially in terms of patient compliance and potential for continuous manufacture. [1]. The three primary methods of strip film manufacture are: hot melt extrusion (HME) [2], solution casting [3] and slurry casting [4-6]. This study investigated the effect of solvent on the critical quality attributes of strip films loaded with BCS Class II drugs when using solution casting. The feasibility of solution casting with a binary solvent system (water and organic solvent mixture) was experimentally examined using hydroxypropyl methyl cellulose E15 (HPMC E15) as a film former, glycerin as a plasticizer, fenofibrate (FNB) as a model BCS class II model drug and three organic solvents: ethanol, methanol and acetone. The critical quality attributes that were systematically explored included: content uniformity, mechanical properties (Youngâ€™s Modulus, Tensile Strength and Elongation at Break), and dissolution rate of films made using the indicated solvents.

Solution casting was found to have many manufacturing problems starting from polymer solution preparation to drying. Each solvent had a unique impact on the viscosity of the polymer solution. As a result, each solvent formulation required different casting conditions to produce a final product with a consistent thickness. Additionally, the maximum drug loading was limited by the solubility of the solvent. Therefore, at drug loadings above the solubility limit for ethanol and methanol films, a slurry was obtained. Interestingly, phase separation occurred in the precursor made with acetone at high drug loadings (60% FNB in dry film).

All the solvents led to different particle shapes and crystallinity in the dry film. At a drug loading of 10% FNB in the dry film the films made with acetone were the strongest, while methanol led to the weakest films When the drug loading was decreased to 2.5% in the dry film, tensile strength of the films made with acetone were decreased while it was observed to increase for ethanol and methanol films. Although all the solution casted films had different properties, similar dissolution rates were obtained, which were improved compared to films loaded with as-received drug particles. Improved dissolution was attributed to presence of amorphous form of drug, although as expected, significant recrystallization was observed, which continued with time. All the solution casted films also had good drug content uniformity for 10% FNB loading in the film (RSD% $<$ 6), despite re-crystallization of the drug. These results suggest that, although solution casting seems to offer some benefits when creating a dosage form for poorly water soluble drugs, the requirement for careful formulation selection (solvent and polymer type) and unpredictable recrystallization, shows that slurry casting [4-6] is a better candidate for practical implementation.

REFERENCES

1. Dixit, R.P. and S.P. Puthli, *Oral strip technology: overview and future potential.* J Control Release, 2009. **139**(2): p. 94-107.
2. Palem, C.R., N.R. Dudhipala, S.K. Battu, M.A. Repka, and M. Rao Yamsani, *Development, optimization and in vivo characterization of domperidone-controlled release hot-melt-extruded films for buccal delivery.* Drug Dev Ind Pharm, 2016. **42**(3): p. 473-84.
3. Garsuch, V. and J. Breitkreutz, *Comparative investigations on different polymers for the preparation of fastâ €dissolving oral films.* Journal of Pharmacy and Pharmacology, 2010. **62**(4): p. 539-545.
4. Sievens-Figueroa, L., A. Bhakay, J.I. Jerez-Rozo, N. Pandya, R.J. Romanach, B. Michniak-Kohn, Z. Iqbal, E. Bilgili, and R.N. Dave, *Preparation and characterization of hydroxypropyl methyl cellulose films containing stable BCS Class II drug nanoparticles for pharmaceutical applications.* Int J Pharm, 2012. **423**(2): p. 496-508.
5. Krull, S.M., R. Susarla, A. Afolabi, M. Li, Y. Ying, Z. Iqbal, E. Bilgili, and R.N. Dave, *Polymer strip films as a robust, surfactant-free platform for delivery of BCS Class II drug nanoparticles.* Int J Pharm, 2015. **489**(1-2): p. 45-57.
6. Zhang, L., Y. Li, M. Abed, and R.N. Dave, *Incorporation of surface-modified dry micronized poorly water-soluble drug powders into polymer strip films.* Int J Pharm, 2018. **535**(1-2): p. 462-472.

531954 Droplet-Coalescence Kinetics for a Non-Newtonian Emulsion Using a Taylor-Couette Shear-Flow Reactor: Characterizing Phase-Separation Risk for a Pharmaceutical Ointment

Abstract:

Topical ointments are complex multiphase fluids, many of which are emulsions comprised of a droplet phase dispersed in a viscous, and often non-Newtonian, continuous phase (e.g., petrolatum or petroleum jelly). The drug substance to be delivered is present in solution within the dispersed droplet phase (propylene glycol in the present case). The physical stability of the ointment after emulsification, i.e., during subsequent processing and packaging, is of great importanceâ€"the growth of droplets and the risk of phase separation must be minimized for a viable commercial product.

Although simpler oil-water emulsions have received considerable attention, the behavior of emulsions containing complex, non-Newtonian fluids (e.g., petrolatum) has not been studied in detail. Specifically, the rate kinetics of droplet coalescence and growth are of interest to characterize the risk of phase separation. In the present work, this problem is addressed by studying an emulsion of white petrolatum and propylene glycol as a model ointment using a vertically-oriented, narrow-gap, Taylor-Couette reactor. This reactor allows control over the temperature and mean strain rate, thus allowing emulsion rheology, droplet coagulation, and phase separation to be investigated under varying shear-flow conditions.

Experimental results on the droplet coalescence mechanisms and rates, as well as phase separation, are generated for varying shear-strain rates using digital cameras and image-processing routines. The results demonstrate that the risk of phase separation increases with increasing strain rateâ€"this is quantified as a characteristic phase-separation time scale. Finally, the experimental results are linked back to the commercial-scale process to show that the operating conditions of the actual manufacturing process are indeed robust with a low risk of phase separation.

537814 Unique Polymorph and Amorphous Dispersion Formation of Suberic Acid Using Monodisperse Droplet Evaporation

Abstract:

In order to effectively deliver an Active Pharmaceutical Ingredient (API), pharmaceutical companies rely on specific particle properties including size, morphology, and structure to achieve desirable drug performance. The structure in particular, be it one of multiple crystalline polymorphs or an amorphous form, directly impacts the solubility of pharmaceutical compounds. As pharmaceutical molecules become ever more complex, their solubility in the body and resulting bioavailability is often decreased. Therefore, methods to understand the formation of different crystalline polymorphs, as well as amorphous dispersions of organic molecular systems is important for pharmaceutical development.

In this work, we use a monodisperse droplet evaporation process technology to form polymorphs that are not obtained with identical chemistry using other formation processes. Monodisperse droplet evaporation is achieved by use of a Vibrating Orifice Aerosol Generator (VOAG), which creates droplets of solution that are well dispersed from one another allowing for the solvent to rapidly evaporate at room temperature. The rapid droplet evaporation leads to high supersaturations, which drives a potential for formation of metastable polymorphs and amorphous dispersions. Suberic acid is the primary small molecule considered in this work first as a standalone system in multiple solvents to study suberic polymorphism. Secondly, it is mixed with polyvinylpyrrolidone (PVP) and dissolved in isopropanol to study the formation of amorphous dispersions. Our results demonstrate the ability of the VOAG to lead to a previously unreported polymorph of suberic acid. Further, we form a variety of amorphous dispersions for which we determine the dispersion stability as well as suberic acid solubility estimates in PVP.

531776 Importance of the Reaction Kinetics of Drug-Bile Micelle Formation in Oral-Drug Absorption Modeling

Abstract:

Understanding the behavior of drug molecules in the gastrointestinal tract is critical to quantifying drug absorption, thus, guiding the development of new drug products. Since in-vivo studies are challenging and expensive, computational modeling of oral drug absorption is a useful tool for guiding development efforts. To obtain accurate predictions, models must account for the drug interactions with bile, which is naturally present in the gastrointestinal tract. The dissolved drug molecules can bind within bile micelles, which form larger structures that cannot be absorbed through the intestinal membrane; only the free, unbound drug can pass through the intestine walls.

The formation/dissociation of drug-bile conjugates happen as reversible reactions, which can be represented as a set of coupled differential equations describing the reaction kinetics. Previously-developed models typically assume that equilibrium between free drug and drug-bile micelles is achieved instantaneously, i.e., infinitely high forward/backward reaction rates. In the present work, this instantaneous-equilibrium assumption is relaxed and the effects of finite rate kinetics of drug-bile micelle formation/dissociation are examined. Using a model-drug molecule (dipyridamole), the set of differential equations incorporating finite drug-bile reaction rates was developed and applied to a setup for in-vitro flux experiments. In combination with these in-vitro flux experiments, the model was used to estimate the forward/backward reaction rate constants for the association/dissociation of drug to bile micelles. Interestingly, it is seen that the reaction rates are sufficiently low that the system deviates substantially from the assumption of instantaneous equilibrium. Therefore, the finite nature of the reaction rates noticeably impacts the absorption rate, hence, the performance of the drug product.

To improve upon the forward/backward reaction-rate estimation method, a second, more detailed model is introduced which accounts for the presence of aqueous boundary layers near the membrane surfaces of the in-vitro experimental setup. This updated model will allow for more accurate determination of the reaction rate constants and, subsequently, their impact on absorption. Establishing the importance of fundamental drug-bile interactions, and subsequent implementation of these mechanisms in oral-absorption software, will aid the development of comprehensive models that can drive rapid, low-cost drug development.

543837 Anti-Static Agent Addition in Excipients â€" Always a Decent Way to Increase Powder Process-Ability in Pharmaceutical Industries?

Abstract:

Granular materials, fine powders and nanostructured powders are widely used in several industrial applications. In particular, excipients are necessary in many dry powder pharmaceutical formulations (dry powder inhaler, tablets, capsules). Thus, any additional information regarding their behaviour and any recommendation about the use of additives could have huge consequences for the optimization of industrial processes or to avoid technical issues (caking, clogging, noncompliance or unconformity of the by-product).

Moreover, to increase powders process-ability, it is commonly admitted to blend excipients and antistatic material, in order to decrease powders triboelectric effect. In this poster, we will show how different talc mass fractions additions affect excipients behaviour. For this purpose, two different lactose grades were selected: Inhalac 400 and Flowlac 90. Five talc mass fractions are studied: 0, 5, 10, 15 and 20%. The blends compaction dynamic (GranuPack instrument), electrical charges (GranuCharge instrument), granulation and dynamic flowability (GranuDrum instrument) are investigated. The additions are found to decrease the electrical charges in contact with Stainless-Steel material (for a specific amount of talc), to limit lactose agglomeration tendency, but also to decrease powder dynamic flowability. Moreover, for the Flowlac 90 it was found that the compaction dynamic is faster with a high fraction of talc.

It will be also shown, that the anti-static agent addition is not every time a decent solution.

544304 CRISPR-Cas9 Plasmid DNA Delivery to Endometrial Cancer Cells for Knockout of PLAC1

Abstract:

Several endometrial cancers overexpress the PLAC1 protein, and overexpression has been linked to more aggressive tumors. The CRISPR-Cas9 paradigm can target the gene locus for this protein specifically and effectively in order to reduce PLAC1 expression and subsequently reduce tumor proliferation.

We have identified and developed three CRISPR-Cas9 platforms encoding for varying degrees of DNA editing at the PLAC1 site. In order to enhance delivery of these plasmids into endometrial cancer cell lines *in vitro*, we have formed nanoplexes by combining plasmid DNA with the cationic polymer polyethylenimine. We varied the ratio of nitrogen in PEI to phosphate in the DNA (N/P) ratio to alter the surface charge and subsequently uptake.

These nanoplexes were approximately 100 nm in diameter with a net positive surface charge. We measured the cell viability and the relative gene expression using MTS and qPCR, respectively, in three different endometrial cancer cell lines. Each of these cell lines has a different p53 mutation, thus allowing us to investigate p53 status as a compounding effect. Our cell viability results indicated that loss of function p53 cells are most susceptible to the nanoplexes. Additionally, the CRISPR sequence that encoded for the least amount of DNA editing was sufficient to reduce PLAC1 gene expression, and was more effective than those that encoded for more editing.

To the best of our knowledge, this is the first investigation into the combinatorial effects of p53 status and PLAC1 expression on endometrial cancers. This research has the potential to improve treatment options for endometrial cancer and offers an alternative to conventional treatment methods such as chemotherapy and surgery.

544305 Sustained Release Polymeric Drug Delivery Systems to Inhibit ERK1/2 Activity

Abstract:

The ERK1/2 signaling pathway is upregulated following heart failure events, and PD98059 has been identified as drug to reduce ERK1/2 signaling. Administering PD98059 shortly after a heart failure event is effective at reducing long-term effects, however, delivering the drug at the necessary concentration for a sustained period of time is difficult.

We have developed nanoparticles, microparticles, and pellets from poly(lactic-*co*-glycolic acid) (PLGA) loaded with PD98059 with the objective of obtaining rapid, sustained release profiles. Loading and release were assessed using high performance liquid chromatography (HPLC). Optimization of formulations that provide early, but long-term sustained release was informed by experimental results.

PLGA particles, both with and without drug, were developed on the nanoscale (200 â€“ 400 nm) and the microscale (20-30 um). The lower limit of detection for drug release measured with HPLC was in the 50-100 ng/mL range. Release from particles began as early as a couple of days, whereas pellet formulations had later initial release and longer sustained release.

In developing various formulations, we were able to assess advantages and disadvantages of each as it pertains to clinical translation. The current treatment option requires that the patient remain in a hospital setting to receive a continuous infusion of soluble drug. However, the formulation presented here offers the potential for out-patient treatment therefore improving the quality of life.

546239 Functionalized Ultra-Thin Shell Microcapsule for Targeted Encapsulation and Release

Abstract:

A continuously growing area of controlled tunable transport and separation of therapeutic has been attracted attention to the stimuli responsive porous hydrogel thin film. The pore of these hydrogel film can be either increase or decrease under stimuli. Such responsive film can be used in the configuration of microcapsules with adjusted pores which has wide applications in filtration, separation and drug encapsulation/delivery. Droplet microfluidic offers unique opportunity for configurate these responsive film into microcapsules that fills the demands. We have fabricated a thin-shell microcapsules encapsulation system by microfluidic technique. This microcapsule with ultra-thin shell that can response multiple stimuli such as ionic strength and osmotic pressure. We also explore their application in controlled therapeutic encapsulation and release.

536541 Synthetic and Compositional Control of Multicompontent Copolymers to Promote Drug Solubility and Bioavailability

Abstract:

Synthetic polymers have enabled amorphous solid dispersions to emerge as a strategy for overcoming poor solubility and bioavailability of intractable drugs via oral administration. In such formulations, polymeric carriers generate and maintain supersaturation of drugs via noncovalent interactions, which serve to increase the apparent drug solubility upon dissolution and enhance gastrointestinal absorption and oral bioavailability. In this work, we use controlled polymerization to create a family of multicomponent copolymers containing numerous functional groups that promote hydrogen bonding and hydrophobic interactions with several drug structures such as phenytoin and nilutamide. Through systematic characterization and screening we show that macromolecular design parameters with comomoners significantly improve efficacy by using: i) one monomer to serve as a precipitation inhibitor that discourages drug recrystallization and ii) another comonomer to provide hydrophilicity and solubility of the hierarchical system. Systems that maintained drug supersaturation in amorphous solid dispersions were identified with molecular-level understanding of noncovalent interactions using NOESY and DOSY NMR spectroscopy. Through this approach, effective polymers have been discovered that promote high drug solubilization and in vivo efficacy such as a Poly(N-isopropylacrylamide-co-N,N-dimethylacrylamide) (poly(NIPAm-co-DMA)). In vitro dissolution tests and in vivo pharmacokinetics of select spray-dried dispersions created via this approach have shown polymer chemical composition and solubilization performance relationships, which will be discussed in detail.

524288 Nanocellulose Gels As a Flexible, High Surface Area Material for Crystallizing Pharmaceuticals

Abstract:

Cellulose is a highly abundant natural polymer that, being a plant-based material, exhibits renewability and biodegradability. It is frequently used in the form of microcrystalline cellulose as a filler and binder in pharmaceutical tablets. A nanoscale form of cellulose can be prepared by removing much of the amorphous cellulose material and leaving small crystalline particles with a large concentration of hydrogen bonding groups on the surface. Due to its strong hydrogen bonding tendency and the presence of charges on the nanocellulose surface, it has the potential to bind with drug molecules and act as a release inhibitor, a stabilizer or a site for heterogeneous crystallization of the drug itself. Beyond hydrogen bonding and charge, however, the nanocellulose surface can be readily altered to display a variety of functional groups, leading to flexibility for rational design of nanocellulose-drug composites. We use nanocellulose functionalization, such as carboxylation and sulfonation, in combination with surfactants such as octadecylamine to prepare gels of nanocellulose. Addition of pharmaceutical compounds to the gels results in changes in drug form, such as polymorph transitions or amorphous form formation due to molecular interactions between the drugs and the nanocellulose gels. The high surface area of the nanocellulose provides a high concentration of interaction sites and the small size of the nanocellulose excipient particles has interesting performance in promoting assembly and packing of the composite in drug products such as compressed tablets and dense free standing films.

528646 Controlling the Particle Morphology of Spray Dried Poly(methacrylic acid-co-methyl methacrylate) (Eudragit L100) Polymer

Abstract:

Title: Controlling the particle morphology of spray dried poly(methacrylic acid-co-methyl methacrylate) (Eudragit L100) polymer

Authors: Kimberly B. Shepard and Michael M. Morgen

Purpose: Poly(methacrylic acid-co-methyl methacrylate) (PMMAMA, or trade name Eudragit L100) is an enteric random co-polymer of methacrylic acid and methyl methacrylate with a high glass transition temperature (~190°C). Spray dried amorphous dispersions (SDDs, left figure) which use PMMAMA as a matrix polymer leverage these properties and often have excellent physical stability, limited gastric release, and fast dissolution rate at intestinal pH. In previous work, we have demonstrated that SDDs of PMMAMA, HPMC and a rapidly crystallizing drug can be formulated to achieve rapid dissolution and sustained supersaturation during *in vitro* dissolution testing. However, during scale-up of PMMAMA SDD manufacture, poor powder properties lead to very low yields, which complicate formulation and process development. In particular, clusters of string-like structures were found throughout the SDDs (see center Figure). In this work, we investigate a process and formulation space in which string formation can be minimized or eliminated.

Methods: PMMAMA SDDs were prepared on small scale (1-100g) or lab scale (100g-10kg) custom spray drying equipment from either 97/3 acetone/water or pure methanol solutions, with solids loading varying from 3-12%wt. Process parameters, such as inlet and outlet temperature (80-140°C, and 35-65°C), liquid atomization pressure (100-600psi), atomization gas pressure (5-30psi), nozzle type (e.g. 2-fluid or pressure swirl), solution temperature, were varied. SDDs were characterized via scanning electron microscopy (SEM).

Results: During a typical pressure-swirl atomization process, droplets are generated when the liquid sheet exiting the nozzle first breaks up to form long one-dimensional "filaments," which in turn de-stabilize and break up to form droplets. During spray drying, liquid droplets dry into particles via a skinning process, i.e. solvent evaporates rapidly from the surface of the droplet, leaving behind a solid-like polymer skin on the surface of the droplet that occurs at a critical concentration. We hypothesize that PMMAMA string structures form when polymer skinning occurs before the filaments fully atomize into droplets. SEM images show examples of "bead on a string" structures, where skinning occurred in the midst of droplet breakup (see right Figure). In these experiments, string formation was reduced by either increasing the time for droplets to skin (e.g. decreasing drying kinetics or decreasing solution concentration) or decreasing the time to droplet formation (e.g. increasing atomization pressure).

We have identified a robust processing space on both small-scale and lab-scale spray dryers where string formation is avoided. Decreasing the outlet temperature from 65°C to 35°C results in adequate drying while eliminating strings. By spray drying from a more dilute solution, the time to skinning can be significantly increased. A solids loading of 3% and 5% in 97/3 acetone/water shows negligible string formation. Increased atomization pressure reduces the time to droplet breakup and subsequent string formation for pressure-swirl and 2-fluid nozzles. Utilizing a slower-drying solvent, such as methanol, or reducing the temperature of the spray solution can also reduce string formation.

Conclusions: By controlling the spray solution solids loading, drying temperature, atomization conditions and other parameters, we identified a process space where PMMAMA SDDs can be manufactured without string formation and with acceptable particle properties (i.e. lack of strings). Stability and performance implications of this processing strategy are currently being evaluated. This process is applicable to other polymer systems in which string formation during spray drying is observed.

537276 Application of Fundamental Relationships and Models to Predict Spray-Dried Dispersion Particle Size

Abstract:

Purpose:

Atomization models can assist spray drying formulators in estimating ranges of droplet size that may be achieved for a given spray solution formulation and drying process. Droplet size can then be related to the final spray-dried dispersion (SDD) particle size by accounting for the fundamentals of droplet solidification and final particle density [1]. Specifically, for film-forming polymeric SDD systems, droplet drying kinetics and thermodynamics must also be considered due to their resulting particle morphological variations [2]. By combining these fundamental relationships with empirical correlations, a rational process development strategy can be employed early in the product lifecycle and give initial insight into future scale-up constraints or risks. Using basic solution properties as model inputs, multiple empirical approaches to predicting droplet size were evaluated. This study showed excellent correlation of predicted droplet size to the resulting final SDD particle size. In addition, use of a simplified porous particle approach that relies on the viscous gel point to predict particle size from droplet size was effectively demonstrated.

Methods:

Viscosity, surface tension and density were evaluated for a range of solids concentrations using two different dispersion polymers, HPMCAS-M (1-12 wt%) and PVP/VA (1-20 wt%) prepared in either acetone or methanol. Additionally, the viscous gel point (skinning concentration) was determined using a gravimetric method. Droplet size was measured sing a phase Doppler particle analyzer (PDPA) to study the effects of solids composition, nozzle geometry, and atomization parameters. Spray-drying experiments were then conducted at lab and feasibility scale to obtain final SDD particle size and evaluate potential impact of this approach on scale-up. SDDs were characterized for particle size distribution using light scattering methods, bulk/tap density, and morphology by scanning electron microscopy.

Results:

Droplet size predictions were conducted using an empirical model developed by Lefebvre:

These predictions correlated well with measured droplet size (PDPA) and measured SDD particle sizes (light scattering) for the formulations studied [3]. HPMCAS-M showed a steeper response of particle size to droplet size than PVP/VA possibly due to higher viscosities observed in solution at similar concentrations. Differences were observed between the relationship of predicted droplet size and SDD particle size based on the spray solvent used for both PVP/VA and HPMCAS-M polymers. A shrinking sphere model was initially applied to understand the relationship between droplet size and final particle size however, this approach was not found to demonstrate the desired predictability. Therefore, a porous particle approach was evaluated that accounts for the differences in skinning concentration of the polymers in various solvent systems. This was found to significantly improve the predictability of final SDD particle size.

Conclusions:

Application of simple models and fundamental relationships was shown to provide a high level of accuracy for predicting both droplet size and final SDD particle size. The utility of this approach can be further expanded to aid in scale-up predictions to rationally choose a droplet size to target a desired SDD particle size during initial process development and optimization on lab-scale spray dryers. Key differences between the dispersion polymer or solvent chosen were explained through either solution properties or relationship of polymer solidification.

[1] Singh, A. and Van den Mooter, G. Spray drying formulation of amorphous solid dispersions. *Adv. Drug Deliv. Rev.* **2015**.

[2] Vehring R. Pharmaceutical particle engineering via spray drying. Pharmaceut Res. 2008;25(5):999-1022.

[3] Lefevbre, Arthur. Atomization and Sprays. CRC Press, 1988.

538532 Improving Spray Drying Processing through Modeling and Characterization

Abstract:

Advances in engineering spray dried dispersion (SDD) for enhancing the oral bioavailability of drugs enable us to develop robust solid dispersion drug products. To this end, research in pharmaceutical spray drying has focused on coupling characterization of particles, solution and the spray with predictive modeling such as using computational fluid dynamics. Here, I will highlight two approaches we have used to gain fundamental understanding of spray drying process and the SDD particles.

To understand the microscopic characteristics of SDDs, we analysed the morphology and chemical distribution of the drug within SDD particles. Our work revealed the existence of local amorphous amorphous phase separation of drug from the polymer within particles. Using theoretical analysis, we elucidated the role of the complex drying kinetics and the thermodynamics of the process resulting in the observed SDD spatial and chemical microstructure. We also highlight the importance of probing the microscopic origins in such a system.

To understand the transport phenomena during spray drying, we turned to computational modelling of spray drying using solvents. We studied the variation in drying kinetics as a function of various processing conditions and the most influential processing parameters. We highlight the similarities and differences in drying when moving from a pilot to a commercial scale dryer. I will also briefly discuss the computational model of polymeric droplets to describe the specific drying trajectories, accounting for the formation of a polymeric skin on the droplet surface, the variation of the solvent diffusivity throughout drying.

533575 Mechanistic Approach to Predict Amorphous Solid Dispersion Thermal Degradation in Spray Drying Processes

Abstract:

Mechanistic approach to predict ASD thermal degradation in spray drying processes

G. Poeiras[1,3], T. Porfírio[1,2*], C. Sá Couto[1], J. Pereira[1], R. Silva[1], I. Duarte[1], M. D. Afonso[3], J. Vicente[1]

[1]Hovione Farmaciência SA, Sete Casas, 2674-506 Loures, Portugal; *tporfirio@hovione.com

[2]LAETA, IDMEC, Mechanical Engineering Department, Instituto Superior Técnico, Universidade de Lisboa, Av. Rovisco Pais 1, 1049-001 Lisbon, Portugal

[3]CeFEMA, Chemical Engineering Department, Instituto Superior Técnico, Universidade de Lisboa, Av. Rovisco Pais 1, 1049-001 Lisbon, Portugal

The number of drugs with solubility limitations is increasing, being the limited aqueous solubility a major challenge in the development of oral-dosage forms, as it may affect oral bioavailability. Several solubilization strategies have been developed, being the amorphous solid dispersions (ASD) one of the most common. Spray drying (SD) process is one of the most widely used technologies for ASD production allowing the control of the product attributes, viz. particle size, density and physical state. SD is a solvent based process that often includes a secondary drying (post-drying stage) to reduce residual solvent content to values below ICH guidelines (International Conference on Harmonization of technical requirements for registration of pharmaceuticals for human use). A key processing phenomena in both equipment is the particles deposition on the walls and the subsequent build-up that affects process operation, safety and may impact the product quality through the deposited particles degradation, Figure 1 [1]. For that reason, it is crucial to assess the product chemical and physical quality, which have to be guaranteed throughout the process.

The degradation is intrinsically dependent on the process conditions in which the product is exposed, such as temperature and saturation. Computational Fluid Dynamics (CFD) may be used to model the temperature profile throughout the chamber and provide an estimation of the product exposure during a manufacturing batch [2]. Moreover, CFD simulations forecast possible zones of wall accumulation. The output of such analysis supports the assessment of the accumulated product quality. Subsequently, the goal is to build a general model capable of predicting the ASD chemical degradation kinetics for any combination of temperature and relative saturation.

Figure 1 – Process exemplification including product accumulation.

As for the physical stability, the thermodynamic properties and molecular mobility inherent to the amorphous state are responsible for the formation of crystalline material over time, compromising the bioavailability and performance of the solid dosage form. Molecular mobility is more or less pronounced according to the process conditions that the product is subjected to, namely temperature, humidity and saturation. The experimental determination of these molecular motions is related to the study of structural relaxation processes, physical aging or annealing, which consist in the continuous evolution of the amorphous product towards a more stable equilibrium state at a given temperature. Structural relaxation can be predicted from the sample thermal history and its degree of non-equilibrium [3].

In sum, this work developed a mechanistic approach to predict ASD stability based on (Figure 2):

· CFD to model and predict temperature and saturation profiles inside the spray drying chamber and/or post drying equipment;

· Estimation of the degree of chemical degradation using a kinetic model;

· Physical stability using structural relaxation studies.

A case study comprising the spray drying and/or post-drying steps using a model drug to form an ASD will be presented following the proposed approach. In this case study, the operating conditions ranges, namely temperature and relative saturation, were defined based on the results provided by the CFD modelling and also aiming to stress the ASD chemically and physically.

Figure 2 – Proposed approach for ASD thermal degradation study.

-

References

[1] S. Keshani, W. R. W. Daud, M. M. Nourouzi, F. Namvar, and M. Ghasemi, "Spray drying: An overview on wall deposition, process and modeling", Journal of Food Engineering, pp. 152–162, 2015.

[2] L. Huang and A. S. Mujumdar, "Simulation of an Industrial Spray Dryer and Prediction of Off-Design Performance", Drying Technology, vol. 25, pp. 703–714, 2007.

[3] R. Hilfiker, Polymorphism in the Pharmaceutical Industry, pp. 260-281, 2006.

Downstream Processing of a Ternary Amorphous Solid Dispersion: the Impacts of Spray Drying and Hot Melt Extrusion on Powder Flow, Compression and Dissolution

Mark T. Davis[1*], Catherine B. Potter[1] & Gavin M. Walker[1]

1. The Synthesis and Solid State Pharmaceutical Centre (SSPC), Bernal Institute, The University of Limerick, Ireland
*email: mark.davis@ul.ie; phone: +353 61 213096

Abstract

Downstream processing aspects of a stable form of amorphous itraconazole exhibiting enhanced dissolution properties were studied. Preparation of this ternary amorphous solid dispersion by either spray drying or hot melt extrusion led to significantly different powder processing properties. Particle size and morphology was analysed using scanning electron microscopy. Flow, compression, blending and dissolution were studied using rheometry, compaction simulation and a dissolution kit. The spray dried material exhibited poorer flow and reduced sensitivity to aeration relative to the milled extrudate. Good agreement was observed between differing forms of flow measurement, such as Flow Function, Relative flow function, Flow rate index, Aeration rate, the Hausner ratio and the Carr index. The stability index indicated that both powders were stable with respect to agglomeration, de-agglomeration and attrition. Tabletability and compressibility studies showed that spray dried material could be compressed into stronger compacts than extruded material. Blending of the powders with low moisture, freely-flowing excipients was shown to influence both flow and compression. Porosity studies revealed that blending could influence the mechanism of densification in extrudate and blended extrudate formulations. Following blending, the powders were compressed into four 500 mg tablets, each containing a 100 mg dose of amorphous itraconazole. Dissolution studies revealed that the spray dried material released drug faster and more completely and that blending excipients could further influence the dissolution rate.

Keywords: Amorphous, itraconazole, flow, compression, tabletting, dissolution

1. Introduction

Amorphous solid dispersions (ASD) have received a surge in research interest over recent decades, due to the large and increasing number of poorly soluble drugs on today's market(Brouwers et al., 2009; Leuner and Dressman, 2000). Although several examples of commercial ASDs exist(Huang and Williams III, 2017), the technology has sometimes lost out to other bioavailability enhancement strategies, often due to problems with recrystallisation or perceived difficulties in downstream processing. Although publications addressing recrystallisation issues are manifold, far less research has been reported addressing the issues of flow and compression(Demuth et al., 2015).

Efficiency of flow and compression can affect processing times and tablet quality. Bridging, arching or rat-holing for instance, can lead to increased processing times, inaccurate die fill and unacceptable dose variation. Compression issues can cause sticking, picking or flaking of tablets and ultimately reduction in patient compliance(Jain, 1999; Leuenberger and Rohera, 1986; Patel et al., 2006). In an age of greater focus on quality by design (QbD), global manufacturing competition and 'right first time' demands, the need to resolve or minimise such issues at the research stage is stronger than ever.

The purpose of this study was to assess the impacts of spray drying (SD) and hot melt extrusion (HME) on the flow, compression and dissolution characteristics of an ASD containing itraconazole (ITZ). Comparisons between properties of amorphous dispersions prepared by SD and HME are popular,(Davis and Walker, 2018) although most of these do not specifically deal with downstream processing issues. We have recently reported the preparation of a novel ternary ASD using both SD(Davis et al., 2017) and HME(Albadarin et al., 2017). Ternary dispersions have become increasingly popular(Davis and Walker, 2018) in the struggle to meet the sometimes competing demands of ASDs: increased bioavailability, adequate stability, efficient processing, the use of existing plant equipment and bearable costs. The formulations, containing drug and excipient polymers Soluplus® and HPMCP, displayed excellent stability and enhanced dissolution in both studies(Albadarin et al., 2017; Davis et al., 2017). The following is an account of the impact of preparation on the downstream processing of one of those formulations.

2. *Materials and Methods*

2.1 Materials

ITZ (>99%) was purchased from Xi'an Lyphar Biotech Ltd. Accurate analytical standard of itraconazole (99.8%) for HPLC calibration was purchased from Sigma Aldrich. Polymers Soluplus® & HPMCP HP-55 (HPMCP) were donated by BASF and Shin Etsu respectively. Excipients Avicel 101 & Avicel PH 200 LM were donated by FMC and Mannogem granular was donated by SPI Pharma. Solvents and chemicals were purchased from Sigma-Aldrich and were of HPLC and reagent grade.

2.2 Spray drying

Spray drying was conducted using a Büchi B-290 mini spray dryer connected to a B-295 inert loop. A 2-fluid pneumatic nozzle (0.7 mm) was fitted and liquid feedstock was pumped at 4 mL/min (15%). Drying nitrogen was set at maximum by fixing the aspirator at 100% (35 m^3/h). A solution of ITZ-Soluplus®-HPMCP (30-40-30 w/w) in dichloromethane-methanol 7-10 (v/v) (1.0 L, 10.0 % w/v) was pumped through the nozzle at inlet temperature 100 °C, producing an outlet temperature of 65 ± 2 °C. High yields (88.6%) were achieved by using high spray gas rates (670 L/h) and attaching the high efficiency cyclone option. Powdered product was immediately transferred into a stainless steel pan and dried overnight in a vacuum oven at 40 °C and 10^{-2} mBar to ensure complete removal of residual levels of solvent. Powder was then stored in a sealed glass bottle in a desiccator over anhydrous molecular sieves.

2.3 Hot melt extrusion and milling

Itraconazole, Soluplus® and HPMCP 30:40:30 (w/w) (50g) were pre-mixed by hand in a polythene bag, having first removed larger particles of HPMCP, by passing through a 435 μm sieve. The pre-mixed powders were then manually fed into a Three-Tec twin-screw extruder (Three-Tec GmbH, Germany) with a barrel diameter of 12 mm and an L/D ratio of 40:1. The set points of the six heating zones, from feed to die, were 80, 110, 120, 140, 150 and 150 °C, respectively and screw speed was maintained at 15 rpm throughout. Following extrusion the material was milled for 1 min at 20 Hz using a 25 mL stainless steel jar attached to a Retsch Mixer Mill MM 400 ball mill (Retsch GmbH, Germany). Milled extrudate was sieved through 435 and 90 μm sieves and the 90 – 435 μm and < 90 μm fractions were stored separately in sealed glass bottles in a desiccator over anhydrous molecular sieves.

2.4 Scanning electron microscopy (SEM)

Drug polymer formulations were attached to double sided carbon tape and sputter coated with a thin layer of gold followed by imaging using a Joel CarryScope JCM-5700 scanning electron microscope. Micrographs were recorded at various magnifications, using a beam acceleration voltage of 5 kV, a spot size of 40 and a working distance of 12.

2.5 Powder Rheometry

Powder flow was analysed using an Freeman FT4 powder rheometer in a manner previously reported(Freeman, 2007). Stability and Variable flow rate was performed in the 25 mm vessel, Aeration in the 25 mm vessel and Shear in the 10 mL and 1 mL vessels.

2.6 Compaction simulation

Powders were tableted using a Gamlen R-series tablet press containing a 500 kg load cell and 6 mm punch and die. The material was subjected to uniaxial compression tests to form flat-face cylindrical tablets with a target weight of 90 mg. Punch load and displacement were measured during compression and the effect of compaction forces from 0.5 to 5 kN, equivalent to pressures of 17 to 173 MPa, was studied. Tablet hardness, defined as the diametral force required to break the compact, was investigated using a PharmaTest hardness tester (Pharma Test Apparatebau AG, Germany) immediately after tabletting.

2.7 True density

The true density of the extruded and spray dried products and powder blends was measured under ambient conditions using an helium pycnometer (AccuPyc II 1340, Micromeritics Inc., Norcross, GA). The volume of the pycnometer was verified with a standard before carrying out analysis. During sample analysis five purge cycles were recorded for samples and the reported result is an average of measurements made on three different days.

2.8 *Formulation and tabletting*

ASD products and excipients were accurately weighed as described in Table 4 and the individual components were vortex mixed for 2 min in 10 mL vessels. The resultant prototype tablet formulations were pressed according to scalable parameters determined through small scale compaction simulation performed on 90 mg material. Blends were weighed into a 0.5 inch (12.7mm) diameter die and then pressed at 1000 kg pressure on a manual hydraulic tablet press to form flat-face cylindrical tablets with a target weight of 500 mg.

2.9 *In-Vitro Dissolution Studies*

Drug release studies were carried out using non-sink conditions, in a USP type II apparatus, PharmaTest dissolution tester (Pharma Test Apparatebau AG, Germany), with 900 mL 0.1 *N* HCl buffer (pH 1.2) per well, a paddle speed of 100 rpm and a temperature of 37 ± 0.2 °C. Formulated tablets of mass 500 mg, equivalent to 100 mg of drug were added to the dissolution vessel (n = 3), with start times staggered by 2 min. 5 mL aliquots of media were withdrawn at predetermined intervals (0.25, 0.5, 0.75, 1, 2, 3, 4, and 6 h), filtered through a 0.45 μm polytetrafluoroethylene (PTFE) syringe filter and then the first 1 mL of filtrate was discarded. 500 μL of the filtrate was subsequently added to 500 μL of HPLC mobile phase in

2 mL HPLC vials. 5 mL of fresh medium was immediately replaced following each withdrawal.

2.10 High performance liquid chromatography (HPLC)

Dissolution samples were analysed using an Agilent (Agilent, Little Island, Cork) 1260 Infinity II high performance liquid chromatography system, comprising of quaternary pump G1311B and diode array detector G1315D set at wavelength 263 nm. The thermostated column compartment G1316A was set at 25.0 °C and equipped with a Kromasil 100 C18 5 µm 250 x 4.6 mm (Kromasil, Mainz, Germany) RP-HPLC column. The system was operated under isocratic flow at 1 mL/min using, a mobile phase of acetonitrile:water:diethanolamine (69.95/30/0.05, v/v/v). Samples were injected from autosampler G1329B in volumes of 20 µL and data was collected and analysed by Agilent OpenLAB CDS Chemstation software. Standards of ITZ (99.8% purity) were prepared from a stock solution of 1 mg/mL in methanol, to quantify the levels of drug in the dissolution media.

3. Results and Discussion

3.1 Preparation of Amorphous Solid Dispersions by Spray Drying and Hot Melt Extrusion

In our previous studies we discovered that it was possible to prepare novel ternary amorphous dispersions of BCS class II drug itraconazole by hot melt extrusion(Albadarin et al., 2017) and by spray drying(Davis et al., 2017). Following rational polymer selection(Davis et al., 2015), these blends proved to have excellent stability and enhanced dissolution profiles. As design of experiments (DoE) had previously been performed during formulation design, only one blend was selected for downstream processing. A ternary ratio of itraconazole-Soluplus®-HPMCP 30-40-30 (w/w) was selected to test a high drug load and simultaneously facilitate HME processing, by choosing a Soluplus® ratio that was slightly higher than that of HPMCP.

Particle size is a critical factor in powder flow. Often the particles from spray drying processes are very small (< 10 µm)(Vehring, 2008), which can be excellent for dissolution, but challenging for powder flow. It is possible to increase particle size by changing nozzle type to a pressure nozzle(Beyerinck et al., 2010), processing parameters, drier geometry or feed stock concentration. Of these options, the most directly available was an increase in feedstock concentration, although it is likely that a pressure nozzle would be more effective. More concentrated feed solutions produce more viscous droplets and hence larger particles(Vehring, 2008). For this reason the feed was prepared at 10% (w/v), following optimisation experiments on the dichloromethane-methanol solvent system(Davis et al., 2017), leading to an optimal ratio of 7-10 (v/v). In addition, the yield of the spray dried powder was optimised by tuning processing parameters. In particular, when the spray gas rate was increased from 25 mm (301 $L.h^{-1}$) to 55 mm (670 $L.h^{-1}$) on the flow meter, the yield increased from 51% to 89%.

With hot melt extrusion, it was less difficult to prepare the quantities of material required (at least 100 g), as the materials are not diluted in liquid before feeding to the equipment. However, it was found that sieving, pre-mixing and manually feeding the ternary materials was favourable, to promote adequate mixing and prevent system overload by excess torque. Early runs lead to the production of opaque amber material and excessive torque. The torque issue was ameliorated by modifying the ternary blend mixture from itraconazole-Soluplus®-

HPMCP 30-35-35 to 30-40-30 (w/w), as the HPMCP was difficult to extrude, but the Soluplus® processed very well. In addition, the adjustment of zone temperatures by increasing zones 4 and 5 from 130 °C and 140 °C to 140 °C and 150 °C respectively, resulted in production of glassy yellow material. After cooling, the glassy extrudate was milled, sieved and stored.

Cleaning of the extruder was also challenging, requiring addition of pure Soluplus® as a 'cleaning polymer', followed by hot removal of thermocouples, hot removal of barrels from screws, cooling, separation of barrel blocks using levers and finally immersion in warm 1% surfactant with sonication.

3.2 Particle size and morphology

The impact of particle size and morphology on powder flow is well known and can drastically effect the efficiency of pharmaceutical powder processes, such as die filling(Gallo et al., 2011; Hancock et al., 2002; Nokhodchi et al., 2007; Sinka et al., 2004; Sivasankaran et al., 2010; Wu and Cocks, 2004; Xie and Puri, 2006). Variation in powder composition or flow can lead to unacceptable deviations in tablet weight and content uniformity, possibly leading to the rejection of a batch in extreme circumstances. Poor flow can potentially also lead to inefficient or even dangerous processes on the plant, caused by bridging or caking of powder masses during storage in containers or silos and the potential for powder flooding. Because spray dried powder contains particles mostly < 10 μm and milled extrudate < 90 μm (or 90 – 435 μm, for the larger sieved extrudate fraction), the cohesion and flow of these powders will differ profoundly and require differing hopper angles, orifices or alternative storage design for safe and efficient handling.

Scanning electron microscopy (SEM) was performed on the materials to provide insight into particle size and morphology. Itraconazole was received as relatively flat, wedge-like crystals (Figure 1a) approximating 40 μm length by 10 μm width, with a mostly smooth surface that nevertheless displayed several brittle shards. Soluplus® was received as much larger, approximately spherical granules (Fig 1b) with a diameter approximating 250 μm. HPMCP (Fig 1c) was received as pod- or cocoon-like elongated particles approximating 350 μm by 70 μm, with a smooth but broken surface. The physical mix was a combination of the above, although the difference in size and shape between the small itraconazole particles and the large HPMCP particles was particularly apparent (Fig 1d). This shaped the decision to sieve the HPMCP before powder blending and extrusion.

The extruded particles were ball milled and sieved prior to SEM analysis. The particles appeared as irregularly shaped granules, comprising considerably more material per granule than the spray dried particles. There were many rough spots on the surface of each individual granule, representing possible areas for particle interlocking and cohesion. Particles were typically 50 μm x 55 μm in the < 90 μm sieved fraction (Fig 1e) and 240 μm x 350 μm in the 90 - 435 μm sieved fraction (Fig 1f). The spray dried sample appeared as a combination of some hollow spheres with diameters of up to 30 μm (Fig 1g), some burst hollow spheres (Fig 1h) of similar diameter, but mostly crumpled spheres of less than 10 μm, that had grouped into agglomerates of various shapes, with 40 - 60 μm in total diameter. The particles size of the spray dried material in this study represented an increase to that of our previous work, where a lower feed concentration of 0.5% (w/v) typically led to particle diameters of 1.5 - 2.4 μm, depending on formulation(Davis et al., 2017). Although increasing the feed concentration did increase the particle size, the spray dried particles were still very small.

While this is advantageous for dissolution, the powders are expected to be quite cohesive during flow analysis.

3.3 Powder Flow

3.3.1 Powder stability during flow

Because unknown powders may be prone to attrition, agglomeration or caking, it is important to begin by measuring whether the powder will change during the course of testing. To discover if the flow properties alter under the stresses imposed during flow,(Iveson et al., 2001; Salman et al., 2002) the 'stability program' runs the powder through a series of identical conditioning and test cycles: if the powder is stable then the results from each measurement are similar. This is observed as a straight line from points 1 through to 7 in Figure 2. The stability index, defined in equation 1, is the ratio of energy in test 1 to energy in test 7:

$$Stability\ index, SI = Energy\ Test\ 7/Energy\ Test\ 1 \qquad (1)$$

Generally, a stable powder will have an SI approximating 1, powders with SI > 1 are unstable, affected by de-aeration, agglomeration, segregation, moisture uptake or electrostatic charge and powders with SI < 1 are unstable, affected by attrition, de-agglomeration, over-blending or coating of the vessel and blade by an additive(Freeman, 2007). The data in Figures 2 and 3 demonstrated that milled extrudate (HME & BM), physical mixture, itraconazole and spray dried ASD were largely stable to flow rate changes, with SI values of 0.96 ± 0.06, 1.08 ± 0.11, 0.97 ± 0.18 and 1.24 ± 0.26, respectively. Only excipients Soluplus® (1.91 ± 0.35) and HPMCP (1.35 ± 0.17) were unstable with respect to flow stress, with the stability results for HPMCP particularly inconsistent (data not shown). The underlying cause for the instability may be connected to static charge accumulation in HPMCP during testing.

3.3.2 Powder variable flow rate during flow

The rate at which powders are handled and transferred varies and the sensitivity of a powder to this is an important factor. Often, cohesive powders are more sensitive to changes in flow rate than non-cohesive or granular materials. During the variable flow rate test (Figure 2), the flow rate is reduced from 100 mm/s to 10 mm/s over tests 8-11. The flow rate index (FRI) is given by equation 2:

$$Flow\ rate\ index\ (FRI) = Energy\ Test\ 11/Energy\ Test\ 8 \qquad (2)$$

Usually, a high flow rate sensitivity (usually for very cohesive powders) is depicted as FRI > 3.0; an average flow rate sensitivity (including most powders) lies between 3.0 > FRI > 1.5; flow rate insensitive materials (with large PSD or surface coating) have FRI ≈ 1.0 and pseudoplastic or Neutonian Flow rates (such as powders containing flow enhancers like magnesium stearate) are found with FRI < 1.0(Freeman, 2007).

As depicted in Figures 2 and 3 the experimental FRI values recorded were as follows: itraconazole 1.89 ± 0.03, Soluplus® 1.28 ± 0.06, HPMCP 1.63 ± 0.24, HME/BM 1.40 ± 0.02; physical mix 1.68 ± 0.11 & spray dried 3.73 ± 0.35. So, most of the materials can be classified as having average flow rate sensitivity, with the Soluplus® and milled extrudate edging towards being insensitive. The clear exception was the spray dried material, with a high FRI value of 3.73, as would be expected for a cohesive powder comprised of very small particles. The difference between the spray dried ASD and the milled extrudate ASD was 3.73 – 1.40 = 2.33 units. It would therefore be reasonable to expect a significantly different sensitivity of these two ASDs to flow rate during processing, despite their identical chemical compositions.

3.3.3 Powder Aeration Studies

The bulk properties of all powders are to some extent influenced by air because the inter-particulate spaces themselves contain air. The volume of air impacts interparticular interactions and flow. Some powders are readily aerated and require only modest quantities of air to transform into a fluidised bed. Others are cohesive and therefore not easily aerated, although the flow properties may still improve with increasing air entrainment. The aeration ratio (AR) is given by the equation:

$$Aeration\ Ratio, AR = Energy\ (Air\ velocity\ 0)/Energy\ (Air\ velocity\ 10) \qquad (3)$$

Powders with AR ≈ 1 are not sensitive to aeration and are generally very cohesive. An average sensitivity to aeration with 2 < AR < 20 is found for most powders. Very sensitive powders with AR >> 20 will most likely become fluidised easily. Referring to Figure 4, the three raw materials and the two ASD powders were measured on the aeration program and produced the following AR results: HPMCP = 1.38 – 1.68 (not sensitive, very cohesive); Soluplus® = 0.45 – 0.54 (not sensitive, special case); itraconazole = 1.62 – 3.51 (cohesive to average); milled extrudate = 45.9 – 58.2 (very sensitive, non-cohesive); spray dried = 1.73 – 2.02 (not sensitive, very cohesive)(Freeman, 2007).

These results can be rationalised by considering the particle size and morphology of the powders (Figure 1 & section 3.2). HPMCP particles are elongated along one axis and typically several hundred μm long, so a preference for horizontal stacking may add to cohesion. Soluplus is spherical and around 300 μm in diameter, the spherical shape is quite aerodynamic and allows a laminar flow around the spheres, leading to the very low AR value of 0.48. This may also explain the unexpected requirement for more energy as the blade flows through the sample at higher aeration velocities. Itraconazole comprises straight, flat crystalline wedges of approximately 10 μm x 40 μm, so it is unsurprising that this material is also cohesive and AR insensitive. The test on the milled extrudate was performed prior to sieving. The particles contain a significant fraction of larger particles (> 300 μm), with various shapes and sizes, so the material was non-cohesive and very sensitive to aeration. Finally, the spray dried material contains hollow, burst and crumpled spheres, with many particles < 10 μm. The material is very cohesive due to the tiny particle size and therefore AR insensitive

The average AR values for the milled extruded and spray dried samples were 50.23 ± 6.89 and 1.88 ± 0.15 respectively, a difference of 48.35 AR units, representing a significant difference in sensitivity towards aeration. The result has implications for understanding Würster coating and other processes that intentionally use air or for flow by gravity (Bruni et al., 2007; Freeman, 2007; Leturia et al., 2014). In particular, die filling during tabletting requires both that the powder fill the die by gravity and also that the air in the die exits easily through the particles as they ingress(Sinka et al., 2004; Wu and Cocks, 2004). As speed of tabletting increases this becomes more important.

3.3.4 Powder shear testing

Shear cell testing is perhaps the most studied and understood flow test performed using powder rheometry(Fitzpatrick et al., 2004; Jenike, 1964; Schulze et al., 2008; Schwedes, 2003; Walker, 1966). The procedures and testing sequences have been reported elsewhere(Freeman, 2007), but briefly consist of powder conditioning, consolidation, vessel splitting, re-consolidation and finally shear testing. The tests were performed in triplicate to increase accuracy. FT4 software enabled automatic construction of Mohr circles, providing valuable data such as cohesion (C), unconfined yield strength (UYS) , major principle stress (MPS), minor consolidation stress (MCS), angle of internal friction (AIF), Peschl's Relative Flow Function (Rel_P) (Peschl and Colijn, 1977) and Jenike's Flow function (FF of ffc)(Ennis, B J; Witt, W; Weinekoter, R; Sphar, D; Gommeran, E.; Snow, R H; Allen, T; Raymus, GJ; Litster, 2007; Freeman, 2007; Jenike, 1964). The equations for FF and Rel_P are given as follows:

$$Flow\ Function, FF = MPS\ /UYS \quad (4)$$

$$Rel_P = (MPS - MCS)/UYS \quad (5)$$

In Figure 5, the yield loci (plot of applied normal stress vs shear stress) of itraconazole, the physical mixture and the two ASD powders were plotted with the corresponding shear parameters reported in Table 1. The most frequently reported value derived from shear cell data to describe flowability is the Jenike Flow Function (FF of ffc)(Freeman, 2007; Jenike, 1964). The FF values for itraconazole (2.03 ± 0.03) and the spray dried ASD (3.45 ± 0.18) indicated that they were cohesive, the physical mixture was very cohesive (1.98 ± 0.25) and the milled extruded ASD was free-flowing (57.01 ± 53.62). The low standard deviation values indicated that repeatability was excellent for the cohesive powders but rather poor for the free-flowing extrudate, demonstrating that this test was more accurate for cohesive powders(Freeman, 2007). Similarly (but with subtle differences in cohesive ranking), the Relative Flowability (Rel_P)(Peschl and Colijn, 1977) values indicated that itraconazole and the physical mix were extremely cohesive, the spray dried ASD was very cohesive and the milled extrudate was cohesionless (Tables 1 & 3).

The highest values of Unconfined Yield Strength (UYS) were found in the physical mixture (11.83 kPa) and itraconazole (10.75 kPa), with much lower values for the spray dried (4.14 kPa) and milled extrudate (2.98 kPa) ASDs, respectively. This was expected, as more cohesive powders will generally have higher values for UYS(Freeman, 2007). In addition, a steeper yield locus often represents a powder that is more sensitive to increasing levels of consolidation. The highest AIF value was for the physical mixture (44.09 ± 8.75°), followed by itraconazole, roughly equal to the extruded/milled ASD and finally the lowest value was for the spray dried ASD (27.75 ± 0.61°). Therefore the pure drug was most sensitive to increasing consolidation and the spray dried ASD was least sensitive overall and less sensitive than the milled extrudate (by 8.64° on average). Consolidation results have implications for packaging, shipping and processing considerations.

3.3.5 A comparison of powder flow testing parameters

To better understand the accuracy and precision of the various parameters used to measure flow, simple mass/volume measurements were taken to calculate the Hausner ratios and Carr's indices for the powders. Although these parameters have little theoretical basis, they are easily obtained and have been used to describe powder flow for many decades, so their inclusion seems worthwhile. The FT4 rheometer conditions the powder prior to recording bulk density and this imparted a high degree of repeatability (accuracy) on the values obtained (Table 2), with standard deviations (n = 3) between 0.01 and 0.08 g/mL for the 5 powders included. Conversely, the 'measured' values were simply taken as a recording of pouring the powders from their containers in whatever state they happened to be resting in (raw materials were stored in 1 kg bags (HPMCP), 5 kg bottles (Soluplus) and 1 kg bottles (itraconazole) and ASDs (100 g) in sealed glass jars in a desiccator) into a pre-weighed graduated cylinder then re-weighing to get the density and this was performed once. Although the repeatability of bulk density values for Soluplus® and HPMCP from the FT4 and direct measurements was excellent (± 0.01 g/mL) and those of spray dried and milled extrudate were comparable (± 0.05 & 0.10 g/mL respectively), the value for itraconazole differed wildly (± 0.19 g/mL, or 37%) (Table 2). The net effect of these differences is demonstrated by Carr index values of 43.39 and 10.71 for itraconazole depending on the bulk density value taken - this clearly demonstrated the requirement for a more consistent method for measuring powder flow, particularly for powders like itraconazole that are both cohesive (C = 2.55 ± 0.17 kPa) and sensitive to consolidation (UYS = 10.75 ± 0.56).

To explore these points further, the various flow parameters obtained from the FT4 rheometer were tabulated with the Hausner ratio and Carr's index in Table 3. For itraconazole, interpretations of values for the Flow Function, Relative Flowability, Flow rate index, Aeration ratio, Hausner ratio and Carr's index were in good agreement with each other: cohesive, extremely cohesive, average flow rate sensitivity, average aeration sensitivity, extremely poor flow and extremely poor flow, respectively. Similar agreement was observed for Soluplus®, HPMCP, physical mixture, spray dried ASD and the milled extrudate. The high level of agreement between results suggests precision in the flow parameters from the instrument. In general, the accuracy of the measurements from the FT4 was excellent. However, while analysing the milled extrudate sample, although the FRI measurement was very accurate (SD ± 0.02), the results for Rel_P and FF were very inaccurate (± 40.07 and ± 53.62 respectively), reflecting the lack of sensitivity of shear testing towards larger particles.

3.3.6 *The effect of additional excipients on ASD powder flow*

To see if flow could be improved by blending with appropriate excipients, the spray dried and the milled extrudate ASDs were blended (separately) with Avicel PH 200 LM (Av200) and Mannogem granular (ManG) to give four formulations (Table 4). In order to enhance powder flow, these excipients had relatively larger particle sizes (on average 200 μm and 400 μm, respectively), although the difficulty in homogenously mixing powders of differing sizes is well documented. Furthermore, both excipients were specially treated commercial forms of MCC and mannitol respectively, both known for their excellent compression characteristics. Thirdly, due to the moisture-sensitive nature of amorphous dispersions, both excipients were selected for their low moisture ('LM') or non-hygroscopic (mannitol) properties. To achieve a 100 mg drug dose in the final tablet (the standard itraconazole dosage) required 333.3 mg (66.6% w/w) of the ASD, allowing only 150 mg (30% w/w) of excipient for the 500 mg total, with disintegrant and lubricant accounting for the remaining 3.4% (w/w).

In order to analyse the effect of these excipients on powder flow, further shear studies were performed using the 1 mL annular shear cell (due to lower powder quantities). The yield locus plots are depicted in Figure 7 with shear data in Table 5. Considering the spray dried ASD and formulations, addition of Av200 reduced cohesion by 0.46 kPa and increased FF by 1.01 units. Conversely, addition of ManG increased cohesion by 0.25 kPa and decreased FF by 0.25 units. Considering milled extrudate the picture was dissimilar: addition of Av200 reduced cohesion by 0.24 kPa and increased FF by 3.56 units, but addition of ManG *further reduced* cohesion by 0.46 kPa and further increased FF by 14.18 units. Clearly the enhancement or hindrance of flow by addition of further excipients is a combination of many complex factors, is specific to the precise set of interparticulate forces acting between the given excipient and dispersion and is therefore difficult to predict or simulate.

3.4 *Compression*

3.4.1 Tabletability and Compactability of raw materials and ASDs

In order to better understand the effects of ASD preparation on tabletability (compaction pressure vs tensile strength) and compactability (solid fraction vs tensile strength), the materials were examined using a Gamlen tablet press (GTP) and a PharmaTest hardness tester. The data was converted into plots according to equation 6:

$$\sigma = 2F / \pi DH \qquad (6)$$

Where, σ is tensile strength, F is the load required to fracture the tablet, D is the tablet diameter and H the tablet height(Banker et al., 2002; Endicott et al., 1961; Fell and Newton, 1970).

The spray dried material produced a stronger compact than milled extrudate at similar compaction pressures (Figure 6a). This was attributed to the bonding area, i.e. the effective surface area taking part in physical interparticulate bonding, which is greater for smaller particles (Olsson and Nyström, 2001). Tablet tensile strength is related to both bonding area and bonding strength with the overall strength due to an interplay between the two (Osei-Yeboah et al., 2016; Sun, 2011). Assuming that a minimal amount of brittle fracture occurs in the powders during the initial stages of compression, the greater mechanical strength of spray dried compacts can be partly attributed to the smaller particle size produced by spray drying relative to milled extrudate.

The strongest compact was formed with pure HPMCP (slightly stronger than pure Avicel 101) and the weakest with Soluplus® (Figure 6a). It is also possible that differences in excipient hygroscopicity impacted compact strength, by introducing moisture mostly during ambient storage or during weighing and compression, causing more hygroscopic materials (such as HPMCP) to form stronger compacts(Malaj et al., 2010; Sun, 2011). Confirmation of this would require further testing.

The compactability (tensile strength vs solid fraction) profiles (Figure 6b) showed a range in tablet tensile stress with solid fraction. Soluplus® was very compacted at the tensile strengths studied, with values over 0.95. Milled extrudate fractions were comparable to values obtainable for a typical filler like Avicel 101 (MCC), but spray dried ASD had lower solid fraction ranges, up to 0.80. HPMCP had the lowest values, up to 0.68, showing the possibility for further compaction at higher pressures.

3.4.2 Tabletability and Compactability of Prototype Tablet Formulations

Formulation of spray dried powders with Av200 (Formulation 1) improved tabletability, however inclusion of ManG (Formulation 2) produced softer tablets (Figure 8a). Neither Av200 nor ManG improved the tabletability of extruded powders, with both Formulation 3 and Formulation 4 producing softer tablets than the milled extrudate alone. Addition of either excipient to the milled extrudate ASD produced a more linear relationship between compaction pressure and tensile strength above pressures of 115 MPa, and this was more marked with the ManG formulation.

Compactability values were higher for all extruded material formulations, although inclusion of both Av200 and ManG resulted in enhanced compactability values for all of the spray dried and extrudate formulations (Figure 8b).

3.4.3 Compressibility of Prototype Tablet Formulations

Tensile strength measurement is carried out on the tablet following compression and is an out-of-die method, however analysis of the powder in-die during compression provides further information, especially on elastic deformation. The non-linear behaviour observed between tensile strength and compression pressure of the extruded powder, Formulation-3 and Formulation-4 provided the opportunity to investigate whether a change in the mechanism of densification from plastic to elastic deformation had occurred. Figure 9 compares the behaviour of extruded powder and extrudate formulations 3 & 4 in-die, illustrating both the change in porosity with compression pressure (dotted line) and the resulting Heckel plot, based on the logarithmic change in porosity (full line). Porosity was calculated by determining the solid fraction using the density of a compact from its mass and volume then dividing by the true density (equations 7 and 8).

$$Solid\ fraction = \frac{Tablet\ Density}{True\ Density} = \frac{Mass/(\pi \cdot \left(\frac{Diameter}{2}\right)^2 \cdot Thickness)}{True\ Density} \quad (7)$$

$$Porosity = 1 - Solid\ fraction \quad (8)$$

For each of the three Heckel plots, an initial non-linear region was observed, typically attributed to particle rearrangement (Heckel, 1961; Shapiro, 1995) or particle fragmentation(Duberg and Nyström, 1982). This was followed by a linear region indicating plastic deformation(Heckel, 1961), then a further non-linear region signifying elastic deformation of the powder bed(Sun and Grant, 2001). While all three regions are present in each of the plots, the nature of the regions varied from powder to powder. The extruded material (Figure 9a) exhibited a shortened plastic deformation region with elastic deformation shown to be the mechanism of compression for pressures greater than 70 MPa. This is in agreement with the plateau in the tensile strength observed for these powders at higher compression pressures (Figure 8a). Similar behaviour is shown by Formulation-3 (Figure 9b), but with a lesser degree of elastic deformation. Formulation-4 (Figure 9c), which maintained a linear relationship between tablet tensile strength and compression pressure for the pressures studied (Figure 8a), behaved close to linearly, even at the highest pressures. Previous studies have shown that while it is difficult to completely separate the contributions of plastic and elastic deformation mechanisms to the strength of a compact following compression, strength increases are mainly due to the formation of physical bonds between particles due to plastic deformation(Hiestand, 1997). Much of the elastic deformation is temporary and disperses once the load is released(Picker, 2001) however it seems that in this case some tensile strength increase due to elastic deformation is permanent, indicated by the fact that the tensile strength of extruded powders in Figure 8 continues to increase at higher pressures and does not completely plateau.

3.5 Dissolution studies

Dissolution profiles of tablets 1, 2, 3 and 4 are plotted in Figure 10. As expected, the formulations containing spray dried dispersions performed better than those containing milled extrudate(Albadarin et al., 2017; Davis et al., 2017). Drug concentrations at 2 h approximated 50 $\mu g.mL^{-1}$ and 30 $\mu g.mL^{-1}$, respectively. After 6 h, average drug concentrations increased to

82.3 μg.mL^{-1} in tablet 1 (spray dried with Av200), 69.1 μg.mL^{-1} in tablet 2 (spray dried with ManG), 44.2 μg.mL^{-1} in tablet 3 (milled extrudate with Av200) and 44.5 μg.mL^{-1} in tablet 4 (milled extrudate with ManG). The highest drug release observed was from one of the spray dried and Av200 tablets, achieving a 91.0 μg.mL^{-1} concentration after 6 h, a drug release of 81.9%.

The most immediate factors facilitating the higher performance of the spray dried material were the smaller particle size, narrower particle dimensions and larger particle surface area, relative to milled extrudate (Figure 1). Only the smaller sieved fraction (< 90 μm) of milled extrudate was brought though for dissolution studies, as this was closer in particle size to the spray dried formulation and therefore more comparable. For an absolute comparison an identical particle size would be best, although this would require a very small sieve fraction from the milled extrudate (or use of a jet mill) and a further enhancement of spray dried particle size by a factor of 5 to 10. This could be achieved using alternative spray drying parameters, such as a pressure nozzle(Beyerinck et al., 2010).

Interestingly, the Av200 tablets performed better than the ManG tablets on average, however the results were not simplistic and at certain times the performance of ManG tablets exceeded Av200 tablets. For example, the Av200 formula out-performed the ManG formula during almost all of the milled extrudate tests, apart from the first half hour, where it was marginally higher and the final test point (6 h), where it was also marginally higher. Additionally, for the spray dried tests, the ManG formula out-performed the Av200 formula during the first 2 hours, but this changed between 2 h and 3 h and the Av200 formulation then released more drug through to 6 h (Figure 10). Depending on the type of formulation required (e.g. immediate or sustained release) either of these results could be seen as a favourable direction for development. It is worth noting that although the ASD and 'filler' are present in each tablet at 66.6% and 30% (w/w) respectively, the impact of 'filler' excipient on each formulation is significant. For example, the highest recorded drug release (at 6 h) for spray dried with avicel was 91.0 μg.mL^{-1} and the lowest for spray dried with mannogem was 67.6 μg.mL^{-1}, a difference of 23.5 μg.mL^{-1}.

Clearly, formulation of either the spray dried or the extruded ASD into an optimised tablet is a very complex task and difficult to predict. Firstly, further studies are recommended for formulation optimisation, such as examining ASD level, filler type, filler level, lubricant level and other factors impacting dissolution rate; including the potential for gelling polymer networks (GPN)(Demuth et al., 2015). Secondly, optimisation of tablet processing factors such as the effect of compression pressure on dissolution, the presence/absence of crystalline drug caused by compression, reduced particle surface area, disintegration time and wettability are also recommended. Finally, optimisation of processing parameters during ASD preparation, such as nozzle type, feed concentration, outlet temperature (for spray drying), or screw size, screw speed and sieve fraction (for extrusion) would further enhance control over critical parameters such as yield, processing time and particle size. Because an exhaustive trial of all possible permutations is impractical, a well-constructed design of experiment (DoE) approach appears the most straightforward and beneficial, given the potential for statistically relevant information based on fewer experimental trials. Optimisation of the formulation aspect of the ternary dispersions themselves using DoE has already been accomplished in this group and was recently reported in two separate publications (Albadarin et al., 2017; Davis et al., 2017).

4. *Conclusions*

The differences in downstream processing properties between amorphous solid dispersions prepared *via* hot melt extrusion or spray drying are significant. Preparation choice substantially impacted flow, compression and dissolution. The fundamental differences can be attributes to variances in particle size and morphology generated from the alternative preparation methods. Spray dried powders had smaller particles sizes, more complex morphology, were more cohesive and had poorer flow than milled extrudate. Good agreement was observed between various rheometric and density based parameters used to measure flow, such as Flow Function (FF), Relative flow function, Aeration ratio (AR), Flow rate index (FRI) and also the more rudimentary methods such as the Hausner ratio and Carr's index. Spray dried materials produced stronger compacts during tabletability studies and had lower solid fractions during compactability studies. Porosity studies revealed differences in the mechanism of densification in extrudate and blended extrudate formulations. It was possible to alter the flow and compression properties of the amorphous dispersions by blending with specific excipients. Dissolution studies on four prototype tablets containing 100 mg of amorphous itraconazole, revealed that drug release was faster and more complete from the spray dried samples and could be further influenced by selective excipient blending.

Acknowledgements

This work was funded under Science Foundation Ireland grant "Synthesis and Solid State Pharmaceutical Centre (SSPC)" - (12/RC/2275). Special thanks are due to our colleagues Dr Barbara Schaller, Dr Patrick Cronin and Dr Javed Iqbal for useful discussions on compression, flow and HPLC, respectively.

References

Albadarin, A.B., Potter, C.B., Davis, M.T., Iqbal, J., Korde, S., Pagire, S., Paradkar, A., Walker, G., 2017. Development of stability-enhanced ternary solid dispersions via combinations of HPMCP and Soluplus® processed by hot melt extrusion. Int. J. Pharm. 532, 603–611. doi:10.1016/j.ijpharm.2017.09.035

Banker, G.S., Siepmann, J., Rhodes, C., 2002. Modern Pharmaceutics, Fourth Edition, Drugs and the Pharmaceutical Sciences. CRC Press.

Beyerinck, R.A., Ray, R.J., Dobry, D.E., Settell, D.M., 2010. Method for making homogeneous spray-dried solid amorphous drug dispersions using pressure nozzles.

Bruni, G., Barletta, D., Poletto, M., Lettieri, P., 2007. A rheological model for the flowability of aerated fine powders. Chem. Eng. Sci. 62, 397–407. doi:10.1016/j.ces.2006.08.060

Davis, M., Walker, G., 2018. Recent strategies in spray drying for the enhanced bioavailability of poorly water-soluble drugs. J. Control. Release 269, 110–127. doi:https://doi.org/10.1016/j.jconrel.2017.11.005

Davis, M.T., Egan, D.P., Kuhs, M., Albadarin, A.B., Griffin, C.S., Collins, J.A., Walker, G.M., 2015. Amorphous solid dispersions of BCS class II drugs: A rational approach to solvent and polymer selection. Chem. Eng. Res. Des. doi:10.1016/j.cherd.2016.04.008

Davis, M.T., Potter, C.B., Mohammadpour, M., Albadarin, A.B., Walker, G.M., 2017. Design of spray dried ternary solid dispersions comprising itraconazole, soluplus and HPMCP: Effect of constituent compositions. Int. J. Pharm. 519, 365–372.

doi:http://dx.doi.org/10.1016/j.ijpharm.2017.01.043

Demuth, B., Nagy, Z.K., Balogh, A., Vigh, T., Marosi, G., Verreck, G., Van Assche, I., Brewster, M.E., 2015. Downstream processing of polymer-based amorphous solid dispersions to generate tablet formulations . Int. J. Pharm. . doi:10.1016/j.ijpharm.2015.03.053

Duberg, M., Nyström, C., 1982. Studies on direct compression of tablets. VI. Evaluation of methods for the estimation of particle fragmentation during compaction. Acta Pharm. Suec. 19, 421–436.

Endicott, C.J., Lowenthal, W., Gross, H.M., 1961. New instrument and method for evaluating tablet fracture resistance. J. Pharm. Sci. 50, 343–346. doi:10.1002/jps.2600500415

Ennis, B J; Witt, W; Weinekoter, R; Sphar, D; Gommeran, E.; Snow, R H; Allen, T; Raymus, GJ; Litster, J.D., 2007. Solid-Solid Operations and Processing, in: Perry, D.W.G.R.H. (Ed.), Perry's Chemical Engineers' Handbook. McGraw-Hill Professional.

Fell, J.T., Newton, J.M., 1970. Determination of tablet strength by the diametral-compression test. J. Pharm. Sci. 59, 688–691. doi:10.1002/jps.2600590523

Fitzpatrick, J.J., Barringer, S.A., Iqbal, T., 2004. Flow property measurement of food powders and sensitivity of Jenike's hopper design methodology to the measured values. J. Food Eng. 61, 399–405. doi:10.1016/S0260-8774(03)00147-X

Freeman, R., 2007. Measuring the flow properties of consolidated, conditioned and aerated powders — A comparative study using a powder rheometer and a rotational shear cell. Powder Technol. 174, 25–33. doi:http://dx.doi.org/10.1016/j.powtec.2006.10.016

Gallo, L., Llabot, J.M., Allemandi, D., Bucalá, V., Piña, J., 2011. Influence of spray-drying operating conditions on Rhamnus purshiana (Cáscara sagrada) extract powder physical properties. Powder Technol. 208, 205–214. doi:10.1016/j.powtec.2010.12.021

Hancock, B.C., Carlson, G.T., Ladipo, D.D., Langdon, B.A., Mullarney, M.P., 2002. Comparison of the mechanical properties of the crystalline and amorphous forms of a drug substance. Int. J. Pharm. 241, 73–85. doi:10.1016/S0378-5173(02)00133-3

Heckel, R., 1961. Density-pressure relationships in powder compaction. Trans Met. Soc AIME 221, 671–675.

Hiestand, E.N., 1997. Principles, tenets and notions of tablet bonding and measurements of strength. Eur. J. Pharm. Biopharm. 44, 229–242. doi:10.1016/S0939-6411(97)00127-6

Huang, S., Williams III, R.O., 2017. Effects of the Preparation Process on the Properties of Amorphous Solid Dispersions. AAPS PharmSciTech. doi:10.1208/s12249-017-0861-7

Iveson, S.M., Litster, J.D., Hapgood, K., Ennis, B.J., 2001. Nucleation, growth and breakage phenomena in agitated wet granulation processes: A review. Powder Technol. 117, 3–39. doi:10.1016/S0032-5910(01)00313-8

Jain, S., 1999. Mechanical properties of powders for compaction and tableting: An overview. Pharm. Sci. Technol. Today 2, 20–31. doi:10.1016/S1461-5347(98)00111-4

Jenike, A.W., 1964. Storage and Flow of Solids, Bulletin of the University of Utah. University of Utah.

Leturia, M., Benali, M., Lagarde, S., Ronga, I., Saleh, K., 2014. Characterization of flow properties of cohesive powders: A comparative study of traditional and new testing methods. Powder Technol. 253, 406–423. doi:10.1016/j.powtec.2013.11.045

Leuenberger, H., Rohera, B.D., 1986. Fundamentals of Powder Compression. I. The Compactibility and Compressibility of Pharmaceutical Powders. Pharm. Res. An Off. J. Am. Assoc. Pharm. Sci. 3, 12–22. doi:10.1023/A:1016364613722

Malaj, L., Censi, R., Gashi, Z., Di Martino, P., 2010. Compression behaviour of anhydrous and hydrate forms of sodium naproxen. Int. J. Pharm. 390, 142–149. doi:10.1016/j.ijpharm.2010.01.036

Nokhodchi, A., Maghsoodi, M., Hassan-Zadeh, D., Barzegar-Jalali, M., 2007. Preparation of agglomerated crystals for improving flowability and compactibility of poorly flowable and compactible drugs and excipients. Powder Technol. 175, 73–81. doi:10.1016/j.powtec.2007.01.030

Olsson, H., Nyström, C., 2001. Assessing tablet bond types from structural features that affect tablet tensile strength. Pharm. Res. 18, 203–210. doi:10.1023/A:1011036603006

Osei-Yeboah, F., Chang, S.-Y., Sun, C.C., 2016. A critical examination of the phenomenon of bonding area - Bonding strength interplay in powder tableting. Pharm. Res. 33, 1126–1132. doi:10.1007/s11095-016-1858-8

Patel, S., Kaushal, A.M., Bansal, A.K., 2006. Compression physics in the formulation development of tablets. Crit. Rev. Ther. Drug Carrier Syst. 23, 1–65.

Peschl, I.A.S.Z., Colijn, H., 1977. NEW ROTATIONAL SHEAR TESTING TECHNIQUE. J Powder Bulk Solids Technol 1, 55–60.

Picker, K.M., 2001. Time dependence of elastic recovery for characterization of tableting materials. Pharm. Dev. Technol. 6, 61–70. doi:10.1081/PDT-100000014

Salman, A.D., Biggs, C.A., Fu, J., Angyal, I., Szabó, M., Hounslow, M.J., 2002. An experimental investigation of particle fragmentation using single particle impact studies. Powder Technol. 128, 36–46. doi:10.1016/S0032-5910(02)00151-1

Schulze, D., Schwedes, J., Carson, J.W., 2008. Powders and bulk solids: Behavior, characterization, storage and flow, Powders and Bulk Solids: Behavior, Characterization, Storage and Flow. doi:10.1007/978-3-540-73768-1

Schwedes, J., 2003. Review on testers for measuring flow properties of bulk solids. Granul. Matter 5, 1–43. doi:10.1007/s10035-002-0124-4

Shapiro, I., 1995. Compaction of powders XII. Evaluation of published compaction equations for modeling purposes, in: Advances in Powder Metallurgy and Particulate Materials. p. 2/97-2/114.

Sinka, I.C., Schneider, L.C.R., Cocks, A.C.F., 2004. Measurement of the flow properties of powders with special reference to die fill. Int. J. Pharm. 280, 27–38. doi:10.1016/j.ijpharm.2004.04.021

Sivasankaran, S., Sivaprasad, K., Narayanasamy, R., Iyer, V.K., 2010. An investigation on flowability and compressibility of AA 6061100-x-x wt.% TiO2 micro and nanocomposite powder prepared by blending and mechanical alloying. Powder Technol. 201, 70–82. doi:10.1016/j.powtec.2010.03.013

Sun, C., Grant, D.J.W., 2001. Influence of crystal structure on the tableting properties of sulfamerazine polymorphs. Pharm. Res. 18, 274–280. doi:10.1023/A:1011038526805

Sun, C.C., 2011. Decoding powder tabletability: Roles of particle adhesion and plasticity. J. Adhes. Sci. Technol. 25, 483–499. doi:10.1163/016942410X525678

Vehring, R., 2008. Pharmaceutical particle engineering via spray drying. Pharm. Res. 25, 999–1022. doi:10.1007/s11095-007-9475-1

Walker, D.M., 1966. An approximate theory for pressures and arching in hoppers. Chem. Eng. Sci. 21, 975–997. doi:10.1016/0009-2509(66)85095-9

Wu, C.-Y., Cocks, A.C.F., 2004. Flow behaviour of powders during die filling. Powder Metall. 47, 127–136. doi:10.1179/003258904225015617

Xie, X., Puri, V.M., 2006. Uniformity of powder die filling using a feed shoe: A review. Part. Sci. Technol. 24, 411–426. doi:10.1080/02726350600934663

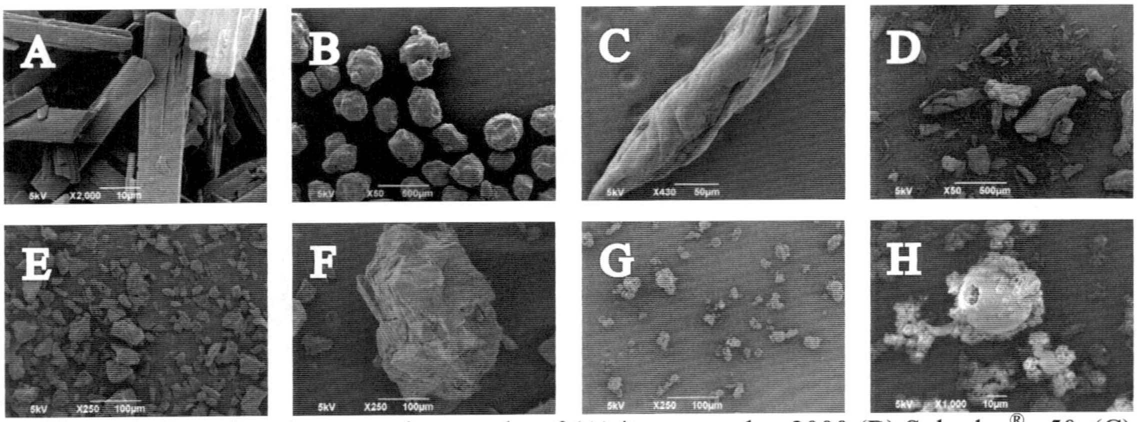

Figure 1: Scanning electron micrographs of (A) itraconazole x2000 (B) Soluplus® x50, (C) HPMCP x430, (D) physical mix x50, (E) milled extrudate < 90 µm sieve fraction x250, (F) milled extrudate 90 – 450 µm sieve fraction x250, (G) spray dried x250 & (H) spray dried x1000.

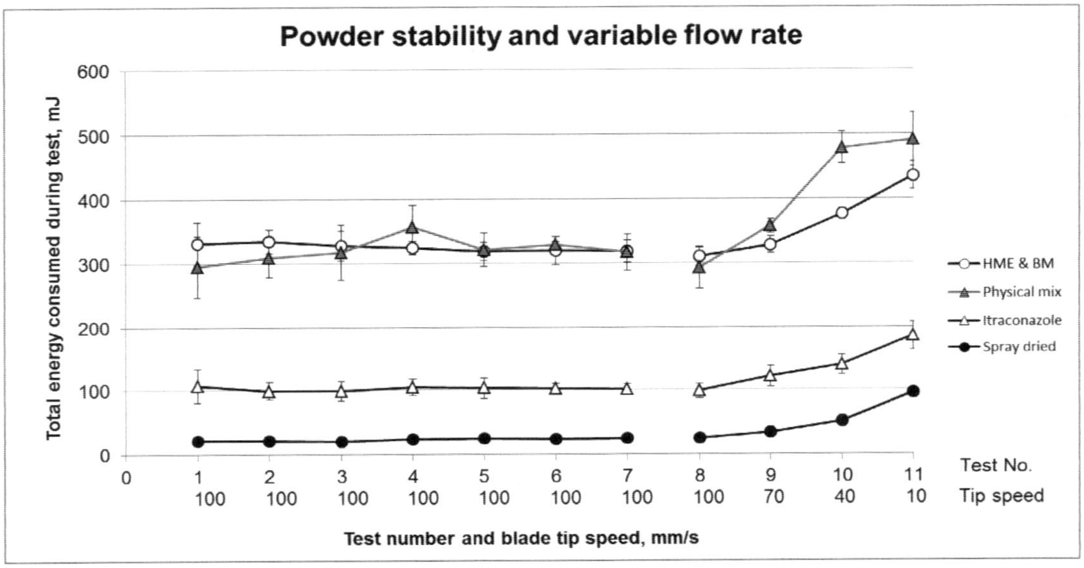

Figure 2: Stability and variable flow rate data recorded on an FT4 powder rheometer for ASD powders of itraconazole-Soluplus®-HPMCP 30-40-30 (w/w/w), physical mix and pure drug. Tests 1-7 are performed at constant blade tip speed and measure stability, a stable powder will show an approximately flat straight line. Tests 8-11 feature a reduction in blade tip speed and measure the sensitivity of powder to flow rate, typically more energy is consumed as the blade passes at a slower speed. Measurements were recorded in triplicate.

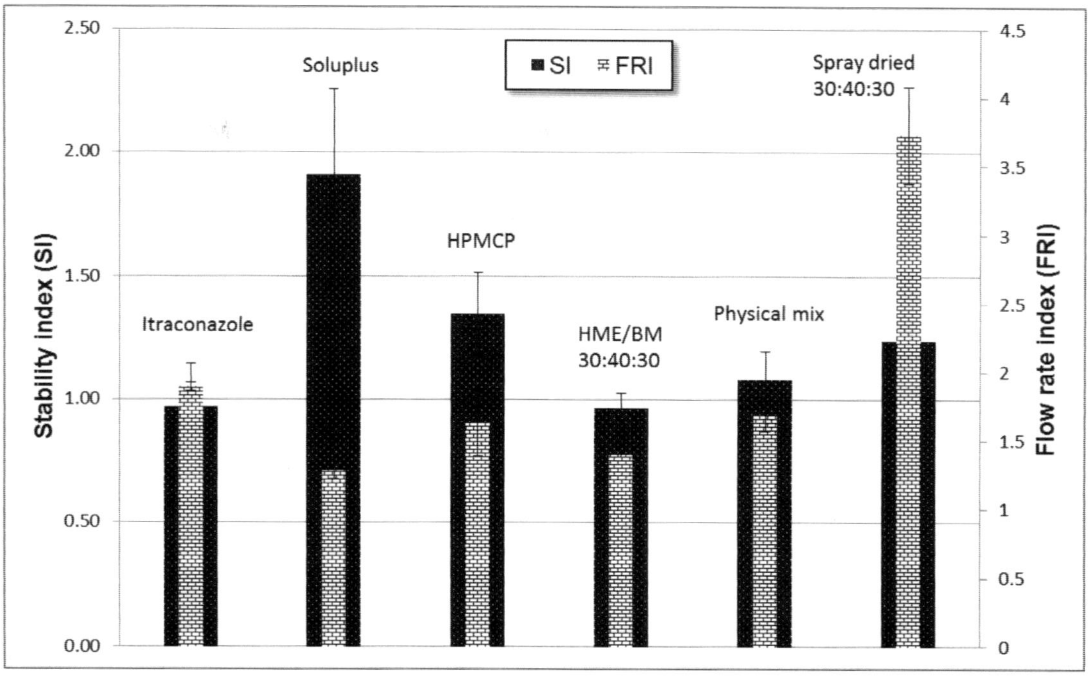

Figure 3: The stability index (SI) and flow rate index (FRI) values recorded in triplicate for itraconazole, Soluplus®, HPMCP, physical mix, HME/BM powder and spray dried powder by the FT4 powder rheometer. An SI value of 1.0 ± 0.1 indicates the powder is not sensitive to flow stress, for example the HME/BM result. An FRI > 3.0 indicates that the powder is highly sensitive to flow rate and most likely cohesive, such as the value of the spray dried sample.

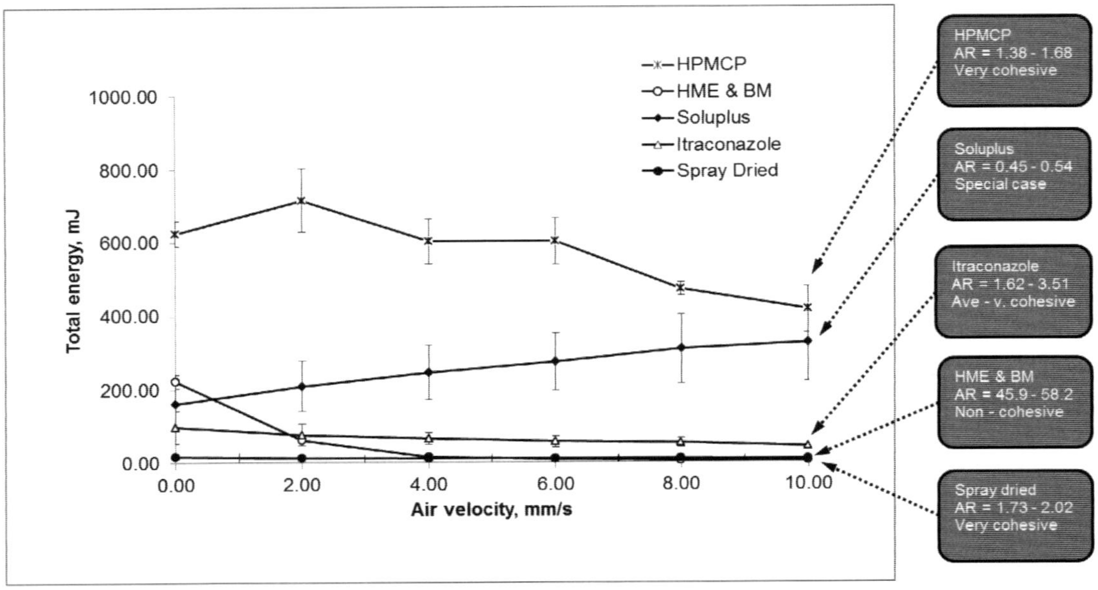

Figure 4: Aeration studies on raw materials itraconazole, Soluplus®, HPMCP and ASD powders prepared by hot melt extrusion then ball milling (HME & BM) or by spray drying. Data was recorded in triplicate on an FT4 powder rheometer. During the test, a blade is passed at 100 mm/s tip speed through a bed of powder while air is introduced at the base at progressive velocities. The extent by which the flow energy is reduced is a function of the physical properties of the powder (such as cohesion) and provides the aeration ratio (AR).

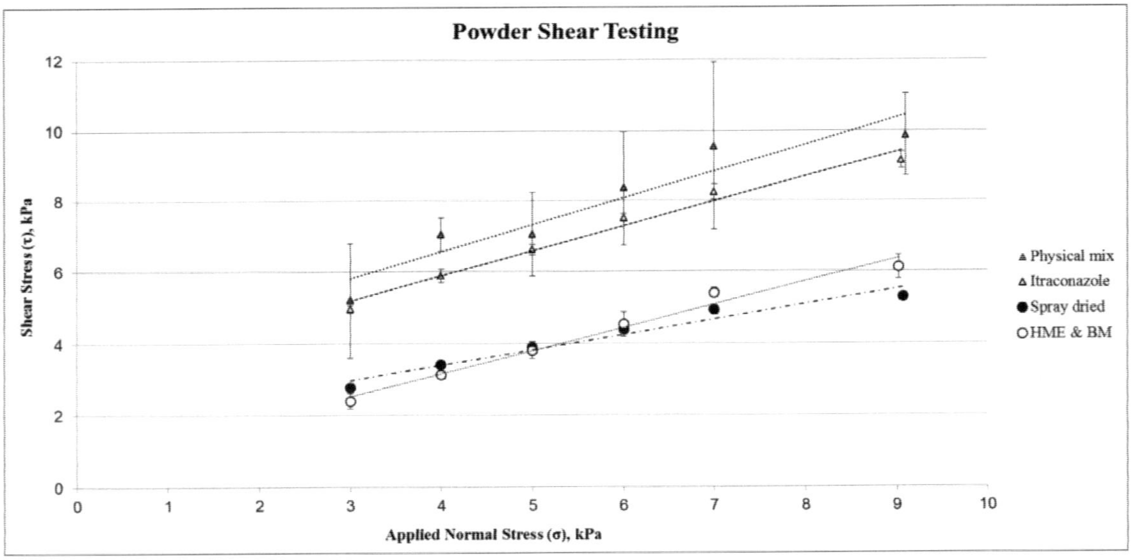

Figure 5: Yield locus plots for itraconazole, physical mixture and ASD powders prepared by spray drying and by hot melt extrusion (followed by ball milling), recorded in triplicate on an FT4 powder rheometer. Corresponding shear parameters are recorded in Table 1.

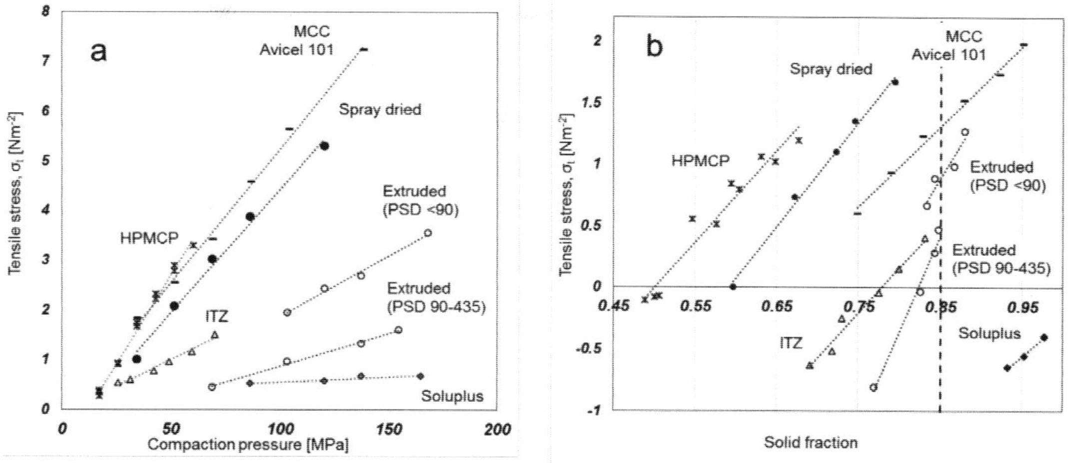

Figure 6: Tabletability (a) and Compactability (b) plots of raw materials, spray dried ASD and milled extrudate. ITZ = itraconazole, MCC = microcrystalline cellulose, PSD = particle size distribution. Two sieve fractions were measured for extrudate: < 90 μm & 90 – 435 μm.

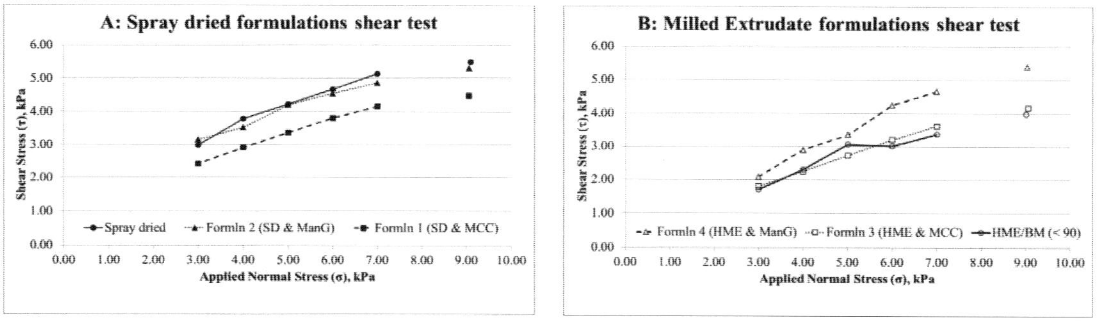

Figure 7: Yield locus plots for prototype formulations, comprising spray dried (SD) powder with Avicel PH 200 LM microcrystalline cellulose (MCC) (Formulation 1) and Mannogem granular mannitol (ManG) (Formulation 2) and milled extrudate (HME/BM <90) with MCC (Formulation 3) and ManG (Formulation 4). For complete formulations see Table 4, for shear data see Table 5 and for discussion see Section 3.3.6.

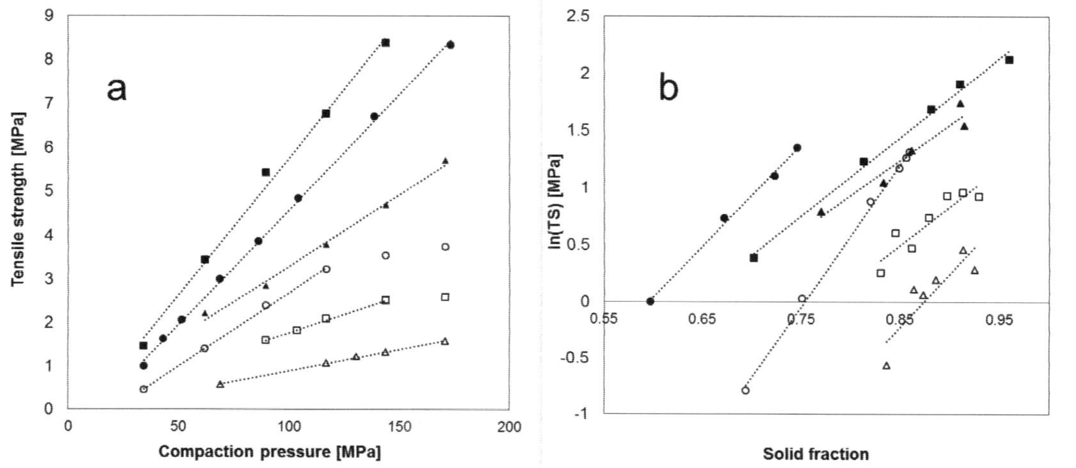

Figure 8: Tabletability (a) and Compactability (b) plots of spray dried (filled circles) and milled extrudate (open circles) & the effect of including Av200 (squares) or ManG (triangles).

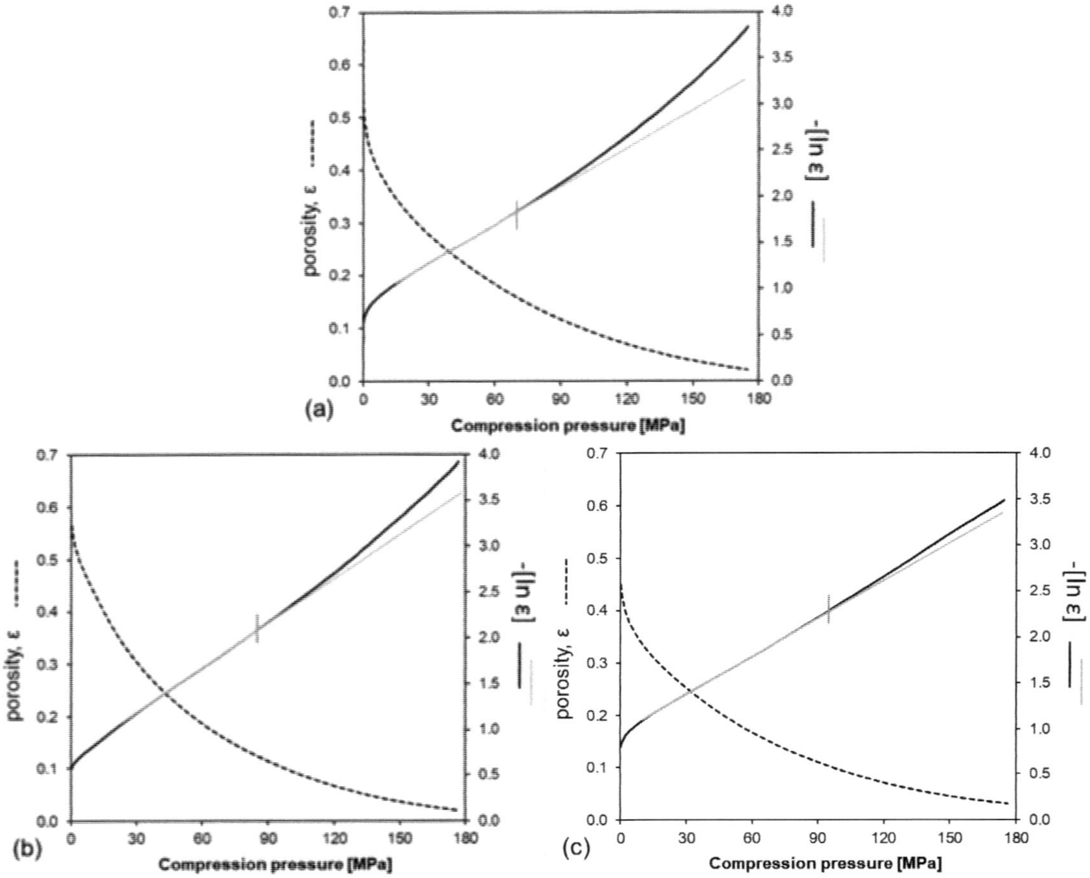

Figure 9: The change in porosity of the powder bed during compression (--- dashed line) of (a) extruded powders, (b) Formulation-3 and (c) Formulation-4 and the resulting Heckel plots, illustrating the logarithmic change in porosity (full line –). The linear portion of the Heckel plot, indicating the region where the powder bed is undergoing plastic deformation, is indicated by a grey line of best fit.

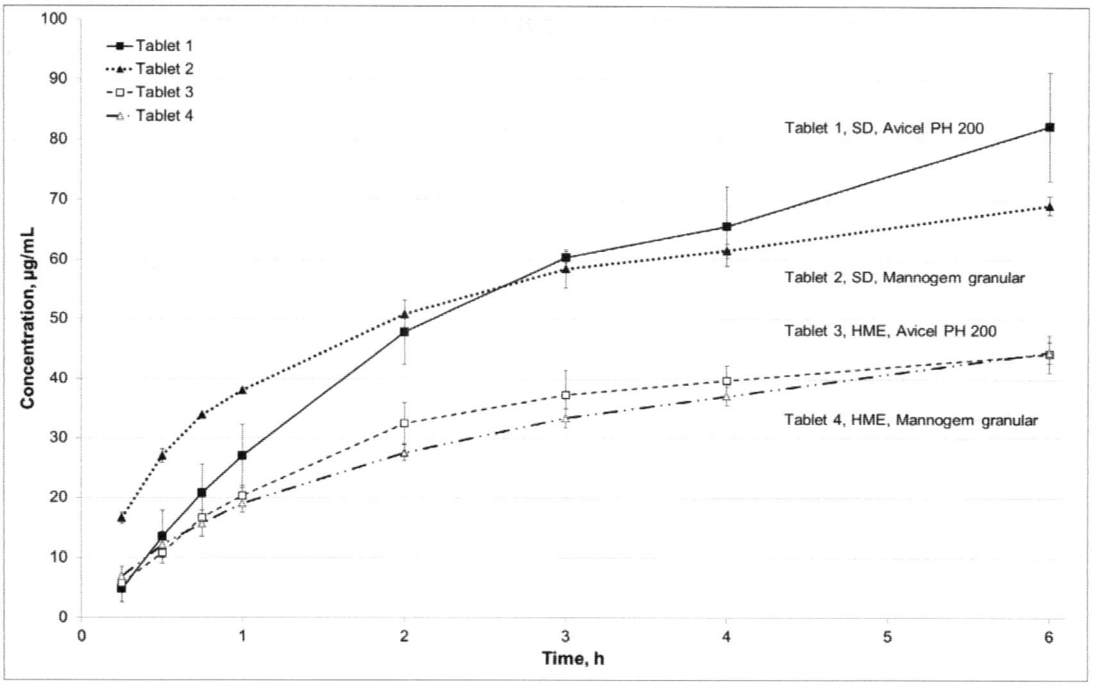

Figure 10: Dissolution profiles of 500 mg prototype tablets, containing a 100 mg dose of amorphous itraconazole prepared by spray drying (tablets 1 & 2) or hot melt extrusion (tablets 3 & 4). ASDs were blended with excipients (Table 4) before compression into spherical disks and dissolution (n = 3) in HCl (pH 1.2), with HPLC analysis at $\lambda = 263$ nm.

540732 Machine Learning Approaches for Enzyme Engineering

Abstract:

Enzyme engineering aims to improve a variety of different properties, such as stability, activity, and binding affinity. A variety of experimental and computational approaches have been used to improve such properties, including directed evolution and structure-guided rational design using various tools (e.g., Rosetta). We have recently developed a machine learning approach (called MLProScape) to model the protein fitness landscape and to guide enzyme engineering efforts. MLProScape numerically captures amino acid based physio-chemical properties as features, builds a machine learning model based on these features, and then identifies synthetic designs with improved enzyme properties. We have applied MLProScape to successfully design novel variants of glycosyl hydrolases with improved catalytic activity for hydrolyzing both cellulose and xylan, two major components found in lignocellulosic biomass. In this case, MLProScape made highly accurate predictions for glycosyl hydrolase activity under 5 different conditions, with pearson correlation coefficients varying between ~0.72 and ~0.90 and root mean squared errors varying between ~0.4 and ~16.2 mmol/min/mg. Examples of other MLProScape protein engineering applications will also be presented. MLProScape is capable of identifying residue changes outside of the active site and can handle complex design criteria that aim to improve multiple enzyme properties, simultaneously. As such, this machine learning based approach will facilitate enzyme engineering efforts aimed at improving various properties, including activity and thermostability.

528003 Simulating the Mechanistic Pathway of Transglycosylation Via a Mutant Glycoside Hydrolase

Abstract:

There is scientific, industrial, and medical value in synthesis of bespoke oligosaccharides, but significant limitations exist to existing laboratory methods. A more desirable approach is to employ highly specific enzymes, but the glycosyltransferases nature employs to do this job are poorly suited to industrial use. An enzymatic alternative to glycosyltransferases is transglycosylation via mutated glycoside hydrolases. The *Thermotoga maritima* Î±-L-fucosidase (*Tm*Afc) mutant D224G has been demonstrated to have good non-specific transglycosylation activity, but the molecular mechanism is not yet understood, hampering engineering efforts. In this talk, we describe transition path sampling simulations to elucidate the transglycosylation pathway of *Tm*Afc D224G, and discuss our proposed mechanismâ€™s implications for engineering improved activity and specificity in this class of enzymes in the future.

534845 Engineering a Multifunctional Family 5 Glycosyl Hydrolase into a Transglycosidase

Abstract:

Glycans are amongst the most abundant biomacromolecules on the planet but we are far from elucidating their role in the design and regulation of biological systems. There is a dire need of tools to systematically synthesize well-defined glycan polymers as is currently possible with other biopolymers like DNA, RNA, and proteins. Glycans synthesis is naturally performed by membrane associated glycosyl transferases (GTs), but GTs can be challenging to express sometimes for large-scale synthesis. Alternatively, transglycosidases can be selected from a large repertoire of available glycosyl hydrolases (GH) and rationally engineered to enable chemoenzymatic synthesis of designer glycans. Various factors play important roles in engineering a glycosyl hydrolase to function efficiently as a transglycosidases; such as, modulating substrate and product binding site affinities, regulating water accessibility, and enzymatic reaction conditions. Here, we focus on engineering a multifunctional family 5 GH from *Clostridium thermocellum* (CelE) into a transglycosidase to synthesize diverse glucan-based polymers. Active site mutations and presence of carbohydrate binding modules (CBM) were all found to have a significant impact on the catalytic rate, specificity, and overall mechanism of transglycosylation. All transglycosylation products were characterized using TLC, HPLC, and mass spectrometry to decode the synthesized glycan structures and reaction specificity. We further use molecular dynamics (MD) simulations and small angle X-ray scattering (SAXS) analysis to study the effect of CBM on the mechanism of transglycosylation to rationally design a highly efficient GH 5 based transglycosidase. In summary, we are developing a broadly applicable tool-kit and workflow to rationally convert glycosyl hydrolases into highly efficient and specific transglycosidases.

535619 Computational Insights into the Catalytic Function of Processive Cellulases

Abstract:

Cellulase enzymes naturally produced by fungi and bacteria are utilized industrially for many purposes, including the production of clean sugar streams from biomass that can be subsequently upgraded to fuels, chemicals, and products. The development of structure-function relationships in these enzymes is a collaborative effort between biochemical experimentation, structural characterization, and computational simulation. Here, we detail recent success stories from our group wherein molecular simulations are able to provide mechanistic insight with atomistic and dynamic resolution not generally attainable via experiment. In the present contribution, we probe different mechanisms for how a processive cellulase may dissociate from a cellulose chain on the surface of cellulose. Under certain conditions, dissociation from a cellulose strand has been demonstrated to be the overall rate-limiting step of catalytic action for cellobiohydrolases. Despite the growing evidence to support this conclusion, the literature is largely silent on the question of *how* dissociation occurs. In the published literature, “dissociation” is defined purely by the beginning (complexed upon a cellulose strand with the complete binding tunnel occupied) and the end (enzyme binding tunnel empty, and perhaps also “lifted off” of the cellulose surface into solution) states of the process. However, several possibilities exist for how the enzyme may proceed from the former to the latter state. Given the importance of this process within the context of cellulase performance (and thus potential cellulase engineering), we formulate several paths for this process (utilizing *Trichoderma reesei* Cel7A) and then compute free energy profiles for each. In so doing, we identify engineering targets for the production of processive cellulases of enhanced efficiency.

THE UNIVERSITY OF KANSAS

Department of Chemical & Petroleum Engineering

Enhanced Lipase-Catalyzed Hydrolysis and Modification of Fats and Oils

2018 AIChE Annual Meeting

Akash Anand

Committee members

Chair: Dr. Laurence R. Weatherley

Dr. Kevin C. Leonard

Dr. Arghya Paul

Dr. Prajnaparamita Dhar

Dr. Edward F. Peltier

1. Introduction

The oils and fats industry is one of the largest industries where chemical engineering techniques have been applied to the food sector. These are obtained from both animal and vegetable sources providing the raw materials for the manufacture of both edible and non-edible products. The products of hydrolysis oils and fats reactions include fatty acids and glycerol, which form the basis for the manufacture of a wide range of modified oils and fats, polyols, polyethers with application for the manufacture of personal care products and other specialty chemicals [1]. However in today's modern world there is an increasing demand for high quality and cheap products. This demand has led to the large scale production of fatty acids and glycerol. Hydrolysis of oils and fats is hindered by solubility issues since oil and water are immiscible. Hence, conducting the reaction at high temperatures increases the solubility of water in oil phase [2]. Currently, superheated steam is used to obtain fatty acids from vegetable oils or/and animal fats and the reaction is carried out at 100 – 260°C and 100 –7000 kPa using a 0.4-1.5 wt% water to oil ratio. This process is called fat splitting [3] and different variations of this technology are being used by various industries to produce fatty acids (e.g., the Twitchell process, the Colgate-Emery synthesis, and the Eisenlohr process). All these processes involves high temperature and pressure equipment [4-7]. With increasing public demand for companies to adopt a cleaner and more efficient attitude towards their processes many scientists and researchers are seeking to develop newer technologies. Of particular interest in this area is the use of enzymes to produce an accelerated and controlled release of fatty acids [8].

2. Lipase – Catalyzed Hydrolysis

Enzymes are basically proteins. Compared with inorganic catalysts such as acids, bases, metals and metal oxides, enzymes are very specific, they limit their action to specific bonds in the compounds with which they react [9]. Lipases are a group of enzymes whose primary function is to catalyze the hydrolysis of fats and oils (Lipolysis) to yield free fatty acids, diglycerides, monoglycerides and glycerol as shown in Figure 1 [10, 11].

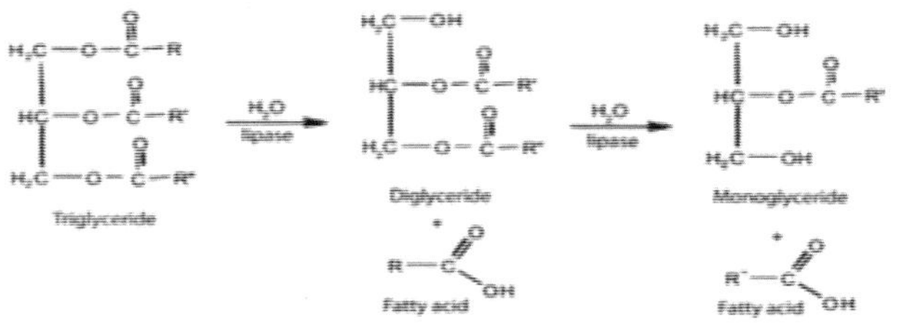

Figure 1. Hydrolysis of triglyceride (oils) catalyzed by lipase

As early as 1958, Sarda and Desnuelle described a sharp increase in lipase activity at substrate concentration exceeding their solubility threshold [12]. Since then the formation of an interface between aggregated substrates and the aqueous solution has been recognized as necessary for the activation of lipases. This behavior known as "interfacial activation" found a structural rationale some years later when an active lipase obtained from the yeast *Candida rugosa* was purified by Yamada and Machida in 1965 and has shown to have a non-specific mode of action [13-16].

3. Hypothesis

The hypothesis of this research is that the enzymatic hydrolysis reaction can be enhanced by immobilization of the enzyme onto a suitable solid support and by the optimization of temperature and pH. It is also proposed that application of an external electric field can enhance both reaction kinetics and reaction selectivity. Furthermore, a mass transfer model in a recirculated continuous reactor with immobilized enzymes could be established with the experimental results and can be used in design of industrial bioreactor and scale up.

4. Enzyme Immobilization

Enzymes, when used in soluble form requires separation from the reaction mixture once the reaction is complete. On account of the complexity of the mixture, recovery and recycling of the enzymes can result in significant loss of total enzyme and specific activity [18]. To overcome these drawbacks enzymes are immobilized onto organic or inorganic support particles by physical adsorption, encapsulation, covalent attachment, or cross-linking. Immobilization can increase the stability of the enzymes and may allow successful performance over a wider range of temperature and pH values [19]. Many supports and techniques for immobilization of lipase have been reported for oil hydrolysis over the years [20-28]. Separation of the immobilized enzyme from the reaction mixture is based on physical separation, allowing continuous use of the enzyme.

5. Glutaraldehyde Cross-linking

Glutaraldehyde cross-linkage is a technique for enhancing the immobilization of the enzymes onto polymeric supports, and in addition shows potential for enhancing the activity [29]. Another potentially significant consequence of applying glutaraldehyde cross-linkage is to enhance retention of the enzyme on the support thereby reducing loss of catalyst activity due to desorption or leakage of the enzyme from the support [30]. Glutaraldehyde has been used extensively as a mild cross-linking agent for the immobilization of enzymes [31, 32]. Another possible use of glutaraldehyde was identified by Haider et al. noting the possibility of it adding a protective layer around the active center of the enzyme [33].

6. Effect of Low Voltage Electric field

Enzymatic splitting of triglyceride esters is limited often by slow kinetics due to mass transfer limitations and by challenges of economic enzyme recycling. Prior work has shown that mass transfer rates and reaction rates can be significantly intensified by creating fine, electrostatic sprays of electrically charged droplets of enzyme solution which are injected into the oil substrate [34, 35]. The principle and sole mechanism for the intensification was assumed to be solely on account of increases in interfacial area and some hydrodynamic turbulence. The possible effect of electrical charge on the lipase activity has hitherto not yet been considered.

7. Experimental Procedures and Results to date

7.1. Materials

The lipase chosen for this study was from Candida rugosa (Type VII, >700 unit/mg solid). Aqueous enzyme solutions used were made up at concentration of 0.5 (%w/v) unless stated otherwise, using 0.1 M phosphate buffer. Vegetable oil and sunflower oil were the main substrate oils. Table 1 and 2 shows the list of support materials which has been used for lipase immobilization. The blended resins create a good balance of hydrophobicity that makes the polymer optimal for immobilization under mild operating conditions. Main advantages of these polymers include high mechanical stability and high enzyme activity at low protein loading. The immobilization criteria were based on finding a low cost method of loading enzyme onto the support. The chosen method of immobilization was adsorption, whereby the enzyme adheres to the surface of the support particles by Van der Waals forces of attraction [35].

Table 1. Properties of supports supplied by Purolite® Corporation

Product Name	ECR8285	ECR1030M	ECR8806M	ECR1090M	ECR1508	ECR1640
Composition	Epoxy/butyl/ methacrylate	Meth acrylic polymer	Octadecyl methacrylate	Macro porous styrene	Macro porous polystyrenic weak base anion	Macro porous polystyrenic strong base anion
Mean pore dia (Å)	400-600	220-340	500-700	900-1100	640	731
Particle size range (μm)	250-1000	300-710	300-710	300-710	-	-

Table 2. Characteristics of Accurel supports supplied by AkzoNobel® Chemicals

Product Name	EP400	EG100	EP100
Composition	Low Density Polyethylene (LDPE)	High Density Polypropylene (HDPP)	High Density Polypropylene (HDPP)
Pore size (μm)	0.1 - 0.5	0.1 - 1.0	0.1 - 1.0
Particle size range (μm)	400 - 1000	3000	200 - 2000

7.2. Immobilization Protocol

The surface of the supports are highly hydrophobic, which indicates that supports surface should be "wetted" before immobilization [36, 37]. The wetting process involved soaking 1 g support particles in 6 ml ethanol. This procedure modified the polymer surface and facilitated adsorption of the aqueous enzyme solution [38]. 10 mL of aqueous enzyme solution were added to 1 g of support particles along with 74 mL distilled, deionized water in a shake flask. The contents of the shake flask were agitated for 20 hours at 25°C in an orbital shaker. Separation of the immobilized support particles and aqueous solution was carried out using vacuum filtration. Immobilized support particles were then washed with 500 mL distilled, deionized water to remove any weakly bound enzymes on the surface of support particles. After washing, the immobilized support particles were dried overnight at 30°C and then assayed for enzyme activity [25, 35, 39]. A graphical summary of this technique is shown in Figure 2.

7.3. Activity Assay (Hydrolysis)

Enzyme activity exhibited by the immobilized enzyme on the polymer support particles was quantified by measuring the rate of hydrolysis of vegetable oil or sunflower oil [34]. All the hydrolysis reactions were carried out by mixing oil, water and immobilized enzyme in 30:10:1 ratio, on a weight basis unless stated otherwise, in a stirred tank reactor operated at 300 RPM [25]. At each and every time interval, 0.5 g of the emulsion was solubilized with 20 ml of 50:50v/v mixture of ethanol and acetone in order to arrest the reaction. The mixture was then titrated against 0.05 M NaOH in the presence of phenolphthalein indicator. Enzyme activity is expressed as the number of moles of free fatty acid produced from the hydrolysis of vegetable oil per minute (molFFA/min) or as moles of free fatty acid (molFFA). A Mettler – Toledo, T5 auto-titrator was used as pH stat.

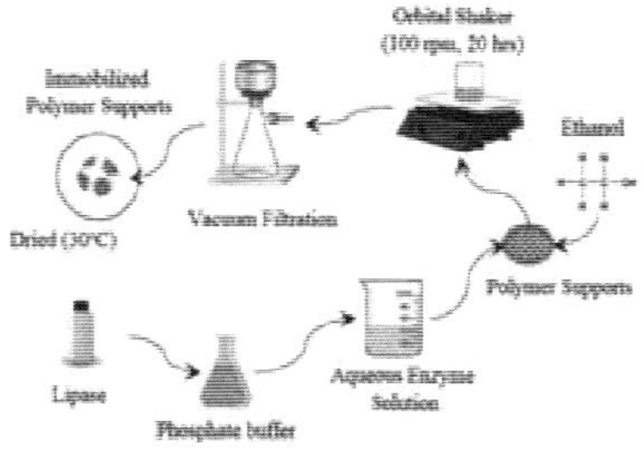

Figure 2. Lipase Immobilization Protocol

7.4. Glutaraldehyde cross-linking

Series of experiments were set up for enzyme immobilized onto polymer supports in the presence of cross-linking agent. The reaction performance in each case was evaluated in order to determine the stability of the cross-linked enzyme. Following a 20-hour contact period of immobilization as previously described, 1 mL of Mc Ilvaine containing 80 μL of 25% glutaraldehyde was added to the immobilization mix and the whole system further contacted for 10 minutes with agitation. 0.2 mL of 20% Sodium hydrogen sulfite ($NaHS_2$) were added and the system was shaken for another 10 minutes. Sodium hydrogen sulfite solution was added to control the activity of the glutaraldehyde. [18, 25, 39].

8. Results for Immobilized Supports

Earlier working comparing the relative stability of the free and immobilized enzyme systems with respect to shelf life, pH and temperature for Accurel supports for the hydrolysis of high oleate sunflower oil, Table 3 and Figure 3 (a and b) showed that: (1) Half-life of the immobilized enzyme was seen to increase up to 7 times that of aqueous enzyme solutions. (2) Lipase immobilized onto EP400 exhibited an increase in enzyme activity over the range of pH 7-9. (3) The temperature stability of lipase was not observed to change on enzyme immobilization and the optimum working temperature of the enzyme was noted at 45°C.

Table 3. Enzyme stability - variation of enzyme activities with time over four week period at room temperature (23°C)

Time (hours)	Enzyme Activity (molFFA)		
	Free Enzyme	EP400 (<400 μm)	EG100 (3 mm)
0	0.10867	0.104	0.00667
168	0.08267	0.1033	0.00533
504	0.00267	0.09333	0.00533
672	0.00133	0.092	0.004

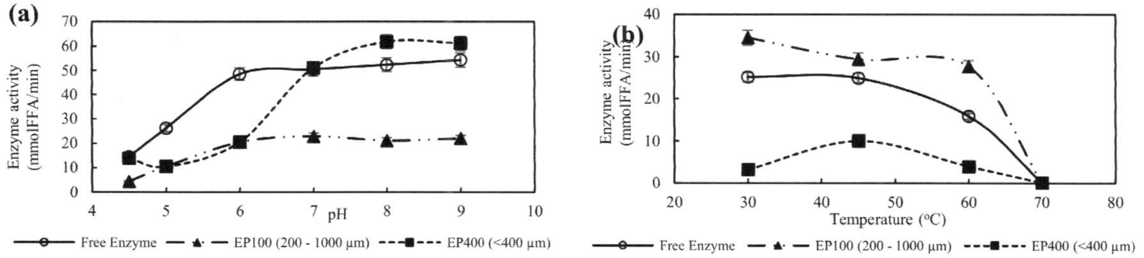

Figure 3. Enzyme activity of free enzyme compared with Accurel EP100 (HDPP) and EP400 (LDPE) immobilized enzyme – (a) varying pH and (b) varying temperature

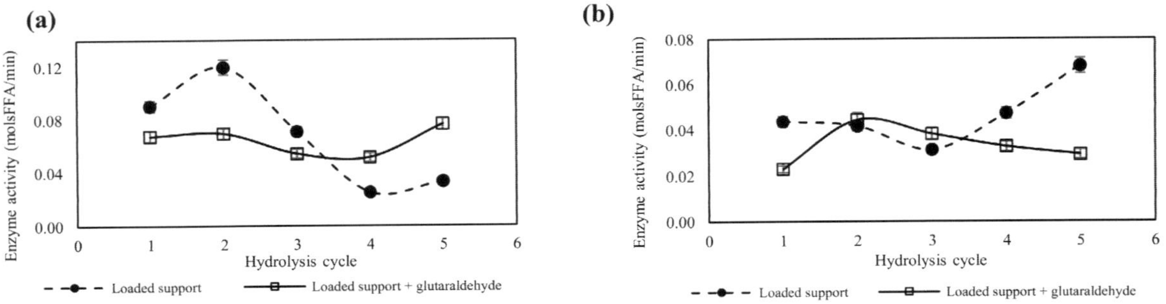

Figure 4. Enzyme activity with respect to different hydrolysis cycle using immobilized support with and without glutaraldehyde cross-linking (a) Accurel EP100 (HDPP) for particle size of 200 – 1000 μm (b) Accurel EP400 (LDPE) for particle size of less than 400 μm

Figures 4 (a and b), shows the enzyme activities measured for enzymes cross-linked onto Accurel EP100 and EP400 supports with glutaraldehyde and comparisons to the loaded supports without glutaraldehyde cross-linking. The activity data are presented for multiple cycles of reaction, with the immobilized enzymes recycled after each activity determination. The results are anomalous, showing significant variation in activity from cycle to cycle. The data for enzyme immobilized onto EP100 with no cross-linking agent, increased in specific activity after one cycle followed by a steep reduction in subsequent cycles. In the case of enzyme immobilized onto EP100 with cross-linking, a much steadier performance from cycle to cycle is observed, suggested improved enzyme stability. Immobilized enzyme, cross-linked onto EP400, showed anomalous behavior from cycle to cycle. In the case of the non-crosslinked catalyst the activity appeared to decrease in the first two cycles, followed by a steep increase in activity after cycle three. Some decrease in activity from cycle to cycle would be expected and the increase in activity after cycle three was unexpected. One possible explanation is based on diffusional effects within the particle structure. It is possible that after each activity experiment, residual reaction was still occurring within the interior surface of the catalysis particle, with fatty acid product for example from cycle three retained in the particle and released slowly during the cycle four activity determination. A further possibility is the continuation of reaction in the inner core of the particle in between cycles. Both of these potential effects, if present, would result in an apparent increase in fatty acid yield during each activity determination beyond cycle three. Figure 4 (a), could be similarly explained by a transient effect. Regardless of the apparent anomalous behavior it is clear for both polypropylene support, and for the polyethylene support that cross-linking imparts a degree of added stability to the immobilized enzyme.

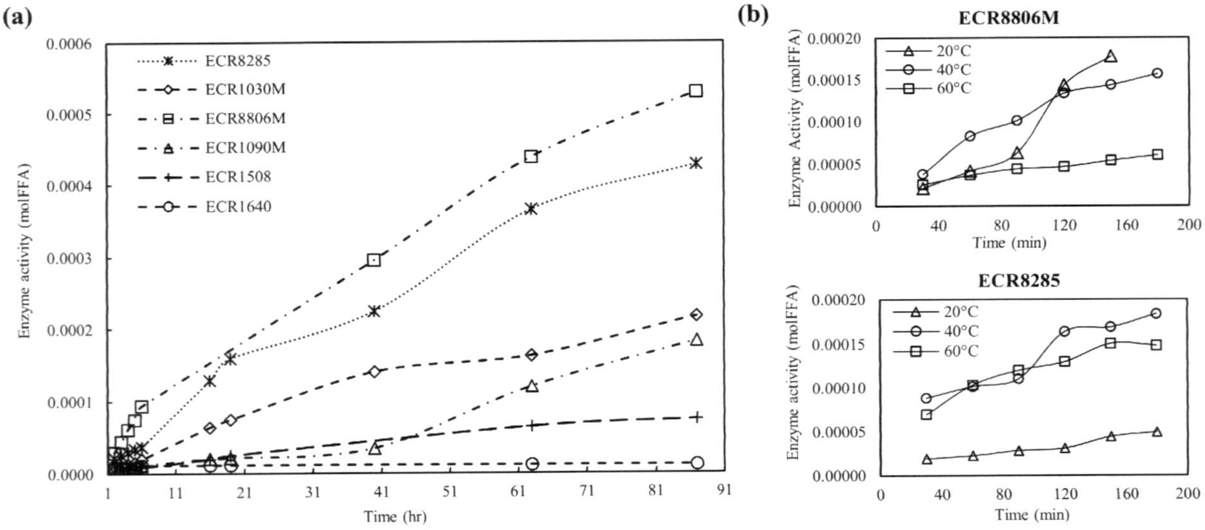

Figure 5. Hydrolytic performance of immobilized lipase on Purolite® supports – (a) Long hydrolysis and (b) Effect of temperature on ECR8806M and ECR8285 resin

Comparison of hydrolytic performance of Purolite supports using vegetable oil hydrolysis shown in Figure 5(a) showed that ECR8806M and ECR8285 performed much better than the all other supports [40]. ECR8806M is an epoxy-supported resin, it shows more stability and activity compared to the other supports because epoxy groups can react with amino or thiol groups of the enzyme and form strong bonds between enzyme and supports which makes it attractive for recycling [36]. Basso et al. showed Octadecyl methacrylate have excellent immobilization capabilities compared to styrene based resin which could explain the poor hydrolytic performance of styrene based resins [37, 41, 42]. From Figure 5(b), ECR8285 and ECR8806M with excellent immobilization capabilities, when examined under three different temperature ranges showed that ECR8285 is more stable at higher temperature range (above 40°C) and ECR8806M at lower temperature range (below 40°C).

10. Electro-Lipase-Activity

Rates of reaction for the enzymatic hydrolysis of vegetable oil in the presence of aqueous enzyme solution were determined in the presence of steady direct current (DC) electrical field. The reactions were conducted in a fixed rectangular geometry batch reactor fitted with planar electrodes as shown in Figure 6(a). The two immiscible liquid phases, oil and aqueous enzyme solution made up at concentration of 1.5 (%w/v), were located between the electrodes with a quiescent rectangular interface of known area. The rates of conversion to the free fatty acid and glycerol were determined for applied DC voltages in the range 0-30 Volts imposed between the two electrodes as shown in Figure 6(b). The results showed clear evidence of significant increase in the rates of reaction in the presence of the electrical field. It was concluded that the electric field has a positive effect on reaction rate, independent of interfacial area [43, 44]. This suggested possible changes in enzyme structure due the electrical field, modification of the nature of the enzyme binding at the liquid-liquid interface, or possibly intensification of local transport rates of substrates and reaction products close to the interface. The latter could involve phenomena such as electro-kinetic transport or electro-osmosis [45-47].

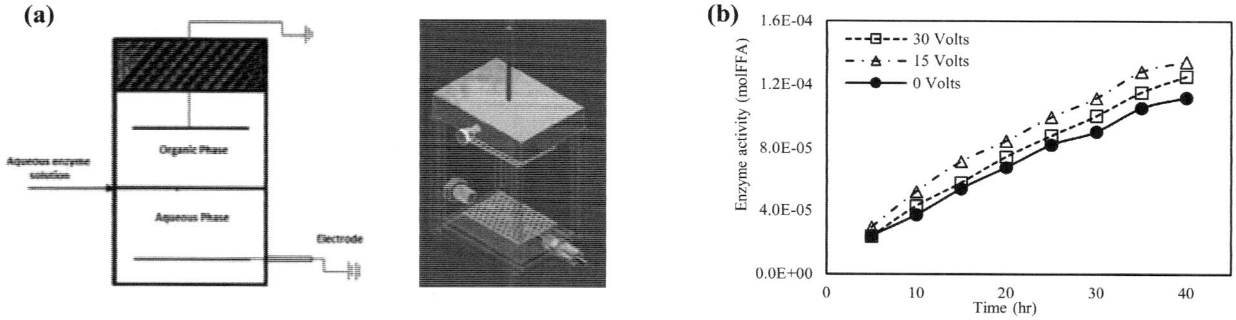

Figure 6. Electrically enhanced enzymatic hydrolysis (EEEH) – (a) Electric field reactor and (b) Comparative performance of free enzymes under electric field

REFERENCES

[1] F.D. Gunstone, R.J. Hamilton, Oleochemical manufacture and applications, CRC Press2001.

[2] K. Ngaosuwan, E. Lotero, K. Suwannakarn, J.G. Goodwin Jr, P. Praserthdam, Hydrolysis of triglycerides using solid acid catalysts, Industrial & Engineering Chemistry Research 48(10) (2009) 4757-4767.

[3] O. Ackelsberg, Fat splitting, Journal of the American Oil Chemists' Society 35(11) (1958) 635-640.

[4] E. Twitchell, Process of decomposing fats or oils into fatty acids and glycerin, Google Patents, 1898.

[5] E.G. W, Process of hydrolyzing oils and fats, Google Patents, 1939.

[6] H. Barnebey, A. Brown, Continuous fat splitting plants using the Colgate-Emery process, Journal of the American Oil Chemists' Society 25(3) (1948) 95-99.

[7] R.L. Holliday, J.W. King, G.R. List, Hydrolysis of vegetable oils in sub-and supercritical water, Industrial & Engineering Chemistry Research 36(3) (1997) 932-935.

[8] W.P. Jencks, Catalysis in chemistry and enzymology, Courier Corporation1987.

[9] A. Zaks, A.M. Klibanov, Enzyme-catalyzed processes in organic solvents, Proceedings of the National Academy of Sciences 82(10) (1985) 3192-3196.

[10] N.N. Gandhi, Applications of lipase, Journal of the American Oil Chemists' Society 74(6) (1997) 621-634.

[11] H. Uhlig, Industrial enzymes and their applications, John Wiley & Sons1998.

[12] L. Sarda, P. Desnuelle, Actions of pancreatic lipase on esters in emulsions, Biochimica et biophysica acta 30(3) (1958) 513.

[13] R. Verger, Enzyme kinetics of lipolysis, Methods in enzymology 64 (1980) 340-392.

[14] Y. Koichi, M. Haruo, High-activity lipase and method of preparation thereof, Google Patents, 1965.

[15] S. Al-Zuhair, K. Ramachandran, M. Hasan, Effect of enzyme molecules covering of oil–water interfacial area on the kinetic of oil hydrolysis, Chemical Engineering Journal 139(3) (2008) 540-548.

[16] M. Hoarau, S. Badieyan, E.N.G. Marsh, Immobilized enzymes: understanding enzyme–surface interactions at the molecular level, Organic & biomolecular chemistry 15(45) (2017) 9539-9551.

[17] R.A. Sheldon, S. van Pelt, Enzyme immobilisation in biocatalysis: why, what and how, Chemical Society Reviews 42(15) (2013) 6223-6235.

[18] C. Mateo, J.M. Palomo, G. Fernandez-Lorente, J.M. Guisan, R. Fernandez-Lafuente, Improvement of enzyme activity, stability and selectivity via immobilization techniques, Enzyme and microbial technology 40(6) (2007) 1451-1463.

[19] C. Laane, S. Boeren, K. Vos, C. Veeger, Rules for optimization of biocatalysis in organic solvents, Biotechnology and Bioengineering 30(1) (1987) 81-87.

[20] H. Noureddini, X. Gao, S. Joshi, P. Wagner, Immobilization of Pseudomonas cepacia lipase by sol-gel entrapment and its application in the hydrolysis of soybean oil, Journal of the American Oil Chemists' Society 79(1) (2002) 33-40.

[21] J.C. Santos, G.F. Nunes, A.B. Moreira, V.H. Perez, H.F. de Castro, Characterization of Candida rugosa Lipase Immobilized on Poly (N-methylolacrylamide) and Its Application in Butyl Butyrate Synthesis, Chemical Engineering & Technology 30(9) (2007) 1255-1261.

[22] Y. Yong, Y.-X. Bai, Y.-F. Li, L. Lin, Y.-J. Cui, C.-G. Xia, Characterization of Candida rugosa lipase immobilized onto magnetic microspheres with hydrophilicity, Process Biochemistry 43(11) (2008) 1179-1185.

[23] M. Monier, A. El-Sokkary, A. Sarhan, Immobilization of Candida rugosa lipase on modified natural wool fibers, Reactive and Functional Polymers 70(2) (2010) 122-128.

[24] M. Basri, W.M. Yunus, Z. Wan, W.S. Yoong, K. Ampon, C.N.A. Razak, A.B. Salleh, Immobilization of lipase from Candida rugosa on synthetic polymer beads for use in the synthesis of fatty esters, Journal of chemical technology and biotechnology 66(2) (1996) 169-173.

[25] L.R. Weatherley, A. Anand, The Performance of Microbial Lipase Immobilized onto Ion Exchange Resins and onto the Natural Zeolite Clinoptilolite, International Journal of Chemical Separation Technology 2(2) (2016).

[26] X.-J. Huang, P.-C. Chen, F. Huang, Y. Ou, M.-R. Chen, Z.-K. Xu, Immobilization of Candida rugosa lipase on electrospun cellulose nanofiber membrane, Journal of Molecular Catalysis B: Enzymatic 70(3) (2011) 95-100.

[27] S.S. Betigeri, S.H. Neau, Immobilization of lipase using hydrophilic polymers in the form of hydrogel beads, Biomaterials 23(17) (2002) 3627-3636.

[28] M.D. Ramos, G.I.G. Gómez, N.S. González, Immobilization of Candida rugosa lipase on bentonite modified with benzyltriethylammonium chloride, Journal of Molecular Catalysis B: Enzymatic 99 (2014) 79-84.

[29] I. Migneault, C. Dartiguenave, M.J. Bertrand, K.C. Waldron, Glutaraldehyde: behavior in aqueous solution, reaction with proteins, and application to enzyme crosslinking, Biotechniques 37(5) (2004) 790-806.

[30] S. Datta, L.R. Christena, Y.R.S. Rajaram, Enzyme immobilization: an overview on techniques and support materials, 3 Biotech 3(1) (2013) 1-9.

[31] F. López-Gallego, L. Betancor, C. Mateo, A. Hidalgo, N. Alonso-Morales, G. Dellamora-Ortiz, J.M. Guisán, R. Fernández-Lafuente, Enzyme stabilization by glutaraldehyde crosslinking of adsorbed proteins on aminated supports, Journal of Biotechnology 119(1) (2005) 70-75.

[32] O. Barbosa, C. Ortiz, Á. Berenguer-Murcia, R. Torres, R.C. Rodrigues, R. Fernandez-Lafuente, Glutaraldehyde in bio-catalysts design: a useful crosslinker and a versatile tool in enzyme immobilization, Rsc Advances 4(4) (2014) 1583-1600.

[33] T. Haider, Q. Husain, Concanavalin A layered calcium alginate–starch beads immobilized β galactosidase as a therapeutic agent for lactose intolerant patients, International journal of pharmaceutics 359(1) (2008) 1-6.

[34] D. Rooney, L. Weatherley, The effect of reaction conditions upon lipase catalysed hydrolysis of high oleate sunflower oil in a stirred liquid–liquid reactor, Process Biochemistry 36(10) (2001) 947-953.

[35] M. Murray, D. Rooney, M. Van Neikerk, A. Montenegro, L. Weatherley, Immobilization of lipase onto lipophilic polymer particles and application to oil hydrolysis, Process Biochemistry 32(6) (1997) 479-486.

[36] M. Mihailović, M. Stojanović, K. Banjanac, M. Carević, N. Prlainović, N. Milosavić, D. Bezbradica, Immobilization of lipase on epoxy-activated Purolite® A109 and its post-immobilization stabilization, Process Biochemistry 49(4) (2014) 637-646.

[37] A. Basso, M. Hesseler, S. Serban, Hydrophobic microenvironment optimization for efficient immobilization of lipases on octadecyl functionalised resins, Tetrahedron 72(46) (2016) 7323-7328.

[38] K. Hernandez, C. Garcia-Galan, R. Fernandez-Lafuente, Simple and efficient immobilization of lipase B from Candida antarctica on porous styrene–divinylbenzene beads, Enzyme and microbial technology 49(1) (2011) 72-78.

[39] A. Anand, L.R. Weatherley, The performance of microbial lipase immobilized onto polyolefin supports for hydrolysis of high oleate sunflower oil, Process Biochemistry (2018).

[40] V.G. Tacias-Pascacio, J.J. Virgen-Ortíz, M. Jiménez-Pérez, M. Yates, B. Torrestiana-Sanchez, A. Rosales-Quintero, R. Fernandez-Lafuente, Evaluation of different lipase biocatalysts in the production of biodiesel from used cooking oil: Critical role of the immobilization support, Fuel 200 (2017) 1-10.

[41] A. Basso, L. Froment, M. Hesseler, S. Serban, New highly robust divinyl benzene/acrylate polymer for immobilization of lipase CALB, European journal of lipid science and technology 115(4) (2013) 468-472.

[42] V.G. Tacias-Pascacio, S. Peirce, B. Torrestiana-Sanchez, M. Yates, A. Rosales-Quintero, J.J. Virgen-Ortíz, R. Fernandez-Lafuente, Evaluation of different commercial hydrophobic supports for the immobilization of lipases: tuning their stability, activity and specificity, RSC Advances 6(102) (2016) 100281-100294.

[43] F. Wu, Y. Jin, D. Li, Y. Zhou, L. Guo, M. Zhang, X. Xu, N. Yang, Electrofluid hydrolysis enhances the production of fermentable sugars from corncob via in/reverse-phase induced voltage, Bioresource Technology 234 (2017) 158-166.

[44] Y. Zhou, Y. Jin, N. Yang, Z. Xie, X. Xu, Electrofluid enhanced hydrolysis of maize starch and its impacts on physical properties, RSC Advances 7(31) (2017) 19145-19152.

[45] C. Albers, L.R. Weatherley, Enzyme reaction kinetics at an electrically charged liquid-liquid interface, (2014).

[46] A. Anand, E. Harmsen, L.R. Weatherley, Free enzyme reaction kinetics at a liquid-liquid interface with an applied electric field, (2017).

[47] A. Anand, J. Hattermer, L.R. Weatherley, Enzyme catalyzed hydrolysis reaction kinetics at a liquid-liquid boundary under effect of an applied electric field, (2017).

[48] F.X. Malcata, H.R. Reyes, H.S. Garcia, C.G. Hill, C.H. Amundson, Immobilized lipase reactors for modification of fats and oils—a review, Journal of the American Oil Chemists' Society 67(12) (1990) 890-910.

[49] Y. Kosugi, H. Tanaka, N. Tomizuka, Continuous hydrolysis of oil by immobilized lipase in a countercurrent reactor, Biotechnology and bioengineering 36(6) (1990) 617-622.

[50] S. Fadiloğlu, Z. Söylemez, Kinetics of lipase-catalyzed hydrolysis of olive oil, Food research international 30(3) (1997) 171-175.

535098 Activity Improvement of D-Psicose-3-Epimerase from *Agrobacterium tumefaciens* CGMCC 1.1488 By Site-Directed Mutagenesis

Abstract:

The enzyme D-psicose-3-epimerase catalyzes the conversion of C3-epimerization from fructose to allulose (D-psicose). Due to the relatively low allulose conversion rate, the improvement of enzymatic activity by site-directed evolution was carried out in this experiment. The DPEase-C58 from *Agrobacterium tumefaciens* CGMCC 1.1488 was supplied to the protein structure comparison and sequence alignment with DTE from *Pseudomonas cichorii*, whose crystal structure and the complex with D-fructose had been determined, to conduct the sites mutagenesis. The two sites of saturation mutation of H228 and T231 were carried out. At first round 5 mutants with 20% relevant higher than wide-type were obtained by high-throughout screening method.

537646 *In Vivo* Characterization of Glucose Responsive Insulin Delivery Systems

Abstract:

***In Vivo* Characterization of Glucose Responsive Insulin Delivery Systems**

Lisa R Volpatti, Morgan A. Matranga, Abel C. Cortinas, Robert S. Langer, Daniel G. Anderson

Koch Institute of Integrative Cancer Research & Dept. of Chemical Engineering, MIT, Cambridge, MA 02139

Motivation: Diabetes mellitus is a disease characterized by poor glycemic control as a result of deficient insulin production (Type 1) or signaling (Type 2). Chronic hyperglycemia often leads to severe complications including retinopathy, cardiovascular disease, kidney failure, and cancer. Therefore, Type 1 and advanced Type 2 diabetic patients are dependent on exogenous insulin to combat hyperglycemia. To improve patient compliance, an attractive diabetes therapy is the self-regulated delivery of insulin to mimic native insulin production or signaling. Many chemically glucose-responsive systems which employ the enzyme glucose oxidase as a glucose sensor have been reported in the literature. However, only ~25% of publications on glucose-responsive insulin release even include *in vivo* studies, and there have been very few reports of true glucose-responsiveness *in vivo*. Here we show extensive *in vivo* results including pharmacodynamic and pharmacokinetic studies of both healthy and diabetic mouse models using our acetalated-dextran based glucose-responsive nanoparticle system.

Results: Acetalated-dextran nanoparticles were formulated containing insulin and glucose oxidase. Glucose oxidase converts glucose to gluconic acid thus reducing the pH of the microenvironment when glucose levels are high. Acetalated dextran rapidly degrades under acidic conditions, releasing insulin on demand. Several *in vivo* studies were conducted to determine the safety and efficacy of these nanoparticles. Safety studies were performed by administering nanoparticles to healthy mice, measuring their blood glucose levels (Fig.1a), and calculating the corresponding hypoglycemic index (Fig 1b). The hypoglycemic index is an indication of the severity of blood glucose depression under normal physiological conditions. We show that two different doses of our nanoparticles have significantly lower hypoglycemic indices than that of naked insulin, indicating that they are safe to inject in the absence of a high blood sugar state.

We further show the efficacy of the nanoparticles using a diabetic mouse model with three simulated meals (Fig 1c) and compare the area under the curve for naked insulin, two doses of nanoparticles, and a dose of nanoparticles with a skipped simulated meal (Fig. 1d). These results indicate that if a diabetic patient were to miss a meal, then he/she would not be at risk of a hypoglycemic episode and would still maintain tight glycemic control. We next compare our nanoparticles to control particles omitting either insulin or glucose oxidase (Fig. 1e) and quantify the area under the curve to show that the enzymes are necessary to achieve normoglycemia (Fig. 1f). We also perform two week experiments by encapsulating the nanoparticles in a hydrogel network to create a long acting insulin depot. Additionally, we perform pharmacokinetic studies by quantifying the amount of insulin in mouse serum samples for both healthy and diabetic models to show glucose responsiveness. Finally, we show by *in vivo* fluorescent imaging that our nanoparticles are glucose responsive.

Conclusions: *In vivo* studies using both healthy and diabetic animal models are key to advancing glucose-responsive insulin therapies. Here we show in a broad range of *in vivo* studies that our optimized acetalated-dextran nanoparticles release insulin in response to high levels of glucose, achieving tight glycemic control, and creating a self-regulated system that could be used as an injectable therapy for diabetic patients.

522263 Silica Nanoparticles Enable Oral Delivery of Insulin

Abstract:

Introduction: Oral delivery of macromolecular drugs is one of the greatest unmet needs in biomedicine, with poor absorption across the intestinal epithelium posing the most significant challenge to clinical translation[1]. Several studies have demonstrated an improvement in bioavailability of macromolecules conjugated onto or loaded into orally administered nanoparticles[2]. However, this work presents the previously unreported ability of anionic nanoparticles to improve intestinal absorption of large molecules without chemically linking them to the particle delivery vehicles.

Methods: A collection of nanoparticles comprising four sizes (20, 50, 100, and 200 nm) and six surface chemistries (bare silica, carboxylated silica, aminated silica, silver, gold, or bare polystyrene) was screened for absorption enhancing effects *in vivo*. Mice received oral gavages of particle suspensions, then either gavages or direct intestinal administration of 4 kDa FITC-labelled dextran (FITC-DX4) or insulin solution. Blood samples were collected and screened for FITC-DX4 or glucose concentration. Mouse intestines were harvested after treatment and processed for histology to assess particle-induced damage. For mechanistic studies, Caco-2 monolayers were used in conjunction with anti-receptor antibodies, enzyme inhibiting peptides, and fluorescent confocal microscopy to investigate tight junction rearrangement.

Results: Nanoparticles with negatively charged surfaces (bare silica, carboxylated silica, silver, and gold) improved oral FITC-DX4 accumulation in the blood, while neutral (polystyrene) and positive (aminated) particles showed little to no effect. Further, silica nanoparticles significantly improved the activity of intestinally administered insulin, matching the total hypoglycemic effect of subcutaneously injected insulin. Intestinal insulin bioactivity exhibited dependence on both nanoparticle dose and insulin dose. Further, orally administered, enterically coated insulin capsules induced pronounced and sustained hypoglycemia at the relatively low oral insulin dose of 40 U/kg, and only when administered alongside silica nanoparticles. Histological analysis of particle-treated mouse intestines showed no evidence of necrosis, inflammation, or changes in tissue architecture. However, *in vitro* mechanistic studies revealed that the nanoparticle treatment causes a substantial rearrangement of the tight junction protein ZO-1, an integral component to the barrier function of the epithelial lining, via a previously-identified, integrin- and MLCK-dependent mechanism[3].

Conclusions and Implications: This is the first study demonstrating the use of co-delivered but unbound nanoparticles for improving adsorption of orally administered protein drugs. These conclusive *in vivo* results underscore the ability of silica nanoparticles to increase epithelial permeability, enabling the oral delivery of macromolecular drugs without encapsulation or conjugation. Future investigations will include more comprehensive toxicity studies in mice, development of an administration vehicle to better co-localize the particles and insulin, and confirmation of safety and efficacy in higher animal models.

Acknowledgements: N. Lamson would like to acknowledge financial support from the NSF Graduate Research Fellowship Program under Grant No. DGE1252522.

References:

1. Hamman, J.H., Enslin, G.M., Kotzé, A.F. Biodrugs. 2005 19(3) 165-177.
2. Banerjee, A., Qi, J., Gogoi, R., Wong, J., Mitragotri, S. J. Control. Release. 2016 (238) 176-185.
3. Walsh, L., Ryu, J., Bock, S., Koval, M., Mauro, T., Ross, R., Desai, T. NanoLett. 2015 15(4) 2434-2441.

536146 Targeted, Systemic Dendrimer-Drug Therapies for Age Related Macular Degeneration

Abstract:

Introduction: Age related macular degeneration (AMD) is a complex, multifactorial progressive disease often implicated with key pathological events such as oxidative stress, inflammation and neovascularization (CNV) in retina and choroid. Current treatments involve frequent, localized intravitreal injections of anti-VEGF agents for CNV suppression that doesnâ€™t work in 1/3 of the patients and there is no viable therapy for early AMD. These treatments only focus on addressing neovascularization and often associated with side-effects such as choriocapillaries atrophy, RPE loss and intraocular toxicity. Hence, targeting and delivering drugs to the key cells involved in disease progression will highly beneficial for both early and late stages, improve compliance and reduce costs. Increasing evidence suggests inflammation (activated microglia/macrophages (mi/ma)) may play a critical role in the development and the progression of AMD. We showed that hydroxyl terminated PAMAM dendrimer can target and localize in activated retinal microglia/macrophages (mi/ma) and RPE upon *intravenous* injections (IV) [1]. In this study, we utilize the intrinsic targeting ability of dendrimers to deliver drugs to the key cells and evaluate is efficacy for early and late AMD in subretinal lipid injected rat AMD model

Methods: Dendrimer-Triamcinolone acetonide (D-TA) and Dendrimer-N-Acetyl cysteine (D-NAC) were designed and synthesized to release drugs intracellularly. Cy5-labelled D-TA and D-NAC were used to evaluate the biodistribution and cellular localization using fluorescence spectroscopy and confocal microscopy. For early AMD, D-NAC monotherapy was administered intravenously on day 3, 5 and 7 post subretinal lipid and evaluated on day 10. For late AMD, a combination of D-NAC + D-TA were administered intravenously on day 11, 13 and 15 and evaluated on day 20. CNV suppression and regression were evaluated by accessing the CNV area and volumes using flat-mount imaging. Attenuation of inflammation were evaluated using macrophage counts and RT-qPCR was employed to evaluate inflammatory cytokines and oxidative stress markers.

Results: Subretinal lipid injection produced blebs that stimulated migration of mi/ma and formation of CNV. Both D-TA and D-NAC were stable in plasma and released drugs under intracellular conditions in a sustained manner. Intravenously administered D-NAC and D-TA were found localized in activated mi/ma and hypertrophic RPE in retina and choroid and were retained for more than 30 days but cleared intact from systemic circulation intact within 24 hrs [2]. D-NAC monotherapy suppressed CNV growth (~78%), reduction in mi/ma accumulation (~63%) and attenuated inflammation by suppressing cytokine production. D-NAC+D-TA combination therapy treatment promotes CNV regression (~72%), reduced CNV volume, enhanced adhesion molecules expression and reduced leakage in late stages of AMD than compared to controls. Dendrimer-drug therapy did not lead to ocular or systemic toxicity.

Conclusions: Targeting and selectively co-localization of dendrimers, in activated mi/ma and RPE, makes them excellent carriers for systemic drug delivery to specific affected cells in the choroid/retina. The efficacy of D-NAC and D-TA in CNV suppression/regression, attenuation of retinal/choroidal inflammation may offer an effective treatment option for early and late stages of AMD and other ocular diseases such as diabetic retinopathy (DR), glaucoma, retinitis pigmentosa (RP) and etc. where inflammation is involved.

Acknowledgements: This study was supported by the funds from NIH-NEIRO1EY025304 (RMK) and Altsheler Durell foundation.

References:

1. Kambhampati, SP., Clunies-Ross, A.J., Bhutto, I., Mishra, MK., Edwards, M., McLeod, D.S., Kannan, RM., Lutty, GA. Systemic and Intravitreal Delivery of Dendrimers to Activated Microglia/Macrophage in Ischemia/Reperfusion Mouse. IOVS. 2015, 56(8), 4413-4424.
2. Kannan, RM., Lutty, G., Kambhampati, SP., Mishra, Mk., Bhutto. I.A. "Dendrimer compositions and their use in treatment of diseases of the eye." U.S. Patent Application 15/307,284, filed February 16, 2017.

535509 Growth Rate Dissipation of Metastatic Triple Negative Breast Cancer Attributed to Slow Tumor-Clearing and Deep Tumor-Penetrating Chemotherapy

Abstract:

Metastatic and/or recurrent Triple Negative Breast Cancer (TNBC) is currently incurable. TNBC accounts for 12-17% of breast carcinomas with the lowest 5-year survival rates among all breast cancer patients due to high proliferation and reoccurrence outside the breast combined with lack of effective targeted therapeutic modalities. Platinum-derived compounds receive extensive clinical use because of their DNA damaging activity to which TNBC tumors frequently show sensitivity. However, clinical efficacy is still limited.

To enhance the efficacy of tumor delivered platinum compounds, we engineered the release and adhesion properties of lipid nanocarriers encapsulating cisplatin with the following two aims: 1) to improve the uniformity in intratumoral drug distributions, and 2) to prolong the exposure of cancer cells to administered therapeutics with the goal of better controlling their growth (and/or spreading).

We have previously demonstrated in 3D multicellular spheroids (used as surrogates of the tumor avascular regions) that uniform intratumoral distributions is facilitated by nanocarriers engineered to release their (rapidly diffusing) therapeutic contents in the tumor interstitium enabling deep tumor-penetration of therapeutics. Key to this approach is to use drug nanocarriers that do not become internalized by cells - so as to maximize the fraction of released drug that may penetrate deeper in the tumor - and, to choose therapeutic agents (for example, cisplatin) which are efficiently transported across the cell membranes independent of the local extracellular milieu.

In this study, to effectively translate this strategy in vivo, the intratumoral residence times of such drug-loaded nanocarriers are now designed to be increased so as to maximize the time-integrated dose delivered at the tumor. To increase the nanocarriers' tumor residence times, we introduce an 'adhesion switch' on the nanocarriers' surface with the aim to slow down their tumor-clearing kinetics. The switch is designed to promote nanoparticle adsorption on cancer cells and/or the extracellular matrix while keeping their internalization by cells to a minimum.

Herein, we present evidence that lipid-based nanocarriers (liposomes) loaded with cisplatin and designed to combine the above two distinct mechanisms, improve the intratumoral drug uniformity and prolong the carriers' tumor residence times. We demonstrate that they result in slow tumor clearance and enable significant decrease of the rate of growth of metastatic TNBC tumors in vivo.

These findings, support the potential of nanocarriers' properties such as the interstitial release of therapeutics by slow tumor-clearing carriers to collectively improve the efficacy of chemotherapeutics and to better control the growth of metastatic TNBC.

537482 Neutrophil–Particle Interactions in Blood Circulation Drive Particle Clearance and Alter Neutrophil Responses in Acute Inflammation

Abstract:

Introduction: Particulate drug carriers for targeted drug delivery have been widely studied for their potential to enhance the therapeutic impact of treatments for a variety of diseases, including cancer, cardiovascular disease, and inflammation. However, the interactions between these drug carriers and circulating leukocytes, particularly neutrophils, remain relatively explored. Such interactions are crucial, as they may impact the ultimate therapeutic efficacy of particulate drug carriers, and may have a therapeutic impact themselves. In this work, we sought to determine what impact, if any, injectable particulate drug carriers have on neutrophil adhesion and transmigration in an inflammatory environment *in vivo*.

Materials and Methods: The impact of injected particulate drug carriers on neutrophil adhesion and migration was evaluated using two different *in vivo* models: a mesentery inflammation model, and an acute lung injury model. For the mesentery inflammation model, female C57BL/6 mice (Jackson) between 3 and 4 weeks in age were used. Neutrophil rolling and adhesion in mesenteric veins were visualized using a 25× oil objective on an inverted fluorescence microscope (Zeiss Axio Observer Z1Marianas microscope) using Slidebook 6 software. Mice were anesthetized, and a tail vein catheter was placed for delivery of reagents. Mice were placed on a custom-made microscopic heated stage at 37 °C, and the mesentery was exteriorized to a glass coverslip through a midline incision. Rhodamine 6G (Rh6G, Sigma, 100 µL of 0.1 mg/mL in PBS) or antiLy6G (Biolegend) was injected IV, and a local injury was induced by topical application of TNF-α (Fitzgerald, 10 µL of 200 µg/mL in PBS). FITC-labeled particles suspended in PBS were injected via IV catheter 3 min following topical TNF-α application and imaged for another 7 min. Analysis was performed using Slidebook 6 and ImageJ by a blinded investigator. Images were recorded continuously in green, red, and bright field channels every 100 ms. Vessels were chosen in each mouse based on size and vessel exposure, with the average diameter of veins ranging from 90 to 190 µm, with an average of 130 µm. The total number of NΦs and particles were counted per frame to obtain average NΦ counts per 3 s of capture in 125 µm length of mesentery vessel, corresponding to 30 frames at 100 ms/ frame.

For the acute lung injury model, male C57BL/6 and BALB/c mice (Jackson) were anesthetized using isofluorane, and then LPS (20 µg of LPS at 0.4 mg/mL) was delivered to the lungs through an orotracheal instillation in a 50 µL volume in PBS. One hour after instillation, mice were injected with a suspension of particles in PBS through the tail vein. Mice received 2 µm particles or 0.5 µm particles in 200 µL injection volume, corresponding to 0.6 mg/mouse, 30 mg/kg. Two or 3 h after LPS instillation, mice were euthanized, and blood was collected via cardiac puncture in acid citrate dextrose (ACD) and perfused with PBS. BALF was collected by inserting a cannula in an incision in the trachea and flushing the lungs with three sequential 1 mL washes of PBS. BALF cells were obtained by centrifugation, separating BALF cells from the supernatant. Following the lavage, organs were harvested and imaged using an Odyssey CLx infrared imaging system (LI-COR), and whole organ scans performed. Following NIR scan, lung, liver, and spleen were homogenized into single-cell suspensions or frozen for histological examination. Frozen samples were embedded in OCT (Fisher) in disposable cassettes and flash frozen with an isopentane slurry; additional liver samples were fixed in 4% paraformaldehyde (PFA) or flash frozen in liquid nitrogen and stored for histology or protein analysis.

Results and Discussion: Circulating blood neutrophils in vivo were found to be capable of rapidly binding and sequestering injected carboxylate-modified particles of both 2 and 0.5 µm diameter within the bloodstream. These neutrophil–particle associations within the vasculature were found to suppress neutrophil interactions with an inflamed mesentery vascular wall and hindered neutrophil adhesion. Furthermore, in a model of acute lung injury, intravenously administered drug-free particles reduced normal neutrophil accumulation in the airways of C57BL/6 mice between 52% and 60% versus particle-free mice and between 93% and 98% in BALB/c mice. This suppressed neutrophil migration resulted from particle-induced neutrophil diversion to the liver.

Conclusions: These data indicate a considerable acute interaction between injected particles and circulating neutrophils that can drive variations in neutrophil function during inflammation and implicate neutrophil involvement in the clearance process of intravenously injected particle therapeutics. Such an understanding will be critical toward both enhancing designs of drug delivery carriers and developing effective therapeutic interventions in diseases where neutrophils have been implicated.

530643 Synthesis and Characterization of pH-Responsive Hydrogels for Oral Delivery of High Isoelectric Point Therapeutic Proteins

Abstract:

Therapeutic proteins are widely researched and offer treatment of cancer, autoimmune diseases such as multiple sclerosis and rheumatoid arthritis, and macular degeneration, as well as many other debilitating diseases, but are currently available to patients via subcutaneous or intramuscular injection. Unfortunately, the frequent injections required for therapeutic dosage are extremely painful, often leading to low patient compliance. Oral delivery of these proteins would be highly preferred, but there are significant barriers to successful delivery to the bloodstream via this route, including the acidic environment of the stomach, enzymatic degradation in the gastrointestinal tract, and the epithelial lining of the small intestine, which tightly regulates the passage of large molecules but is where most drugs are absorbed. Bioavailability of protein drugs after oral administration is extremely low, less than one percent. Environmentally-responsive anionic hydrogels containing methacrylic acid (MAA), which protect the protein under acidic conditions but swell to release the protein at the higher pH of the small intestine, have been developed as drug delivery vehicles for therapeutic proteins and have demonstrated successful delivery of proteins such as insulin. However, higher isoelectric point (pI) proteins, suffer from limited release using these systems due to electrostatic binding resulting from their positive charge at neutral pH. Of the over 150 currently FDA-approved protein therapeutics, at least 46 exhibit high isoelectric points (likely more, as not all of the approved drugs have accepted pI values available). These therapeutics are used to treat a wide variety of diseases: breast, lung, colorectal, and head and neck cancers; rheumatoid arthritis; osteoporosis; Crohnâ€™s disease; infertility; non-Hodgkinâ€™s lymphoma; multiple sclerosis; macular degeneration; anemia; and growth failure, among others. Because these diseases together account for many millions of patients who may benefit from repeated administration of protein therapeutics, it is imperative that any oral delivery strategy be able to accommodate the high pI exhibited by the proteins. The aims of our work were to explore different strategies for achieving high bioavailability of therapeutic proteins delivered via the oral route, with emphasis on seeking innovations that could allow the previously studied systems to be expanded or modified to function with high pI proteins. Copolymeric systems containing itaconic acid have been synthesized and possess similar pH-dependent swelling behavior, but show potential for delivery of high pI proteins when compared to the MAA-based systems.

538390 Zwitterionic Polymer Coatings to Limit Protein Adsorption to Nanocarrier Surfaces

Abstract:

The formation of a protein corona on drug carriers significantly reduces carrier penetration into cells. The goal of this work is to design zwitterionic polymers that can be readily attached to the surface of nanocarriers and limit protein adsorption from biological fluids. Zwitterionic polymers containing positively and negatively charged groups in close proximity can imbue surfaces with high levels of protein resistance (similar to virus coatings). In the current work, polymethacryloyloxyethyl phosphorylcholine (MPC) was synthesized using photoinitiated radical addition-fragmentation chain transfer (RAFT) polymerization. Polymer conversion was confirmed by 1H-NMR and monodispersity in molecular weight by GPC analysis. MPC, containing thiol end groups, was attached as a self-assembled monolayer to the surface of 100 nm spherical gold particles, synthesized using a modified version of the Turkevich protocol. DLS, TEM, laser Doppler anemometry, and XPS confirmed polymer coating on the gold surfaces. The level of proteins adsorbed from serum and bronchoalveolar lavage fluid (BALF) onto the nanoparticle surfaces was determined via standard protein assays. Significant inhibition of protein adsorption was observed with both serum and BALF, however complete inhibition was not achieved with BALF likely due to differences in the protein composition between these two biological fluids.

533721 High-Throughput Synthesis and Characterization of Rapidly Eroding Polyanhydride Nanoparticle Libraries for Drug Delivery

Abstract:

Oral administration is a preferred route for drug delivery as it is simple, well-tolerated by patients, and doesnâ€™t require a medical professional for needle handling. However, the acidic and enzymatic degrading environment of the gastrointestinal tract can limit the bioavailability of proteins and some small-molecule therapeutics. Oral delivery of hydrophobic drugs, such as estradiol, can be particularly challenging due to low aqueous solubility and a high first pass metabolism, further limiting drug bioavailability [1]. For other drugs, such as chemotherapeutics for cancer, the gut may need protection from toxic drug activity [2]. Controlled drug release nanoparticles could help meet these challenges by protecting their drug payload from the acidic environment of the stomach and releasing their payload in the more pH-neutral environment of the small intestine to allow transport into systemic circulation.

Polyanhydrides are a class of biodegradable, biocompatible polymers that have been used for controlled release of small-molecule drugs and proteinaceous payloads. They display hydrolytic degradation and surface erosion, resulting in a tunable release rate when hydrophobic and hydrophilic monomers are co-polymerized. Sebacic acid (SA) is a hydrophobic polyanhydride monomer that contributes favorable thermal properties for room-temperature synthesis of nanoparticles and a high rate of internalization by phagocytic cells, but its high local acidic microenvironment can negatively impact drug or protein stability. In contrast, 1,8-bis(*p*-carboxyphenoxy)-3,6-dioxaoctane (CPTEG) is an amphiphilic monomer that has been shown to stabilize protein antigens and also facilitates internalization by immune cells. In this work, we created a new family of CPTEG:SA copolymers with a rapid rate of degradation that could be ideal for oral delivery. Since polyanhydride degradation is base-catalyzed, drug release is expected to be delayed during transport through the stomach, and then progress rapidly in the intestines.

We synthesized CPTEG:SA copolymers of varying compositions via both conventional and high-throughput methods using an automated robot. Number average sequence length analysis indicates that SA-rich chemistries had a high SA sequence length, indicative of weakly segregated block-like microstructure. Chemistries closer to equimolar amounts of CPTEG and SA had similar, moderate sequence lengths, yielding an alternating copolymer. Compositions between 50:50 and 15:85 CPTEG:SA had reactivity ratios close to unity. Thermal characterization of CPTEG:SA polymers indicated semicrystalline behavior in SA-rich chemistries (i.e., at or below 35 mol% CPTEG). All polymers within this range had glass transition temperatures and melting points above room temperature, which facilitates nanoparticle synthesis via flash nanoprecipitation. Film erosion studies showed a faster rate of degradation as SA content increased.

The effect of polymerization time (and thereby molecular weight) on nanoparticle synthesis and morphology was investigated using the high-throughput methodology. A reaction time of five hours, corresponding to number-average molecular weight of approximately 20 kDa, yielded particles with sub-micron diameters. Nanoprecipitation conditions were tested a high-throughput manner, and rhodamine B, a hydrophobic small-molecule model drug, was encapsulated into the particles to investigate release kinetics. CPTEG:SA nanoparticles released rhodamine B rapidly, within three days. Drug release kinetics showed a chemistry-dependent trend, as increasing SA content increased the rate of release, consistent with the erosion kinetics data. Cell internalization assays in human monocyte/macrophage cell line THP-1, murine macrophage cell line J774, and murine bone-marrow derived dendritic cells (BMDCs) revealed chemistry-dependent internalization, with SA-rich 5:95 and 10:90 CPTEG:SA nanoparticles demonstrating the highest internalization by all cell lines. Altogether, the CPTEG:SA copolymers display a high tunability of degradation rates, drug release kinetics, and immune cell internalization at short time scales. This family of biodegradable carriers could facilitate the rational design of controlled release drug delivery and nanovaccine carriers by combining the favorable thermal, protein stabilization, and internalization properties of anhydride monomers.

Selected References

[1] Sahana D K, Mittal G, Bhardwaj V and Kumar M N V R 2008 PLGA Nanoparticles for Oral Delivery of Hydrophobic Drugs: Influence of Organic Solvent on Nanoparticle Formation and Release Behavior In Vitro and In Vivo Using Estradiol as a Model Drug *J. Pharm. Sci.* **97** 1530â€“42

[2] Sharpe L A, Daily A M, Horava S D and Peppas N A 2014 Therapeutic applications of hydrogels in oral drug delivery. *Expert Opin. Drug Deliv.* **11** 901â€“15

537702 Development of an Automated Multi-Stage Continuous Reactive Crystallization System with in-Line Pats for High Viscosity Process

Abstract:

An automated multi-stage continuous reactive crystallization system with in-line PATs for a high viscosity process was developed in the present study. Data Acquisition (DAQ) hardware and Labview software were used as the local control system. A â€œforward-backwardâ€ burst pumping strategy was developed to smoothly transfer the highly viscous hot slurry from one vessel to the next. In addition, a comparative analysis between a PFR and CSTRs in series revealed that to achieve the same conversion, the latter would require more volume than the former, but less than a single CSTR. For a second-order reaction, the value of the DamkÓ§hler number necessary to achieve conversion of 90.0% in a single CSTR is 90. Thus, it is reasonable to approximate a PFR using CSTRs in series to obtain a high yield with a smaller reaction volume (compared with a single CSTR). In-line FBRM and React IR were used to monitor the crystal size and reactant concentration in the vessel containing cold slurry, as the probes could not be positioned in the hot slurry due to fouling issues. E-factors of batch and continuous processes were also compared, and the continuous reaction could obtain a lower E-factor because less waste was generated.

532114 Disturbance Studies of Continuous Cooling Crystallization of Carbamazepine

Abstract:

Continuous manufacturing (CM) is an emerging technology in the pharmaceutical manufacturing sector. Continuous crystallization is important in an integrated CM process and may significantly affect the drug substance critical quality attributes (CQAs), such as particle size distribution and polymorphism. However, few publications have been focused on potential manufacturing disturbances, the impact on drug substance quality and how the risks can be mitigated. Here, we designed and built an automated two-stage Mixed Suspension Mixed Product Removal (MSMPR) crystallization platform for a model compound (Carbamazepine, CBZ) that exhibits multiple polymorphs. The crystallization process includes integration of Process Analytical Technology (PAT) tools (on-line Raman microscopy and Focused Beam Reflectance Microscopy, FBRM) for real-time monitoring. We performed continuous runs and evaluated the effects of typical manufacturing disturbances, such as temperature fluctuations, agitation rate change, and feed concentration hikes, on process dynamics. This talk will discuss our results and findings.

532752 Interface Nucleation in Segmented Continuous Crystallization Process

Abstract:

The continuous crystallization behavior of segmented tubular crystallization process was investigated using glycine as a model compound. In our previous work, air-liquid slug flow crystallization was studied systematically through experiments coupled with a Realizable kâ€“Îµ model in a computational fluid dynamics platform and obtained optimal results. In order to alleviate the phenomenon of wall crystallization occurred in the air-liquid segmented tubular crystallization system, liquid-liquid slug flow crystallization was investigated using dodecane as the oil phase. Process parameters such as surface active agent, ultrasonic were studied experimentally for their effect on the growth rate, size distribution, and morphology of the produced crystals as well. Static nucleation experiment under microscope of air-liquid interface and liquid-liquid interface was specially conducted and compared, and the results indicated that the oil restrained the nucleation compared to gas. The binding energy between glycine and different tubing materials were simulated using the material studio software and found that the binding energy between glycine and silicon tube was the strongest.

531977 Model Based Process Design on Continuous Cooling Crystallization

Abstract:

The use of first-principle models in the pharmaceutical industry presents great promise, as the models may significantly help on process design and control. Models can enhance scientific understanding of the process dynamics, and predict the behavior of a system at various conditions. As part of the effort to understand emerging technologies such as continuous manufacturing,[1-2] the development and verification of a population balance model (PBM) for the continuous cooling crystallization of Carbamazepine (CBZ) is studied. A systematic experimental design will be followed to estimate the kinetic parameters for the stable and metastable form (CBZ form III and II, respectively); a series of batch cooling crystallization experiments were performed to determine the parameters for nucleation, growth and dissolution kinetics.

In addition, we will perform a design of experiment (DOE) to systematically determine the design space to produce products of a certain size and polymorphic composition range. The population balance model (PBM) will be used to simulate our continuous crystallization processes. Five factors considered in the determination of the design space are: feed concentration, initial concentration, seed loading, seed crystal size distribution, and residence time. Various points within the design space and at the boundaries were evaluated experimentally.

1. Lee, S. L.; Oâ€™Connor, T. F.; Yang, X.; Cruz, C. N.; Chatterjee, S.; Madurawe, R. D.; Moore, C. M.; Lawrence, X. Y.; Woodcock, J., Modernizing pharmaceutical manufacturing: from batch to continuous production. *J Pharm Innov* **2015,** *10* (3), 191-199.
2. Yang, X.; Acevedo, D.; Mohammad, A.; Pavurala, N.; Wu, H.; Brayton, A. L.; Shaw, R. A.; Goldman, M. J.; He, F.; Li, S.; Fisher, R. J.; Oâ€™Connor, T. F.; Cruz, C. N., Risk Considerations on Developing a Continuous Crystallization System for Carbamazepine. *Organic Process Research & Development* **2017**.

534345 High Throughput Continuous Crystallization of Lysozyme in an Oscillatory Flow Baffled Crystallizer with Real Time Process Monitoring

Abstract:

High throughput continuous crystallization of lysozyme in an oscillatory flow baffled crystallizer with real time process monitoring

Biopharmaceuticals are increasingly used for medicinal therapies and now make up roughly 20% of the pharmaceutical market share. While most (small molecule) pharmaceutical products are purified with at least one crystallization step, biopharmaceuticals (specifically proteins) are very difficult to crystallize due to extraordinarily slow growth kinetics[1]. Moreover, large protein chains require subsequent folding operations in order to reach a desired activity level. Until now, oscillatory flow strategies have primarily been used in batch mode to improve the process kinetics of biopharmaceuticals through enhanced micromixing and heat transfer[2]. Known to have high surface area to volume ratios, oscillatory flow baffled reactors (OFBR) generate turbulence by superimposing piston driven oscillatory flow onto the net flow generated by a pump. This combination approach allows for adequate solid suspension with greatly improved micromixing compared to the traditional stirred tank approach[3]. Herein, lysozyme was chosen as a model compound for both crystallization and folding applications, as its kinetic parameters are relatively well studied in literature[4].

Using the commercially available DN6 (Alconbury West Ltd) with the new Blaze probe (Blaze 900, Blaze Metrics), the oscillatory system was able to significantly reduce the time required to crystallize lysozyme to several sizes (varying between 50-200μm). The Blaze Metrics probe allowed for real time monitoring of the system, both in terms of steady/transient state analysis as well as monitoring the critical quality attributes of the crystals (size and shape). These dynamic measurements allow for real time feedback control strategies to be implemented on this new commercially available platform.

In a separate application, the DN6 was also used to continuously fold the newly purified lysozyme. Proper folding of proteins is essential for their activity, and, for enzymes, activity can be measured using a colorimetric assay utilizing fluorescently tagged substrates. To ensure the activity of lysozyme was not affected by the crystallization process, renatured lysozyme was examined for its ability to degrade fluorescein-labeled *Micrococcus lysodeikticus* cell walls and compared to pre-crystallization lysozyme. It was found that the recrystallization step had little to no impact on the lysozyme's activity. Continuous oscillatory flow systems with proper monitoring offer advanced approaches to generating quality crystalline proteins for high throughput applications.

References

1. Gavira, J. A. (2016). Current trends in protein crystallization. *Archives of Biochemistry and Biophysics*, *602*, 3–11. http://doi.org/10.1016/j.abb.2015.12.010

2. Abbott, M. S. R., Harvey, A. P., Perez, G. V., & Theodorou, M. K. (2013). Biological processing in oscillatory baffled reactors: operation, advantages and potential. *Interface Focus*, *3*(1), 20120036.

3. McDonough, J. R., Phan, A. N., & Harvey, A. P. (2015). Rapid process development using oscillatory baffled mesoreactors - A state-of-the-art review. *Chemical Engineering Journal*, *265*(1), 110–121. http://doi.org/10.1016/j.cej.2014.10.113

4. Liu, Y., Wang, X., & Ching, C. B. (2010). Toward further understanding of lysozyme crystallization: Phase diagram, protein-protein interaction, nucleation kinetics, and growth kinetics. *Crystal Growth and Design*, *10*(2), 548–558.

525444 Dynamic Modeling of a Continuous Reactive Crystallization Process

Abstract:

Pharmaceutical and fine chemical companies often employ reactive crystallization to make intermediates and final products. The particle formation rate, particle size and shape, and growth rate are influenced by two temperature and concentration dependent phenomena: A) the reaction, via temperature and concentration dependent kinetic and reaction rate and activation energy; B) crystallization, via temperature and concentration dependent supersaturation ratio and nucleation and crystal growth rate. Therefore, the critical process parameters in productivity (reaction yield) and control of particle size distribution could be manipulated to achieve desirable quality and economics.

Determining the impact of operating temperature on the particle size distribution could be challenging when higher temperature would advance the reaction rate (more solute generation/concentration), but may impact supersaturation ratio and crystal growth rate (more solubility at higher temperature). Beside temperature, the reagents concentration (flow-rate) would impact the reaction rate and generating solute molecule that would lower supersaturation. Effect of micromixing and local supersaturation is not studied in this project.

Most of the reactive crystallization processes have been in batch mode. The reactive crystallization can be used in continuous fashion for API synthesis. The continuous reactive crystallization in a MSMPR (CSTR) method has been less studied and published. In this work, a dynamic process simulation was developed for a continuous reactive crystallization, where the effect of temperature and concentrations (reagentsâ€™ flow-rates) were studied on the system response and particle size distribution. The dynamic mode case studies cover a variation of CPPs disturbances and demonstrate system dynamic in terms of the particle size distribution and response rate in comparison to the RTD of the continuous reactive crystallizer.

527889 Convergent Synthesis of Peptides at Large Scale

Abstract:

Solid Phase Peptide Synthesis (SPPS) is the primary approach used in manufacturing to produce synthetic peptides. SPPS provides certain advantages over full solution phase chemistry approaches as the peptide on resin can be synthesized more rapidly and isolated with a higher purity. However, as a peptide chain grows longer the cost of SPPS increases disproportionately due to the use of excess reagents, the large volume of solvents, and decreasing coupling efficiency related to mass transfer resistances and steric hindrance.

Convergent peptide synthesis can be a very effective alternative to SPPS alone, when synthesizing long synthetic peptides. The method combines SPPS and the solution phase chemistry, where the fragments of a peptide chain are first synthesized linearly on resin, and then coupled in solution. The synthesis of short peptide fragments reduces the risk of single deletion impurities in the synthesis. Coupling fragments in solution also may significantly reduce the solvent use, resulting in greener synthetic approaches. Additionally, convergent approaches to peptide synthesis are far more amenable to alternative reactor designs such as flow infrastructure given the low concentration, high potency and high value of the peptide assets.

Control strategies for convergent chemistry approaches can also be significantly simplified given that peptides are routinely terminally purified through chromatographic means. Synthesis impurities that are markedly different in molecular weight are often more effectively separated in preparative chromatographic systems due to the differences in retention time. Synthesis impurities resulting from full linear builds are often characterized by single deletion impurities, with very similar amino acid sequence and therefore much more challenging to separate without sacrificing significant yield.

The focus of this presentation will be on our efforts to develop a platform for peptide synthesis focused on linear fragment builds, convergent fragment couplings, and enhanced process intensification and efficiency through incorporation of continuous processing strategies.

521342 Continuous N-Hydroxyphthalimide Mediated Electrochemical Aerobic Oxidation of Benzylic C-H Bonds

Abstract:

Electroorganic chemistry has emerged as an environmentally benign tool for synthetic chemists to achieve efficient transformations that are challenging with traditional reagent-based methods. Continuous flow chemistry brings pharmaceutical industry numerous advantages, but implementing electroorganic synthesis in flow is challenging, especially for electroorganic reactions with coupled electrode reactions and slow chemical reactions. Here, we present an electrolysis flow cell engineered for the N-hydroxyphthalimide (NHPI) mediated electrochemical aerobic oxidation of benzylic C-H bonds. Measuring the kinetics of the chemical reaction with cyclic voltammetry gives insights into effects of molecular structure on reaction rate and explains the need for recycling to achieve good yields.

521459 Future Manufacturing Platforms

Abstract:

Within R&D at Eli Lilly, a new capability team was created to identify, design, develop and tests new and/or improved manufacturing platforms into internal and external manufacturing network for both drug substance (DS) and drug product (DP). Future Manufacturing Platforms (FMP) goal is to ensure that chemistry, engineering and analytical solutions can meet future manufacturing needs, exploit new/improved technologies and help accelerate commercialization of medicines.

Responsible for prioritizing areas of investment, the recent journey of FMP will be highlighted. This includes the identification of technical gaps, development of external partners and academic collaborations, and the understanding of the regulatory environment to influence the adoption of new/improved manufacturing platforms. A few case studies will be presented.

537176 Harnessing Power of Membranes to Enable Process Intensification and Improvement

Abstract:

Continuous processing in biopharmaceutical industry could be very well enabled by the appropriate application of continuous membrane technology, which has been used in its limited capacity.

The presentation will focus on the appropriate membrane selection and associated process design to not only the removal of the inorganic salts but also the undesired solvent and/or reagents. It will also delve in the engineering methodology to develop a membrane process and touch upon scale-up considerations and challenges for successful technology implementation in a cGMP chemical pilot plant setting.

Manish S. Kelkar, Daniel Weyant, Raja Reddy, John Bellettini, and Nandkishor K. Nere are employees of AbbVie. The design, study conduct, and financial support for this research was provided by AbbVie. AbbVie participated in the interpretation of data, review, and approval of the publication.

526243 Nanofiltration: Integration in Medium Molecule Processes

Abstract:

Nanofiltration is a pressure and shear driven process for separating components of a process stream based on molecular size differences. Membrane separations using both ceramic and polymeric membranes have been around since the 1960s for industrial processes, with their use largely limited to aqueous systems. While membranes compatible with organic solvents have been used in manufacturing for many years, recently developed membrane materials allow their use to be extended to many other chemical processes and solvent systems, extending from small molecule development to peptide synthesis. Additionally, the semi-continuous nature of the nanofiltration system and ability to perform impurity removal at low temperatures makes nanofiltration well suited for incorporation into continuous processes. This talk will highlight the exploration of nanofiltration technology to help with: a) concentration, b) impurity purification, and c) solvent exchanges using different case study examples ranging from small molecules to medium size molecules (peptide) separation.

539794 Liquid Phase Oligonucleotide Synthesis with Membrane Separation for Efficient Large Scale Manufacturing

Abstract:

Liquid Phase Oligonucleotide Synthesis with Membrane Separation for Efficient Large Scale Manufacturing

Danilo Cuccato[a], Jack H. J. Cordrey[a], Piers R. J. Gaffney[a], Patrizia Marchetti[a], Jeong F. Kim[a], Daniela Negru[b], Mike S. Anson[b], Andrew G. Livingston[a].

[a]Department of Chemical Engineering, Imperial College, London, SW7 2AZ, UK

[b]GSK, Medicines Research Centre, Gunnels Wood Road, Stevenage, Herts, SG1 2NY, UK

Abstract

Oligonucleotides (oligos, 18 to 25-mers) are difficult to prepare at high purity using solid phase oligo synthesis (SPOS) on scales greater than 1 kg, and extensive chromatography of the crude full-length oligo is required to reach acceptable purity. We have developed a liquid phase oligo synthesis (LPOS) strategy that allows in-line monitoring of coupling efficiency throughout the synthesis, so that any incomplete reactions so detected can be driven to completion. After each chain extension the growing oligo is separated from reaction debris using a two-stage nanofiltration apparatus. To enhance the separation, three oligos are linked to a central hub in a three arm oligo-star (**3*n***). The purified oligo-star is then concentrated, again using molecular sieving, ready for the next chain extension. This process enables LPOS in standard chemical process plant, and for building block usage to be minimised. A new class of membranes was developed that efficiently retains oligo-stars but permeates large reactant debris (MW *ca.* 600 Da) in strong organic solvents over more than 19 synthesis cycles. The synthesis cycle was repeated up to a 20-mer (**4**) using a 5'-methoxyisopropyl protecting group (**2**, ***P*** = Mip). Current work, now using commercial phosphoramidites (**2**, ***P*** = Dmtr), seeks to reach greater than 90% crude purity. This technology is being commercialised by EXACTMER, a new start-up from Imperial College London.

522346 Mammalian Cell Culture Broth Clarification and Affinity Capture of Secreted Antibodies in a Novel Single Integrated Process Using a Compact Settler

Abstract:

Enhanced sedimentation on inclined surfaces has been exploited successfully to separate the larger live and productive mammalian cells (>12 microns in diameter) from the smaller (<8 microns diameter) dead cells and cell debris. The selective recycle of live and productive Chinese hamster ovary (CHO) cells to the continuously fed or perfused bioreactor results in a high cell concentration/density inside the bioreactor, while the dead cells and cell debris are continuously removed from the top of the "inclined settler". This selective cell retention device has been scaled up initially in a multiplate rectilinear design and used by several biomanufacturing companies, such as Bayer, Biomarin, Eli Lilly, Roche, etc. This original lamellar design requires a large footprint for the large production scale (>3000L) perfusion bioreactor and is quite inefficiently scaled up.

Sudhin Biopharma Company has now scaled up the inclined surfaces in a more compact cylindrical and conical design, which houses over 6 x more inclined surfaces on a given footprint, compared to the original lamellar inclined settler design. With its more efficiently scalable patent-pending design, this compact settler has been initially demonstrated to settle live yeast Pichia pastoris cells (3 - 5 microns in size) from the fermentation broth and return them to the continuous bioreactor, achieving very high cell densities (O.D. >800) over two months long culture duration. In mammalian cell cultures applications, this compact settler again demonstrates selective retention of live and productive CHO cells while the dead cells and cell debris are continuously removed along with the secreted proteins in the harvest stream.

A novel application for the compact settler is the clarification of mammalian cell culture broth at the end of a fed-batch bioreactor culture, which is more commonly used in the biomanufacturing industry. This device is currently being evaluated at a collaborating large biopharmaceutical manufacturing company to potentially replace the primary clarification device of a continuous centrifuge and reduce the requirement of depth filtration membrane used as a secondary clarification device. Current results from our own cell culture broth clarification experiments carried out in a single use disposable plastic settler will be presented in this talk.

Another novel extension of the compact settler's usage is to contact the mammalian cell culture broth containing the secreted antibody product directly with protein A coated affinity capture beads inside the settler, without any need for prior clarification or filtration of the cell culture broth. After sufficient contacting time for antibody molecules to be captured by the protein A beads suspended inside the settler, all the cells, cell debris, and unbound host cell proteins can be washed out of the settler by increasing the top effluent rate from the settler. After adequate washing, bound antibodies can be released by the elution buffer and the pure antibodies can be readily harvested in the top effluent stream. Subsequently regeneration of protein A beads are carried out inside the settler by pumping the equilibration buffer into the settler. Latest experimental results from the sequential batch operations will be presented.

It is relatively simple to convert the above sequence of batch affinity capture, wash, elution and regeneration steps inside a single compact settler into a continuous flow process of protein A beads undergoing capture, wash, elution and regeneration in a series of settlers, with the beads flowing from one settler to the next in a truly continuous cross-flow contacting pattern with different buffers. This continuous affinity capture on protein A beads suspended inside a series of compact settlers will be a more efficient usage of the more expensive protein A beads than the current packed protein A bead affinity column chromatography, commonly used in the industry today. The relative advantages and disadvantages of the novel integrated process of affinity capture of antibodies inside compact settlers will be discussed.

534991 Examples of Model Informed Drug Development in Drug Design and Delivery

Abstract:

Modeling and simulation in the pharmaceutical industry is getting more diverse and influencing more phases of drug development. In this talk examples of the impact of modeling in various phases of drug development will be shown. A metric for understanding the properties of a target, a monoclonal antibody and the disease which conveys important insights for dose and regimen will be presented. This has been very useful in explaining to research scientists the interplay between target and drug and the effect of rug concentration on target suppression.

Another important part of drug development is targeting a specific organ. Finite element based modeling of organ function and drug delivery have helped understand and influence development of drugs targeted to specific organs. Examples of inhaled drug development and drug delivery for spinal injury will be shown.

538412 Drug Loading into and Drug Release from pH- and Temperature-Responsive Hydrogels

Abstract:

Hydrogels that undergo deformation upon appropriate changes in pH or temperature have considerable promise as drug delivery vehicles. Uptake of drug macromolecules by swelling and non-swelling hydrogels and release of drug macromolecules from these into a target fluid are investigated here. A mathematical model for hydrogel-solution composite, a composite of a distributed parameter system (cylindrical, slab or spherical hydrogel) and a lumped parameter system (surrounding solution), is developed. The polymer network displacement in a swelling/deswelling hydrogel is described by a stress diffusion coupling model. The analytical solution for network displacement is used to predict solvent intake by swelling gels, solvent efflux from deswelling gels, and changes in pressure, porosity and effective drug diffusivity resulting from network displacement. These in turn influence drug uptake during and after hydrogel swelling and drug release from hydrogel during and after deswelling. Numerical results illustrate benefits of hydrogel swelling for drug loading and merits of different modes of drug release. Drug uptake and drug release by temperature-responsive hydrogels are compared to those by hydrogels not subject to deformation.

527521 Modelling Simultaneous Saccharification and Fermentation of Natural Polymers: Population Balance Interlinked with Cybernetic Modelling

Abstract:

The generation of important and useful products (e.g. ethanol, lactic acid etc.) through microbial fermentation often involves the breakdown of complex polymeric feedstock such as starch and cellulose through enzymatic scissions followed by subsequent metabolic conversion. The interplay between the kinetics of enzymatic depolymerization and the response of the microbes towards changes in the abiotic phase is critical for the adequate description of such a complex process. In this work, two unrelated frameworks, i.e. the Population Balance Modelling (PBM) and the Cybernetic Modelling (CM) were interlinked to model such a system. Specifically, the PBM technique was used to describe the enzymatic depolymerization whereas the CM framework was used to model the microbial response toward complex environmental changes. The interlinked PBM and CM framework was implemented on two case studies involving the Simultaneous Saccharification and Fermentation (SSF) of starch by two recombinant yeast strains capable of excreting glucoamylase alone or together with Î±-amylase. The simulation results revealed that the proposed framework captured features not attainable by existing approaches. Examples of such include the ability of the model to indicate (in case study one) that an appropriate amount of glucose mixed with starch as initial substrates yielded an optimum productivity of ethanol. Not only that, the model showed (in case study two) that SSF is indifferent to the type of starch when both enzymes are present as opposed to when only glucoamylase is present. Thus, the effect of various enzymatic actions on the temporal evolution of the polymer distribution and how the microbes respond to the initial molecular distribution of the polymers can be studied. Such a framework also enables a more molecular and fundamental study of a complex SSF system, a feat which heretofore was unattainable by existing SSF models.

531047 Transport of Paramagnetic Nanoparticles in Magnetic Field for Drug Delivery

Abstract:

Paramagnetic particles can be used as carriers to effectively transport drugs through cerebrospinal fluid and deliver to the targets using a magnetic field. Since the *invivo* experiments are expensive and difficult to conduct for such systems, a mechanistic model that can predict the dynamics and distribution of paramagnetic particles is necessary for the optimal design of drug delivery systems. The existing models are applicable to dilute solutions of paramagnetic particles and cannot be extended to concentrated solution, which is generally created near the magnetic targets. In this talk, I will present a framework to model magnetophoretic diffusion and convection of paramagnetic particles in a concentrated solution. An analytical solution to the magnetic field-driven transport of particles in 1-D will be discussed with application to drug delivery through cerebrospinal fluid.

532510 Making Tangible Impact through Mathematical Modeling in Bio-Pharmaceutical Industry

Abstract:

Chemical engineers play a vital role in the development of various bio-pharmaceutical processes that achieve consistent product quality, predictable scale-up, and efficient operation. Achieving these goals require in-depth understanding of both the scientific and engineering aspects of the processes providing ample opportunities for chemical engineers to add value. Due to constrained timelines and intrinsic cost considerations around the drug substance and drug product development of novel, â€œfirst in the worldâ€ molecules, any attempts to reduce expensive and time consuming experimentation is highly desirable. To this end mathematical modeling can play an important role in conjunction with the chemical engineering principles and conventional wisdom.

The presentation will introduce some of the examples of tangible impact made through pragmatic modeling approaches employing computational fluid dynamics (CFD), discrete element method (DEM), population balance modeling (PBM) and applied mathematics. The cases include some of the key process equipment and unit operations that are employed as workhorses in the bio-pharmaceutical industry, namely crystallizer, agitated filter dryer; bioreactor and lyophilizer.

Nandkishor K. Nere, Kushal Sinha and Natarajan Ramasubramanyan are employees of AbbVie. The design, study conduct, and financial support for this research was provided by AbbVie. AbbVie participated in the interpretation of data, review, and approval of the publication.

534116 Photochemical Models for Drug Substance Stability

Abstract:

The mathematical models generally utilized in process design of photochemical reactions can be readily implemented in the development of controls strategies for drug substances prone to photodegradation. We present here the development of two photochemical models used to quantify the risk to photodegradation of solid and liquid drug substance. A key feature of the model development is streamlined estimation of the relationship of the absorption spectra and the quantum yield to the wavelength of the incident light. The predictive power of the model enabled the design of a control strategy to limit the formation of the photodegradant. A methodology to minimize the experimental burden in the development of a mathematical model is proposed. Also presented are insights obtained from quantum mechanical modeling of the electronic transitions associated with the UV absorption spectra.

530232 A Study of Photon Transport in Gas-Liquid Flow: Scalability of Photooxidations from the Micro- to the Milli-Scale

Abstract:

The use of light energy constitutes a green activation mode of organic molecules. A photoreaction of increasing interest is the photosensitized addition of singlet oxygen, as it is an atom-economic method of functionalization and is applied in the synthesis of commercial chemical products as fragrances and pharmaceuticals [1]. Many singlet oxygen oxidations were applied to organic substrates in flow photo microreactors due to their small penetration depth and the presence of segmented flow which leads to enhanced interfacial mass transfer [2,3]. However, as microreactors lack the throughput required by industry, scale-up could be realized using milli-scale reactors which combine the advantages of microreactors with the throughput of conventional batch reactor systems [4]. However, scale-up is usually preceded by an extensive parameter study (e.g. variation of the light source intensity, concentration of the sensitizer, transport gas fraction) realized for each reaction system on the small and large scale [5,6]. If the light absorption and scattering phenomena in the employed gas-liquid photoreactors are understood at a fundamental level, they can be predicted and the scale-up procedure becomes faster and cost efficient.

This study aims to investigate the photon transport in gas-liquid flows on the micro- and milli-scale. The microreactor consists of a glass plate with a serpentine channel characterized by a volume of 0.6 mL and a diameter of 1 mm. The milli-scale reactor is a Corning® Advanced-Flow™ G1 Photo Reactor which is composed of heart-shaped elements and is characterized by a volume of 8 mL and a channel height of 1.1 mm. Both reactors are irradiated by green Light-Emitting Diodes (LEDs). Firstly, the photon flux and the optical pathlength are experimentally determined by visible-light actinometry following a methodology previously reported [7]. The actinometric measurements are performed in single liquid phase and in gas-liquid two-phase flows. Nitrogen is used as inert gas with the volumetric gas transport fractions comprised between 0.2 and 0.5. An example of the different gas-liquid flow patterns obtained in the microreactor and G1 Photo Reactor is shown in Figure 1. Secondly, a modelling tool based on the ray tracing technique implemented in Matlab is used to support the experimental quantification of the optical pathlength in single and two-phase flow.

As application, the gas-liquid photoxidation of 2-methoxyfuran in the presence of molecular oxygen and Rose Bengal in ethanol is scaled-up from the microreactor to G1 Photo Reactor. The optimal concentration of the photosensitizer on micro- and milli-scale is correlated to the optical pathlength determined by actinometry. It is found that the photoreaction is not limited by the gas-liquid interfacial mass transfer on both micro- and milli-scale. Therefore, the maximum conversion of 2-methoxyfuran occurs for a gas fraction where the liquid receives the highest photon flux.

This study will contribute to the understanding of the light absorption and scattering phenomena in different gas-liquid flow patterns at the micro- and milli-scale. This knowledge is crucial for making the scale-up process more efficient, but also for designing improved multiphase flow photochemical systems.

Figure 1. Example of gas-liquid flow patterns in the studied microreactor (left) and Corning® Advanced-Flow™ G1 Photo Reactor (right).

References

1. J. P. Knowles, L. D. Elliott, and K. I. Booker-Milburn, Flow photochemistry: Old light through new windows. Beilstein Journal of Organic Chemistry, 2012. 8: p. 2025-2052.

2. D. Cambié, C. Bottecchia, N. J. W. Straathof, V. Hessel and T. Noël, Applications of continuous-flow photochemistry in organic synthesis, material science, and water treatment. Chemical Reviews, 2016. 116: p.10276–10341.

3. A. Gavriilidis, A. Constantinou, K. Hellgardt, K. K. (Mimi) Hii, G. J. Hutchings, G. L. Brett, S. Kuhn, and S. P. Marsden, Aerobic oxidations in flow: opportunities for the fine chemicals and pharmaceuticals industries, Reaction Chemistry & Engineering, 2016. 1: p. 595-612.

4. A.Woitalka, S.Kuhn and K.F.Jensen, Scalability of mass transfer in liquid–liquid flow, Chemical Engineering Science, 2014. 116: p. 1–8.

5. C. Mendoza, N. Emmanuel, C. A. Páez, L. Dreesen, J.-C. M. Monbaliu and B. Heinrichs, Transitioning from conventional batch to microfluidic processes for the efficient singlet oxygen photooxygenation of methionine, Journal of Photochemistry and Photobiology A: Chemistry, 2018. 356: p. 193-200.

6. N. Emmanuel, C. Mendoza, M. Winter, C. R. Horn, A. Vizza, L. Dreesen, B. Heinrichs, and J.-C. M. Monbaliu, Scalable photocatalytic oxidation of methionine under continuous-flow conditions, Organic Process Research and Development, 2017. 21: p. 1435–1438.

7. A. Roibu, S. Fransen, M. E. Leblebici, G. Meir, T. Van Gerven and S. Kuhn, An accessible visible-light actinometer for the determination of photon flux and optical pathlength in flow photo microreactors, Scientific Reports, 2018. 8: p. 1-10.

528094 Photoredox Iridium-Nickel Dual Catalyzed Decarboxylative Arylation Cross-Coupling: From Batch to Continuous Flow Via Self-Optimizing Segmented Flow Reactor

Abstract:

Photoredox decarboxylative cross-coupling via iridium–nickel dual catalysis has emerged as a valuable method for C(sp^2)–C(sp^3) bond formation. Herein we describe the application of a segmented flow ("microslug") reactor equipped with a newly designed photochemistry module for material-efficient reaction screening and optimization. Through the deployment of a self-optimizing algorithm, optimal flow conditions for the model reaction were rapidly developed, simultaneously accounting for the effects of continuous variables (temperature and time) and discrete variables (base and catalyst). Temperature was found to be a critical parameter with regard to reaction rates and hence productivity in subsequent scale-up in flow. The optimized conditions identified at microscale were found to directly transfer to a Vapourtec UV-150 continuous flow photoreactor, enabling predictable scale-up operation at a scale of hundreds of milligrams per hour. This optimization approach was then expanded to other halide coupling partners that were low-yielding in batch reactions, highlighting the practical application of this optimization platform in the development of conditions for photochemical synthesis in continuous flow.

533728 Studies Toward Scalable Photochemical Reactions in Flow

Abstract:

Studies Toward Scalable Photochemical Reactions in Flow

Emily B. Corcoran, François Lévesque, Jonathan P. McMullen, and John R. Naber

Process Research and Development, Merck & Co., Inc., Rahway, New Jersey 07065, USA

In recent years, there has been a significant growth in the reports of reactions promoted by visible and UV light, often detailing highly selective transformations not feasible through non-photochemical, manifolds. Due to the unique advantages of these reactions, we foresee an increasing use of light-promoted transformations in the synthesis and production of API. The implementation of bimolecular, non-chain photochemical reactions on large scale remains a key challenge. The implementation of these transformations on large scale is most amenable to flow, due to most photochemical reactions requiring a high surface area to volume ratio for good efficiency. This presentation will highlight key parameters that need to be taken into account for an effective transition from lab scale to preparative scale, namely reactor geometry, selection of irradiation source, pathlength, and residence time. As part of the discussion, ongoing efforts by the flow chemistry team at Merck & Co., Inc. toward developing scalable photochemical reaction in flow will be presented in the form of several short case studies about our experiences with scaling photocycloadditions and multi-catalytic photochemical transformations.

529446 Improving Reactor Design for Scaling-up Photoredox Reactions in Flow

Abstract:

Improving Reactor Design for Scaling-Up Photoredox Reactions in Flow

Eric G. Moschetta,* Kaid C. Harper*, Shailendra V. Bordawekar, and Steven J. Wittenberger

AbbVie, Inc. Process R&D, 1401 Sheridan Road North Chicago, IL 60064

*co-corresponding authors: eric.moschetta@abbvie.com, kaid.harper@abbvie.com

Photoredox catalysis has emerged as an innovative methodology for constructing organic molecules in selective fashion. These one-electron chemistries rely on special transition metal and organic photoredox catalysts that absorb photons from sources of visible light to promote the catalysts into electronically excited states. Once energized, the catalysts undergo single electron transfer with organic substrates to afford products. These reactions are attractive from an industrial viewpoint because they operate at mild conditions, use green energy sources in the form of inexpensive visible light, and offer exceptional selectivity and unique reactivity in chemical transformations. Such transformations are especially appealing in the pharmaceutical industry. In spite of these advances in synthetic methodology, reactor technology does not yet exist to execute photoredox reactions on a commercial scale. One promising solution is to implement flow chemistry in photoredox processes. Flow reactors offer the benefits of increased throughput, improved heat and mass transfer properties, and potentially smaller reactor volumes. For photoredox applications, flow reactors also offer improved photon fluxes throughout the volume of the reactor, especially when compared to batch reactors. The major challenge in developing scalable flow reactor technology for photoredox chemistry in the pharmaceutical industry is understanding the fundamental kinetics of these reactions and the other necessary scale-up parameters for rational design of flow reactors to achieve production on the order of kilograms per day.

We present an example of a successful kilogram-per-day scale reaction in a flow reactor for a dual metallaphotoredox Câ€"N coupling using an iridium photoredox catalyst and a nickel cross-coupling catalyst. We show how we collected kinetic data in batch and used simple guidelines for transitioning from the batch reactor to a flow reactor. These guidelines should apply to a broad range of photoredox reactions as well as photoreactors. We also address how the intensity of the light source apparently affects photoredox reactions in different fashions and how that impacts the design of the flow reactor. Finally, with photoreactor-specific scale-up principles established, we reintroduce more conventional principles of chemical reaction engineering to address how flow reactor design may be approached at commercial scales to maximize throughput.

The design, study conduct, and financial support for this research was provided by AbbVie. AbbVie participated in the interpretation of data, writing, reviewing, and approving the publication. All authors are AbbVie employees.

531790 Photon Mediated Coupling Reactions Using Copper Nanocatalysts - Molecular Mechanisms on Homo Coupling and Hetero Coupling Activity

Abstract:

Metal catalyzed coupling reactions are widely used reactions in the pharmaceutical industry. These reactions are traditionally executed using expensive, rare-earth palladium (Pd) based homogeneous catalysts. Copper nanocatalyst catalyzed cross-coupling reactions will have several advantages as they can be used as a platform to develop energy efficient and environmentally friendly flow processes. Recently, promising results have been reported in the literature on the use of metal nanocatalysts for coupling reactions. However, identifying the catalytic pathway (i.e., heterogeneous versus homogeneous pathway) remains a herculean task.

In this contribution copper nanocatalysts are used to perform cross-coupling and homo-coupling reactions. We demonstrate whether the catalyst utilizes the homogenous or heterogeneous pathway under different conditions. Molecular reaction mechanisms are proposed for both hetero and homo coupling activity and molecular structures for the intermediates are proposed. We will show how the reaction conditions determine whether the reaction is catalyzed by leached homogeneous complex or heterogeneous surface. Furthermore, by using a tunable light source with copper catalysts, we show an enhanced rate of reaction compared to the conventional thermally driven catalysis. We will discuss different possible mechanisms responsible for the photocatalytic activity.

542620 The Impact of Shale Gas and Oil on the Chemical Industry

Abstract:

Lower alkanes and aromatics are very useful as both fuels and chemical raw materials. Occasionally supply disruptions constrained availability but also encouraged investigation of alternative feedstocks and chemistries. Recent new technologies for the production of unconventional gas and oil sources from tight shale formations significantly increased availability of both methane and condensate, radically altered process economics, resulted in the construction of much new intermediates capacity, and multiplied the ways traditional and alternative chemistries might be exploited. This presentation will outline historically how natural gas components became useful as chemical feedstocks, how government regulations occasionally led to perturbations in availability and price, how supply crises led to the development of alternative chemistries, catalysts, and ultimately the exploitation of shale resources. More specifically, will be discussed how excess coproducts, especially condensate, from increasing shale gas and oil production for energy are having significant unexpected impacts on the chemical process industries and opportunities for new process chemistry research.

541186 Sustainable Energy and Chemicals: Past, Present, and Future

Abstract:

The last 40 years has seen the U.S. energy situation vacillate from gasoline rationing and concerns over Peak Oil, to projected energy and feedstock prices that were noncompetitive in global markets for chemicals production, to the economic revolution of frack gas and light tight oil. The importance of low cost energy and feedstocks in driving economic opportunity will be discussed relative to this history, along with the importance of sustainability in decision making to gain stakeholder buy-in and acceptance. Strategies for managing career paths and competencies despite changing targets and high uncertainty will be addressed.

542618 Disruptions: What the Future May Hold

Abstract:

Disruptive forces are all around us. They change the way we communicate and work. They change the way we develop goods and services. They also change the types of products that are manufactured. Examples such as the way in which we produce and consume energy or the feedstocks we use to produce product have seen many changes over the course of the last few decades, and these disruptions are expected to continue. The chemical engineering discipline and the expectations for a chemical engineerâ€™s skills has also changed and adapted as a result of these disruptive forces. A discussion of the broad themes associated with these forces and the implications that they may have to chemical engineers in the future will be discussed.

542621 Geopolitical Factors Influencing the Evolution of the Chemical Industry

Abstract:

The basic chemicals and plastics industry is built on a simple principle: find a cheap feedstock and convert it into something more valuable. However, the right feedstock depends on place and time. Certain policies and actions of various governments have influenced the evolution and profitability of the chemical industry during the past 50 years, sometimes in surprising ways. For example, high oil prices and US government funding encouraged the development and commercialization of coal gasification technology during the late 1970s and early 1980s. This enabled China to develop a competitive basic chemical industry using coal as a feedstock during the last 25 years. Similarly, an abrupt increase in oil prices during the first decade of the 21st century encouraged the present shale gas revolution, causing a major shift in the economy of producing basic chemicals and polymers in the US. If history is our guide, then what governments choose to fund and regulate may be as important as market forces in determining the shape of our industry and profession in the future.

542622 Agility & Resilience: How to Maintain Career Competitiveness in the Changing Chemical Industry

Abstract:

The world around is changing. The shift to protectionism, global instability and price volatility of raw materials and feedstocks has the potential to wreak havoc on the chemical industry. With the revolution of shale gas in North America and the sudden drop in feedstock costs, the chemical sector in the US has reasserted its historic dominance. Learn how the role of a chemical engineer can contribute to industry growth and resilience, while adapting to allow for personal agility throughout the career life cycle.

535092 Leveraging Deep Learning for Pharmaceutical Discovery Lead Profiling

Abstract:

As part of small molecule drug discovery, potential lead compounds are studied to ensure that they have a low risk of adverse behavior. Cytochrome P450s (CYP) are a family of enzymes that are an important drug metabolism pathway; drug candidates that inhibit the activity of these enzymes can lead to dangerous drug-drug interactions. Once synthesized, candidate compounds are submitted to high throughput *in vitro* assays to screen for CYP inhibition. By prioritizing more promising options and measuring the effect of changes to the core structure, the ability to virtually screen compounds for CYP inhibition would save teams time and effort during drug design.

This work describes the development of a model that can predict CYP inhibition from chemical structure using in-house data from BMS. Such quantitative structure activity relationship (QSAR) models have been successfully developed from publicly available data sets, however, the limited amount of public CYP inhibition data hinders their practical utility. As part of an innovative multidisciplinary team at BMS, analytics experts from different functional areas contributed to exploring and analyzing the BMS CYP assay database. Using this much larger set of private data (>1M entries) combined with recent advancements in machine learning algorithms, a model suitable for virtual screening was developed. Of the many modeling approaches that were explored, >85% classification accuracy was achieved using a deep learning approach. Important caveats for handling real-world data as well as a comparison of different modeling methodologies will be presented.

528297 Learning to Design and Validate Small-Molecule Synthetic Routes from Historical Reaction Data

Abstract:

Advances in laboratory automation promise to decrease the manual effort of synthesis, but determining *how* to synthesize a compound currently requires time and effort investment from expert chemists. To achieve full autonomous chemical synthesis, one must have robust synthesis planning software that can propose fully-specified synthetic routes to target molecules.

In this talk, we will describe our recent efforts to develop such software. The overarching theme of our work is how to most effectively leverage historical reaction data to inform decision-making in small molecule pathway design.

The overall synthesis planning workflow contains a number of interconnected modules. We focus on two critical aspects of computer-aided synthesis planning and how machine learning and other data-driven techniques have enabled new approaches to both challenges. First, we discuss the problem of retrosynthetic planning (i.e., identification of suitable starting materials) and how the recursive expansion and search strategy are both conducive to machine learning approaches. Second, we discuss the challenge of *in silico* reaction validation, which can be addressed by solving the inverse problem of forward reaction prediction. We summarize neural network-based approaches we have taken to develop models that can anticipate the products of a chemical reaction after being trained on previously published reactions. Here we use the model for reaction validation, but its utility extends to prediction of side products and impurities. Finally, we describe how these techniques for retrosynthesis and forward prediction are integrated into an overall workflow that, for a given molecular target, predicts a rank ordered list of reaction paths that connect the target to purchasable starting materials via a series of plausible reaction steps. The integrated program offers additional features for excluding specific reactions or chemicals, e.g., for IP or toxicity concerns.

530970 Using Machine Learning to Recommend Reaction Conditions and Quantifying Similarity of Catalysts, Solvents, and Reagents

Abstract:

Using Machine Learning to Recommend Reaction Conditions and Quantifying Similarity of Catalysts, Solvents, and Reagents

Hanyu Gao, Thomas J. Struble, Connor W. Coley, William H. Green, Klavs F. Jensen

One essential element for synthetic planning is the reaction conditions. Properly designed reaction conditions can promote desired reactivity, and sometimes small changes in reaction context can lead to drastically different reaction outcomes. While other aspects of computer-aided synthetic planning has seen rapid development, including retrosynthesis[1] and evaluation of reaction outcomes, [2,3] the suggestion of reaction conditions is primarily considered a human task and relies heavily on chemists' knowledge and experience, and computer aided condition recommendation remains a challenging and under-explored problem. [4–6]

In this work we develop a neural-network based model to predict suitable reaction conditions for organic transformations. The model is trained on roughly 10 million examples from the Reaxys[7] database in order to predict the exact chemicals used as catalysts, solvents, reagents, and an appropriate temperature for the reaction. Prediction results are evaluated both quantitatively, using a variety of accuracy metrics, and qualitatively, using multiple sets of representative examples. The model is able to predict a closely similar context combination, based on strictly defined criteria, to recorded catalyst, solvent and reagent in the top-ten predictions with accuracy of 67.4%, with accuracies for individual elements reaching 80%~90%. Empirical evaluation shows that many of the failed predictions are potential alternatives to the true conditions, or due to data quality issues. Temperature is predicted to be within 20K deviation from the recorded temperature in 60%~70% of the test cases, with dependence on the correctness of chemical context prediction.

Besides making predictions, the model can also generate information that can be used to quantify the similarity of catalysts, solvents and reagents. We discovered that the weight matrix in the last layer for predicting catalysts/solvents/reagents can be effectively used to calculate a score that represents the functional similarity between catalysts/solvents/reagents. The identified near neighbors of common catalysts, solvents and reagents agree well with established knowledge. This work demonstrates the possibility of inferring functional similarity of chemicals purely from reaction data, without relying on property calculations.

The reaction condition suggestions and similarity information provided by this tool can be used to aid experimental design, improve the accuracy of in silico evaluation of reactivity, and pathway-level assessment and improvement of chemical synthesis processes.

Reference

(1) Segler, M. H. S.; Preuss, M.; Waller, M. P. Planning Chemical Syntheses with Deep Neural Networks and Symbolic AI. *Nature* **2018**, *555* (7698), 604.

(2) Jin, W.; Coley, C.; Barzilay, R.; Jaakkola, T. Predicting Organic Reaction Outcomes with Weisfeiler-Lehman Network. In *Advances in Neural Information Processing Systems*; 2017; pp 2604–2613.

(3) Coley, C. W.; Barzilay, R.; Jaakkola, T. S.; Green, W. H.; Jensen, K. F. Prediction of Organic Reaction Outcomes Using Machine Learning. *ACS Cent. Sci.* **2017**, *3* (5), 434–443.

(4) Marcou, G.; Aires de Sousa, J.; Latino, D. A. R. S.; de Luca, A.; Horvath, D.; Rietsch, V.; Varnek, A. Expert System for Predicting Reaction Conditions: The Michael Reaction Case. *J. Chem. Inf. Model.* **2015**, *55* (2), 239–250.

(5) Segler, M. H. S.; Waller, M. P. Modelling Chemical Reasoning to Predict and Invent Reactions. *Chem. Eur. J.* **2017**, *23* (25), 6118–6128.

(6) Lin, A. I.; Madzhidov, T. I.; Klimchuk, O.; Nugmanov, R. I.; Antipin, I. S.; Varnek, A. Automatized Assessment of Protective Group Reactivity: A Step toward Big Reaction Data Analysis. *J. Chem. Inf. Model.* **2016**, *56* (11), 2140–2148.

(7) Reaxys https://new.reaxys.com/.

537077 Applying Data Science Techniques to Solubility Data for Synthetic Compounds: An Expedited End-to-End Workflow from Data Collection to Crystallization Process Design

Abstract:

Crystallization is an important process often employed several times within synthetic routes for drug substances. The development of robust crystallization processes can be separated into two tasks: (1) the selection of a solvent-system, and (2) the design of a process within that solvent system that affords materials with desired chemical/physical purity and meets yield requirements. An end-to-end workflow relying on data science techniques was developed for capturing, visualizing, and interpreting solubility data to facilitate the consistent and rapid execution of these tasks.

This workflow starts with the collection of solubility data using standardized equipment sets and approaches into templated tables. These tables contextualize the solubility data by joining each measured value (concentration, temperature, composition, etc.) with relevant meta-data (solute purity, x-ray diffraction results, equipment, date, etc.). The contextualized solubility data is ingested within a database – providing a single source for all solubility data. A templated visualization then consumes the data from this source. Further it can be filtered within the visualization as necessary (e.g., limited to a specific solute and limited to only data collected for a specific lot of solute).

To facilitate solvent selection as the first task in crystallization process development, this visualization automatically applies a decision tree to collected solubility data to classify solvents with regards to crystallization as solvents that are likely: "good for a thermal process", "solvent within antisolvent driven crystallization", or "good antisolvents". Based on this classification the scientist working with the system can either begin the task of process design or apply a predictive solubility model that has been integrated within the visualization to determine other solvent systems that may be worth investigating. The application of this model uses simple R scripts with open source libraries (e.g., non-linear optimization packages) doing the "heavy-lifting".

Once a solvent system has been selected, the task of process design begins using an automated script to fit and select the best-model for solubility data across ranges of temperature/antisolvent ratios within that system. The contextualization of the solubility data allows for visual identification and rapid exclusion of outliers within the model fitting step. Once a solubility model has been selected and fit for a given system – a constrained optimization algorithm is applied to determine the process that affords the highest yield given user supplied constraints (which are modified to ensure the process meets desired chemical/physical purity needs). This initial crystallization process conditions are then attempted and refined as necessary.

This end-to-end workflow has resulted in significant time savings, and allowed for the setting of consistent expectations for initial designs across projects. Further, it is an early demonstration of the integration of modelling and data science techniques within process development.

534161 Optimizing Drying Profile of Polymeric Drug Products Using Machine Learning and First Principle Modeling

Abstract:

Drying is one of the critical steps for manufacturing of polymeric drug products. The key quality attributes for the dried product can be the residual solvent content, and the flowability. The flowability of the powders can be dependent on the glass transition temperature which in turn depends on the solvent content. Hence it requires special precautions to keep the temperature of the drug product during drying below a critical transition temperature (Tg). This work shows a case study where such type of issues can be tackled by employing a hybrid model combining machine learning and first principles to optimize the drying profile in minimal number of experimental runs.

Typically the drug product is dried in an agitated Nutsche filter type of a dryer. This drying requires special care to ensure that at all points of drying; the temperature of the product is below the critical value (Tg). While a simple heat transfer model (like in Figure 1, section A) can be used to link the jacket temperature to the product temperature, the presence of solvent and its evaporation needs additional considerations as detailed below. In the model building step, Model 1 was a Support Vector Machine (SVM) based model to predict the Tg for different residual solvent levels (based on experimental data). The model predicted with accuracy over 98%. Model 2 captures the mass transport of the solvent from the drug product, accounting for the temperature dependent mass transport of the solvent.

Here K_i and C_i are the diffusivity and concentration of solvent i, and T is the product temperature. is calculated as a function of temperature of each solvent in the system, based on the regression of the experimental data.

Figure 1: Modeling approach for drying

To predict the optimal jacket temperature profile for drying to meet the product requirements, the heat transfer model, Model 1 and Model 2 are solved using an algorithm like the one shown in Figure 2. For a given temperature profile, and initial solvent levels, Model 2 predicted the time evolution of the solvent concentration. At each time, Model 1 calculated the Tg. The optimal input temperature profile was chosen such that the product temperature at any time was lower than the predicted Tg at that time.

Figure 2: Modeling algorithm for predicting the optimal temperature profile

The above model building approach of combining machine learning and first principle modeling can be extended to other unit operations as well where only part of the underlying physics is understood. In combination with a mixing type of a model, this approach can be applied to scale-up drying (Figure 1). The presentation shows specific example cases of such applications.

525894 A Reaction Database for Small Molecule Pharmaceutical Processes Integrated with Process Information: Future Developments

Abstract:

To resolve process engineering issues during the development of synthetic routes for new drugs at an early stage of process development, access to chemical information is needed, which can be provided through knowledge databases, experience, literature review and/or computer-aided tools [1,2]. The available data (knowledge) is used for similarity search, reaction data analysis, synthesis route planning, drug discovery-development and prediction of physicochemical properties [3]. A computer-aided tool for reaction synthesis based on efficient retrieval of chemical information that are organized in chemical reaction databases have been developed. The reaction data in the developed databases consist of individual reactions and chemical information for the associated molecules involved in the reactions.

At the current stage, the database covers 350 chemical and biochemical reaction systems with the data-information mentioned above. Extension of the database is an-going objective and involves several tasks. One current task is to develop and use algorithms for efficient collection of reaction data integrated with process information. This task is crucial and very demanding and requires tests to verify the accuracy of the data and check for their consistency, in the same way phase equilibria data are checked through established consistency tests. Another task is to create an efficient interface for data import-export integrated with tools for synthesis-design of pharmaceutical processes.

The overall objective is for the reaction database to cover a very wide range chemical and biochemical reaction systems and integrate it with generic model-based prediction tools that combines reaction pathways synthesis algorithms, solvent selection tools [5] consistency tests for reaction kinetics data, prediction of missing thermodynamic properties, process synthesis/design algorithms [6, 7], process analysis [6] and optimization tools [8]. Such an integrated computer aided tool will enhance the amount of data-information available, thereby reducing time and cost for problem solution. Also, the tool would provide useful insights for multi-objective optimization involving decision making related to screening of different alternatives for potential active ingredients candidates at the early stage of drug/process development.

This presentation will give a description of the developed reaction database where the objective is to collect-store data for reactions involved in the synthesis of small molecules in pharmaceutical processes together with a search engine to retrieve the data when required. For example, in studies involving the synthesis-design of reaction-separation schemes and/or the role of organic solvents in reaction performance improvement. The goal is to provide a data rich environment by making useful process information available during the early stage synthesis of pharmaceutical products. The database is structured based on a specially developed ontology in terms of reaction classification (reaction types); compounds participating in the reaction; need-use of organic solvents and their function; information for single step and multistep reactions; target products; reaction conditions and reaction kinetic data. Information related to reactor scale-up together with information on downstream separation and other relevant information for each reaction are also available in the database, including the references for the collected data. Additionally, the information available in the database allows the calculation of well-known "green" sustainability metrics [4]. The application of the database will be illustrated through the synthesis of ibuprofen, for which data on different reaction pathways are available whose performances are compared using "green" sustainability metrics.

[1] Caron, S.; Thomson, N.M. Pharmaceutical process chemistry: Evolution of a contemporary data-rich laboratory environment. J. Org. Chem. 2015, 80, 2943–2958.

[2] Gasteiger, J. Chemoinformatics: Achievements and challenges, a personal view. Molecules 2016, 21, 151.

[3] Warr, W.A. A short review of chemical reaction database systems, computer-aided synthesis design, reaction prediction and synthetic feasibility. Mol. Inform. 2014, 33, 469–476.

[4] Papadakis, E.; Anantpinijwatna, A.; Woodley, M.J.; Gani R. A reaction database for small molecule pharmaceutical processes integrated with process information. Processes. 2017, 5(4), 58.

[5] Gani, R.; Jiménez-González, C.; Ten Kate, A.; Jones, M.; Powell, L.; Atherton, J.H.; Cordiner, J.L.A modern approach to solvent selection product formulations. Chem. Eng. 2006, 113, 30-43

[6] Tula, A.K.; Eden, M.R.; Gani, R. Process synthesis, design and analysis using a process-group contribution method. Comput. Chem. Eng. 2015, 81, 245-259

[7] Emenike, V. N.; Schenkendorf, R.; Krewer, U. A systematic reactor design approach for the synthesis of active pharmaceutical ingredients. Eur. J. Pharm. Biopharm. 2018, 126, 75-88.

[8] Diab, S.A.; Jolliffe, H.G.; Gerogiorgis, D. I. Plantwide technoeconomic analysis and separation solvent selection for continuous pharmaceutical manufacturing: Ibuprofen, artemisinin, and diphenhydramine. Comput. Chem. Eng. 2018, 41, 85-120

536023 Single-Molecule Characterization of Cazyme Protein Modules Adsorption to Multivalent Glucan Polymers like Cellulose

Abstract:

Protein adsorption to multivalent carbohydrate (or glycan) derived ligands plays a critical, but poorly understood, role in both fundamental cellular (e.g., glycocalyx and viral protein binding) and industrial processes (e.g., cellulosic sugar-based biofuels production). However, we currently have a poor understanding of the molecular mechanisms driving protein adsorption to glucan polymers like cellulose. We also lack suitable experimental tools and predictive adsorption models that can suitably explain the heterogeneous and varying affinity multivalent protein-carbohydrate binding interactions. Here, we study the adsorption-desorption of a family 1 carbohydrate-binding module (CBM), and its conserved aromatic amino acid mutants, to model glucan polymers like cellulose using a combination of bulk ensemble and single-molecule protein-ligand binding methods. Our single-molecule force spectroscopy method allows us to directly probe the structure-function relationships driving CBM adsorption to glucan polymers. Analysis of the CBM-cellulose unbinding rupture forces and total bond lifetimes dataset has provided a molecular basis for distinct binding orientations for CBMs on ordered glucan polymer surfaces like cellulose and the corresponding binding affinity for each orientation. Molecular dynamics simulations provide further insights into the possibility of multiple binding orientations for CBM on cellulose surfaces, which is an often-neglected feature of these interactions. We also report ongoing work with the aromatic amino acid mutants that are critical to stabilizing specific binding orientations and how these binding orientations could vary for various Type-A and Type-B CBMs on cellulose allomorphs. Our findings allow us to generate a comprehensive picture of non-productive binding interaction mechanisms of CBMs with glucan polymers that will ultimately lead to the development of more efficient enzymes for cellulosic biomass deconstruction into fermentable sugars.

535107 Combinatorial Experimental and Computational Approach for the Effective Entrapment of Glucose Oxidase in Hyaluronic Acid Nanogels

Abstract:

Biocatalysis offers important advantages over traditional chemical catalysts when sustainable technology pathways are being considered. The advantages of high specificity, activity, mild reaction conditions, and high product selectivity of the biocatalysts are however outflanked by the limitations that an enzymatic approach will have in commercial-scale processes, namely loss of enzyme stability over time, narrow process operating conditions, and requirement of intensive and costly separation processes from products. Much work is currently done in the field of enzyme immobilization for enzymatic activity retention and reusability, as well as for improvement of product yield, all at low implementation costs. In particular, the immobilization of enzymes onto renewable biopolymers has recently gained attention for both its sustainable approach and its abundant, inexpensive starting materials. Herein we evaluate the suitability of a naturally occurring polysaccharide and main component of the extracellular matrix as both a support as well as an active and protective encapsulator during enzyme immobilization. For this, we combined hyaluronic acid (HA) with model enzyme glucose oxidase (GOx) and evaluated both the enzymeâ€™s functionality at the interface with the polysaccharide as well as its ability to catalyze direct electron transfer controlled reactions. The reaction kinetics at the interface have been elucidated both experimentally using spectroscopic and atomic force microscopy assays and computationally using molecular dynamics simulations to offer both a viable mean for understanding transfer reactions when product formation is being considered as well as for evaluating the effectiveness of immobilization. Our combinatorial approaches provide a feasible strategy to control enzyme-based interface reactions that allow for fast product formation while preserving and extending enzyme functionality and stability. Further, our results provide a viable approach to implement enzyme-based immobilization for different applications ranging from glucose biosensors for the control of diabetes to stabilizers for food preservatives and colors.

523575 Spatiotemporal Dynamics from a Classic Enzyme Cascade with Self-Governing Substrate Competition

Abstract:

The construction of enzymatic reaction networks *in vitro* with programmable and predictable dynamics is a central goal of synthetic and systems biology. Biomacromolecules such as genes and enzymes (proteins) are the basic building blocks in the fabrication of such dynamic systems. Despite the abundance of complex dynamics realized with genetic circuits, and even though enzymes play vital roles in the regulation of metabolic pathways and signal transduction processes, the demonstration of enzyme-based networks *in vitro* has only been explored very recently.[1,2]

In this presentation we will introduce a two-enzyme reaction network consisting of a FAD-dependent oxidoreductase and a peroxidase that can generate tunable complex dynamics.[3] In a classic enzyme cascade of glucose oxidase (GOx) and horseradish peroxidase (HRP), we discovered a new reaction that GOx can take $ABTS^{+\bullet}$ as a competitive substrate with oxygen to oxidize glucose. This reaction becomes dominated when the dissolved oxygen is depleting. Based on this, we can generate multiple pulsed responses including the typical charging/discharging, rectangular and parabolic waveforms in a closed system. The waveforms can be well tuned by changing the concentrations of substrates and enzymes. We also demonstrated a safer version of the green bottle experiment with this network, in which the system toggles between colorless and green when the solution is shaken.

In an open system, this reaction network can spontaneously form and display reaction-driven Rayleigh-Bénard convective patterns. The convection is initiated by the accumulating hydrodynamic instability at the air-solution interface where the enzymatic reactions create a denser layer of product. The patterns manifest themselves in green resulting from the interfacial formation and convective flow of $ABTS^{+\bullet}$. The spatiotemporal patterns are highly dependent on the depth of the solution and will become increasingly unstable with greater depth.

In summary, we demonstrated the smallest enzymatic reaction network with tunable output waveforms and spatiotemporal pattern formation. The complex dynamic behaviors of this network are based on self-governing substrate competition, rather than activation or inhibition. In addition, the ease of direct observation of convective patterns may also encourage further research on reaction-driven Rayleigh-Bénard convection.

References:

1. Semenov, S. N. et al. Rational design of functional and tunable oscillating enzymatic networks. *Nature Chemistry* **2015**, 7, 160–165.
2. Nijemeisland, M., Abdelmohsen, L. K. E. A., Huck, W. T. S., Wilson, D. A. & van Hest, J. C. M. A compartmentalized out-of-equilibrium enzymatic reaction network for sustained autonomous movement. *ACS Central Science* **2016**, 2, 843–849.
3. Zhang, Y., Tsitkov, S. & Hess, H. Complex dynamics in a two-enzyme reaction network with substrate competition. *Nature Catalysis* **2018**, 1, 276–281.

528491 Uncovering the Quantum Mechanical Origins of Enzymatic Catalysis with Systematic QM/MM Methods and Accelerated, Large-Scale Electronic Structure

Abstract:

The large size of most enzymes necessitates hybrid quantum mechanical-molecular mechanical (QM/MM) simulation for mechanism elucidation. Numerous convergence studies of QM/MM methods have revealed over the past several years that key energetic and structural properties approach asymptotic limits with only very large (ca. 500-1000 atom) QM regions. This slow convergence has been observed to be due in part to significant charge transfer between the core active site and surrounding protein environment, which cannot be addressed by improvement of MM force fields or the embedding method employed within QM/MM. I will discuss our recent development of systematic methodologies that enable atom-economical determination of optimal QM regions (typically ca. 200-300 atoms) and provide insight into the crucial interactions in enzyme active sites captured only in large QM regions. The two approaches, based on measuring the change in chemical environment around an active site or the degree of charge transfer between core substrates and the surrounding environment, produce robust, consistent, and readily validated results on test case enzymes (e.g., catechol O-methyltransferase, cytochrome P450cam, and hen eggwhite lysozyme) that asymptotically approach results from very large QM regions. These systematically selected QM regions enable identification of charge-transfer-mediated prolonged electrostatic attraction as a previously overlooked component of enzyme rate enhancement in group transfer across a wide family of methyltransferases. I will present how pairing our efforts with graphics-processing-unit-accelerated QM/MM simulation has enabled our collection of over a nanosecond of enhanced sampling dynamics with range-separated hybrid density functional theory to reveal the magnitude of charge transfer and free energy barriers in QM/MM with QM regions up to 600 atoms in size. These simulations reveal dramatic fluctuations in the charge density mediated through hydrogen bonding interactions in the greater protein environment as a key component of enzyme action that is apparent only with large, systematically selected QM regions.

532386 Towards Scalable Production of Enantiomerically Pure Amines: Enzyme Mechanism and Kinetics

Abstract:

Towards scalable production of enantiomerically pure amines: enzyme mechanism and kinetics

Robert D. Franklin‡, John M. Robbins‡, Joshua A. Whitley‡, Andreas S. Bommarius‡#

‡School of Chemical and Biomolecular Engineering, Georgia Institute of Technology,

Parker H. Petit Institute for Bioengineering and Bioscience,

Engineered Biosystems Building, 950 Atlantic Drive N.W., GA 30322, USA

#School of Chemistry and Biochemistry, Georgia Institute of Technology,

950 Atlantic Drive, Atlanta, GA 30332-2000, USA

Amine dehydrogenases (AmDHs) catalyze the reductive amination of prochiral ketones to produce chiral amine compounds[1-3]. The reaction requires ammonia as the nitrogen source and a hydride transfer from an NADH cofactor, as seen in the reaction scheme below. Chiral amines are frequent intermediates in high value active pharmaceutical ingredients. AmDHs were developed by engineering the substrate binding pocket of amino acid dehydrogenases (AADHs) to eliminate activity towards keto acids and introduce activity toward ketones. Determining kinetics is a necessary step toward to designing a reactor for processing on scale. While significant work has been done to determine the kinetics of AADHs, the same treatment has not been given to AmDHs. In the present work, we demonstrate the first detailed study into the kinetic mechanism of amine dehydrogenases, in both the reductive amination and oxidative deamination directions. Of particular interest are the differences and similarities between the mechanisms and kinetics of AmDHs and their parent AADHs, as well as the inhibition patterns. We also employed isothermal titration calorimetry to determine binding order and relative substrate affinity. Lastly, we investigated temperature- and pH-dependence of the rate. Understanding of enzyme kinetics, and determination of a rate law are key steps toward developing a viable API manufacturing strategy using enzyme catalysts.

References:

1. Bommarius, B. R.; Schurmann, M.; Bommarius, A. S., A novel chimeric amine dehydrogenase shows altered substrate specificity compared to its parent enzymes. *Chem Commun (Camb)* **2014,** *50* (95), 14953-5.

2. Abrahamson, M. J.; Vazquez-Figueroa, E.; Woodall, N. B.; Moore, J. C.; Bommarius, A. S., Development of an amine dehydrogenase for synthesis of chiral amines. *Angew Chem Int Ed Engl* **2012,** *51* (16), 3969-72.

3. Abrahamson, M. J.; Wong, J. W.; Bommarius, A. S., The Evolution of an Amine Dehydrogenase Biocatalyst for the Asymmetric Production of Chiral Amines. *Advanced Synthesis & Catalysis* **2013,** *355* (9), 1780-1786.

522125 Construction of Artificial Metalloenzyme Catalyst and the Study of Size Effect

Abstract:

The incompatibility between enzyme and metal catalyzed reactions is a critical issue in integrating chemoenzymatic cascade processes. In this study, we report the synthesis of temperature-responsive lipase-palladium nanohybrids (Pluronic-CALB-Pd) with the size of Pd NPs ranging from 0.8 nm to 2.5 nm. As the Pd NPs downsizing to sub-nanoclusters, the catalytic performance of Pluronic-CALB-Pd in racemization of (S)-1-phenylethylamine was dramatically improved and even exhibited efficient catalytic activity at 55 °C which perfectly matched the optimal temperature of CALB. The yield of dynamic kinetic resolution (DKR) of racemic 1-phenylethylamine catalyzed by Pluronic-CALB-$Pd_{1.61}$ was up to 96 % and 99 e.e. at 55 °C for 12 h. The catalytic performance was greatly enhanced compared to the combination of commercially available Novozym 435 and Pd/C. In addition, the Pluronic-CALB-Pd nanohybrids can be homogeneously dispersed in organic solvents at relatively high temperature (>50 °C) and be precipitated at ambient temperature for an easy reuse after reaction.

Figure 1. Catalytic performance of Pluronic-CALB- Pd nanohybrids in the racemization of (S)-1-phenylethylamine

539823 Functionalized Magnetic Graphene Oxide Sheets for Efficient Cota Laccase Immobilization

Abstract:

In the recent decades, massive industrial revolution of textile dyes brought a serious environmental pollution, especially for water. Synthetic dyes gave warning not only for the environment but also for human health because of their existence in wool, textiles, papers, and leathers. As environmentally friendly biocatalysts, laccases can degrade a broad range of industrial toxic contaminants. However, the industrial applications are limited due to the low stability and poor reusability of free laccases.

Enzyme immobilization can reduce the restrictions by increasing the stability, durability and realizing continuous operations. Up to now, laccase has been successfully immobilized on various kinds of carriers, such as microspheres, nanoparticles, nanofibers and membrane. Among these carriers, the specific magnetic particles can be produced by immobilization of an affinity ligand on the surface of prefabricated magnetic beads, which can be quickly separated from the reaction medium and controlled by applying a magnetic field; then the catalytic efficiency and stability properties of the enzyme can be greatly improved.

In the current study, functionalized magnetic graphene oxide (MGO) support attached with NÎ±,NÎ±-Bis (carboxymethyl)-L-lysine hydrate (NTA-NH_2) and chelated with Cu^{2+} were synthesized. The Cu^{2+}-chelated MGO (MGO-NTA-Cu^{2+}) exhibited the highest adsorption capacity of 177 mg/g-support for CotA laccase among all synthesized nano-composites. The maximum activity recovery of laccase using MGO-NTA-Cu^{2+} was 114%. The catalytic properties of MGO-NTA-Cu-CotA laccase were significantly improved in comparison with those of free laccase. MGO-NTA-Cu-CotA laccase showed efficient decolorization rate for Congo red (CR) reached 100 % after 5 h reaction at 60 Â°C and pH 8. MGO-NTA-Cu-CotA laccase retained 89.4 % of its initial activity after 10 consecutive cycles. These MGO-NTA-Cu^{2+} supports provide promising potential for large-scale laccase immobilization in practice.

530222 Multidimensional Dynamic Experiments in Flow As a Kinetic Platform for Data Rich Experimentation in Pharmaceutical Process Development

Abstract:

Continuous flow processing continues to garner interest in the pharmaceutical industry for the production of small molecule drug substances given the advantages associated with improved mass and heat transfer, more facile scale-up, and smaller process footprints. However, to date, in-depth understanding and development of reactions can be hindered by limited availability of key raw materials and short timelines during the early stages of drug development. A need exists for the development of scalable reaction platforms enabling extensive reaction characterization in a data-rich manner using minimal material and time. Transient reaction experiments provide a viable process development intensification option to fill this gap. One common misconception of flow chemistry is that steady-state needs to be achieved in order to extract meaningful information. Recent studies have utilized dynamic conditions to rapidly obtain reaction profiling data for the generation of kinetic models [1,2,3,4]. In this presentation, we will discuss the development and application of an automated continuous flow reactor operated transiently for both process optimization and kinetic investigations in support of drug substance development.

[1] Mozharov, S.; Nordon, A.; Littlejohn, D.; Wiles, C.; Watts, P.; Dallin, P.; Girkin, J.M. *J. Am. Chem. Soc.* 2011, 133, 3601-3608.

[2] Hone, C.A.; Holmes, N.; Akien, G.R.; Bourne, R.A.; Muller, F.L. *React. Chem. Eng.* 2017, 2, 103-108.

[3] Moore, J.S.; Jensen, K.F. *Angew. Chem., Int. Ed.* 2014, 53, 470-473

[4] Aroh, K.C.; Jensen, K.F. *React. Chem. Eng.* 2018, 3, 94-101.

538491 Autonomous Reaction Platform for Continuous Chemical Synthesis

Abstract:

During all stages of the pharmaceutical development, the ability for synthetic chemistry to create and modify new molecules is crucial for library generation, toxicology, and pharmacokinetics optimization; however, the field has struggled with the slow pace of synthetic route planning and synthesis resulting in long project timelines in the development pipeline. Advances in laboratory automation can address this bottleneck by increasing expert chemists productivity when developing new molecules. Automated reaction platforms require the integration of route planning with a library of process units, beyond simple mixing and stirring, capable of carrying out multi-step syntheses, inline reaction monitoring, and easily reconfigurable between molecules.

Our research has focused on the development of a continuous flow chemistry platform capable of multi-step syntheses, and robotic reconfigurability. Our talk will discuss, the integration of a library of flow chemistry modules capable of handling solids, aggressive reagents, inline separations and the intense reaction conditions required for organic synthesis. The system is designed to assemble modules and fluidic connections into a continuous flow process. The assembly of the system is coordinated through a web-based interface which generates reaction recipes and can be connected to a synthetic route planning software. Our platform has been used to rapidly synthesize a variety of active pharmaceutical ingredients (API) and dyes requiring stereoselectivity, site-selectivity, library generation, and convergent synthesis. Through the development of flexible process modules, our integrated robotic platform aims to decrease the time required to synthesizing new molecules while increasing synthetic repeatability and lab-to-lab transferability. Automation of synthetic chemistry can decrease the time for molecule development and allow chemists to focus on pathway refinement, reaction optimization, and process analytics.

537591 Advancing Flow Chemistry Portability: Approach to Crossing the Chasm

Abstract:

Advancing Flow Chemistry Portability: Approach to Crossing the Chasm

François Lévesque, Nicholas J. Rogus, Glenn Spencer, Plamen Grigorov, Jonathan P. McMullen, David A. Thaisrivongs, Ian W. Davies, John R. Naber

The manufacturers of active pharmaceutical ingredients and their intermediates have recently rediscovered flow chemistry and continuous processing.[i] This renewed interest in these technologies[ii] has arisen from the anticipated benefit in supply chain economics and regulatory pressure in addition to the obvious opportunity for improved control including heat and mass transfer,[iii] process safety, access to high pressure and high temperature conditions,[iv] and use of supported catalysts[v] and biocatalysts[vi].

In some parts of the community there has been an assumption of ease and straightforward scalability for flow chemistry. However, when a flow step is embedded between two batch operations the flow step would ideally be completed within the same amount of time as a typical batch operation, which in our experience is usually 8 - 24 hours. To achieve this requirement, productivity obtained in a laboratory scale reactor would need to be increased by several orders of magnitude, precluding the simple scale out or numbering-up approaches. In the case of reactions that require optimal mixing for selectivity and yield, the need to preserve the same characteristics at production scale is essential. Achieving this goal would depend on our capacity to characterize mixers and define the minimum requirements to maintain the same mixing intensity.

We report mixing characterization of five lab scale and eight production scale static mixers using a modified 4th Bourne reaction. An efficient inline method relying on UV-visible spectroscopy was developed to streamline analysis of product distribution. As a result of these studies we have designed, 3D-printed and characterized a stainless steel static mixer across four orders of magnitude. This approach allowed evaluation of different configurations and ensures efficient scale-up across development and commercial facilities that should allow for enhanced portability of mixing sensitive processes.

[i] Cole, K. P.; McClary Groh, J.; Johnson, M. D.; Burcham, C. L.; Campbell, B. M.; Diseroad, W. D.; Heller, M. R.; Howell, J. R.; Kallman, N. J.; Koenig, T. M.; May, S. A.; Miller, R. D.; Mitchell, D.; Myers, D. P.; Myers, S. S.; Phillips, j. L.; Polster, C. S.; White, T. D.; Cashman, J.; Hurley, D.; Moylan, R.; Sheehan, P.; Spencer, R. D.; Desmond, K.; Desmond, Gowran, O. Kilogram-scale prexasertib monolactate monohydrate synthesis under continuous-flow CGMP conditions, *Science*, **2017**, *356*, 1144-1150.

[ii] Plutschack, M. B.; Pieber, B.; Gilmore, K.; Seeberger, P. H. The Hitchhikerâ€™s Guide to Flow Chemistry, *Chem. Rev.* **2017**, *117*, 11796-11893.

[iii] Yoshida, J.-I.; Takahashi, Y.; Nagaki, A. Flash chemistry: flow chemistry that cannot be done in batch, *Chem. Comm.* **2013**, *49*, 9896-9904.

[iv] Gutmann, B.; Cantillo, D.; Kappe, C. O. Continuous-flow technologyâ€"a tool for the safe manufacturing of active pharmaceutical ingredients, *Angew. Chem. Int. Ed.* **2015**, *54*, 6688-6728.

[v] Kirschning, A.; Solodenko, W.; Mennecke, K. Combining Enabling Techniques in Organic Synthesis: Continuous Flow Processes with Heterogenized Catalysts, *Chem. Eur. J.* **2006**, *12*, 5972-5990.

[vi] Sheldon, R. A.; Woodley, J. M. Role of Biocatalysis in Sustainable Chemistry, *Chem. Rev.* **2017**, asap. DOI: 10.1021/acs.chemrev.7b00203

521032 Development of a Fixed Bed , Gas Liquid Flow Reactor for Pharmaceutical Applications

Abstract:

Development of a Fixed Bed , Gas Liquid Flow Reactor for pharmaceutical applications

Jason Mustakis Pfizer Worldwide R&D, Groton, CT 06340, USA

Hydrogenation is a very atom efficient, green transformation frequently encountered in the synthesis of modern pharmaceutical molecules. Operating large batch hydrogenators introduces a number of challenges due to the multiphase nature of the reaction and the specialized infrastructure required. These reactors rely on mechanical means for the dispersion of hydrogen which frequently leads to scale up challenges, while frequent handling of potentially pyrophoric catalyst leads to safety concerns. Trickle bed reactors provide an alternative solution, the gas dispersion is controlled by the flow rate and reactor geometry, the handling of the catalyst is also reduced and the flow set up provides an easy way to terminate the reaction during a safety event. Additionally, the geometry and size of the reactor allows access to significantly greater operating pressure which can open the space for alternative chemistry. Fixed bed reactors also provide a way to intensify the hydrogenation operation, providing production flexibility and reducing the size of the reactors enabling flexibility of operations. This is in line with the new small volume specialized medications found in modern pharmaceutical development pipelines.

Trickle bed technology is widely developed in the chemical industry however, its adaptation by the pharmaceutical industry has been very slow and it application introduces a series of additional challenges. Our goal is to design a trickle bed reactor that is flexible enough to support a project throughout its development lifecycle, starting with clinical supplies and enabling small scale manufacturing capability without the need for specialized infrastructure. The relative low production rates required by this application are dictating that such trickle bed operates on conditions that have not been well studied in literature before and in flow regimes that are controlled primarily by surface tension. This reactor will be used either as a standalone batch process replacement or in coordination of other flow capable unit operations.

Operating a trickle bed also offers a number of challenges due to the high localized catalyst loading that leads to interesting heat and mass transfer issues. The complexity of pharmaceutical molecules of requires catalysts to have high activity and selectivity. Catalyst activity is of special interest, as it is important for achieving the column size reduction, while temperature variations may lead to selectivity challenges. Localized starvation of catalyst from hydrogen can lead to side reactions, as well as, catalyst deactivation which will affect the productivity goals. Powder carbon supported precious metal catalysts (PCPMC) are commonly applied in pharmaceutical batch process due to their large surface area as they combine high activity and robustness. However, due to their particle size, they cannot be directly applied in trickle bed applications (pressure drop, bed cracking, fine migration). The smaller size of the column also does not allow us to utilize structure supports found in the classical TBR reactor. Due to chemistry considerations we have the desire to stay with carbon supported catalyst and we are exploring a wide range of morphology from large powders to granular and spherical particles with narrow distribution. Overall, we are exploring catalyst in the range of 80 and 800 micron. Mechanical strength is a consideration, as large pressure drops can affect typical activated carbon, generating fines which then can contribute to pressure increases. Spherical carbon based catalyst offer a number of advantages such as mechanical strength and predictable flow characteristics while maintaining the high surface area that is characteristic of carbon supported PMC. We will describe the journey of the development of a TBR suitable for pharmaceutical application from lab scale to a cGMP environment. We will be examining the main factors affecting the mass and heat transfer performance of the reactor, as well as, catalyst selection as we develop methodologies to scale from batch operations to fixed bed a cGMP environment.

525426 Multiscale Modeling of a Plug Flow Reactor for a Continuous Drug Substance Manufacturing Process

Abstract:

A simple tabular reactor (at different sizes) for a Friedel-Craft acylation reaction, conventionally being used in continuous manufacturing of drug substance (flow-chemistry), is simulated at macro and micro level. The effect of reactor dimensionality and flow rate on dispersion and deviation of models from ideal plug flow regime is investigated. A Finite Element Method model, with radial and axial discretization, and a Distributed Lumped Model with axial discretization are used to simulate the reactorâ€™s transport phenomena (at low Reynolds number) and the reaction (endothermic). The reaction modeling accuracy is compared with reported experimental values for model validation analysis. The model demonstrates the RTD and dispersions in the reactor and effects of process parameter on deviations from ideal plug flow regime.

In Quality by Design framework, mathematical models can be utilized at every stage of product development and manufacturing including risk management. Predictive models can aid process development by estimating the impact of process and equipment parameters, and material attributes (i.e., the model inputs) on product attributes (i.e., the model outputs) thereby providing a quantitative framework for assessing risk and evaluating risk mitigation approaches. The emerging technology, continuous manufacturing, is a potential driving force for the utilization of process modeling and simulation. Mechanistic model can describe process nonlinearities and can be used to control the process and mitigate quality risks. Process modeling and simulation and given that process design, integration, evaluation, scale up, and optimization can be performed for the overall process once individual process unit have been characterized mathematically using models.

516568 Integrated Quality By Design for Continuous Pharmaceutical Manufacturing: Accounting for Dynamics and Feedback

Abstract:

Design spaces are sets of critical process parameters (CPPs) that guarantee that critical quality attributes (CQAs) of a manufacturing process are within specifications [Q8-Q10]. The design spaces can be constructed through extensive experimentation, mathematical models, or a combination of the two. For individual continuous pharmaceutical processes, design spaces are typically determined by assuming steady-state operation. The full design space from drug substance through drug product can be calculated using a plantwide approach or a unit-by-unit approach. Common inner approximations of the design space (e.g., hyper-rectangles) can result in significant conservatism, especially when a unit-by-unit approach is employed. Approximation of the design space by polytopes [e.g., Q11, page 25], however, is an effective means for the construction of design spaces for end-to-end continuous manufacturing facilities by combining the design spaces for individual unit operations [LF]. This approach means that the design spaces for individual processes constructed by individual research and development teams can be integrated to construct the design space for the entire continuous manufacturing facility at scale. In particular, design spaces across different value streams and between R&D and manufacturing can be integrated together in a "plug-and-play" manner in the same way as dynamic models [AM, MH, RL] without introducing any additional constraints on the process operations.

The effects of feedback control should be considered during the construction of design spaces. Due to the presence of lower level feedback control systems, design spaces for closed-loop processes can often be calculated using low-order Taylor series approximations, which results in simple expressions for the full design space (e.g., polytopes). Feedback control loops also tend to enlarge design spaces, sometimes by more than an order of magnitude. Unfortunately, disturbances, noise, and uncertainties will prevent real manufacturing processes from ever reaching "steady state". Therefore, design spaces calculated at steady state should not be used to guarantee quality specifications. This presentation will describe how to take feedback control and dynamics into account during the assurance of critical quality attributes for drug substance, intermediates, and drug product. The approach is to explicitly incorporate information on the dynamics and control systems directly into the analysis of the full integrated manufacturing facility. The approach is demonstrated for both a model predictive control and a classical plant-wide control strategy.

Although the focus of the presentation is on continuous manufacturing processes, most of the observations and approaches are also applicable to batch processing, and dynamics and feedback are as relevant to batch processing as there are to continuous manufacturing.

References

[AM] A. T. Myerson, M. Krumme, M. Nasr, H. Thomas, and R. D. Braatz. Control systems engineering in continuous pharmaceutical processing. Journal of Pharmaceutical Sciences, 104(3):832-839, 2015.

[LF] Lucas Charles Foguth, "Integration of Quality-by-Design into Control Systems Design for Continuous Pharmaceutical Manufacturing." Ph.D. thesis, Massachusetts Institute of Technology, Cambridge, 2016.

[MH] Moo Sun Hong, Kristen A. Severson, Mo Jiang, Amos E. Lu, J. Christopher Love, and Richard D. Braatz. Challenges and opportunities in biopharmaceutical manufacturing control. Computers & Chemical Engineering, 110:106-114, 2018,

[Q8] Q8 (R2) Pharmaceutical Development. U.S. Department of Health and Human Services, Food and Drug Administration, Silver Spring, Maryland, November 2009.

[Q9] Q9 Quality Risk Management. U.S. Department of Health and Human Services, Food and Drug Administration, Silver Spring, Maryland, June 2006.

[Q10] Q10 Pharmaceutical Quality System. U.S. Department of Health and Human Services, Food and Drug Administration, Silver Spring, Maryland, April 2009.

[Q11] Q11 Development and Manufacture of Drug Substances. U.S. Department of Health and Human Services, Food and Drug Administration, Silver Spring, Maryland, November 2012.

[RL] R. Lakerveld, P. L. Heider, K. D. Jensen, R. D. Braatz, K. F. Jensen, A. S. Myerson, and B. L. Trout. End-to-end continuous manufacturing: Integration of unit operations. In Continuous Manufacturing of Pharmaceuticals, edited by P. Kleinebudde, J. Khinnast, and J. Rantanen, Wiley, New York, Chapter 13, pages 447-483, 2017.

522556 Material Tracking in a Fully Continuous Drug Substance Process

Abstract:

This work summarizes a recent experience that uses a residence time distribution (RTD) model to track material across 12 unit operations of a drug substance process. The process features 48 feed tanks, several PAT points, 3 divert stations, and 3 surge points. This also corresponds to the first GMP campaign run in the new small volume continuous (SVC) facility and it resulted in 200 Kg of active pharmaceutical ingredient produced.

Application of flowsheet modeling to simulate the aforementioned continuous process, as well as the importance of event tracking (e.g., startups, diversion, stoppages, flow rate changes, feed composition changes, heel in surge vessels, complex schedule, etc.) will be highlighted. Of major importance were model linkage to process data (historian, batch records), modular approach, and challenges faced during development.

532159 Kinetic Study for Comprehensive Understanding of Solid State Phase Transitions of Nicotinamide/Pimelic Acid Co-Crystals

Abstract:

For many material, understanding of solid-state phase transformation (SSPT) is beneficial not only for the fundamental reasons but in the case of pharmaceuticals, for the avoidance of product issues due to the change in physicochemical properties. Despite the fact that this topic has been regarded as an interesting research subject, the mechanism of SSPT is still not fully understood. It is especially surprising that most co-crystal studies have been biased to identify novel polymorphs of co-crystals and to reveal their crystal structures with very little effort on understanding the kinetics. Herein we presented the comprehensive kinetic study of the polymorphic system of 1:1 Nicotinamide (NA)-pimelic acid (PA) co-crystals with the combination of different experimental techniques. The bulk kinetic modeling with powder x-ray diffraction (PXRD) indicated that polymorphic transition in the solid state of NA/PA co-crystals occurs by nucleation and growth mechanism, and no phase separation of components in the co-crystals was observed during the transitions. In-situ atomic force microscopy (AFM) observation was performed to monitor the surface morphology change during transitions at the nanoscale to supplement the bulk kinetic analysis.

Differential scanning calorimetry (DSC) thermal analysis was also conducted to obtain insight into the dependence of the crystal sizes and the defects on the kinetics of transitions. The deviation in the activation energies of SSPT for both powder and polycrystalline form were achieved respectively. In addition, the change in the tendency of transition point with increase in the number of transition recycling was observed (initially decrease then after increases), which might be attributed to the change in crystal sizes and defect inclusions during the transition process.

533858 Case Study: Modifying API Crystal Habit to Improve Flowability

Abstract:

Case Study: Modifying API Crystal Habit to Improve Flowability

Daniel A. Green, Jennifer L. Tansey, & Yan T. Sun â€“ GlaxoSmithKline, Upper Merion, PA, USA.

Abstract

An Active Pharmaceutical Ingredient (API) precipitates as acicular crystals from the original crystallization solvent mixture: water and isopropyl alcohol (IPA). Extremely poor flowability of the solid API was noted after shipping a sample from the US to the UK. Our hypothesis was that the crystal habit exacerbates the observed poor flowability upon shipping. To test this hypothesis, we investigated altering the crystal habit by crystallizing from different solutions. We varied the ratio of water and IPA in the original water/IPA solvent system, and investigated the acetone/water system. We compared bulk density and powder flow properties for several cases. We find both crystal habit change, density, and flowability improvement. Specifically, the average aspect ratio of the crystals is significantly less in acetone/H_2O compared to IPA/H_2O. The aspect ratio depends on the acetone/H_2O ratio as well: increasing as the fraction of acetone increases over the range studied (50 - 90% acetone), while the dependence on the IPA/H_2O ratio is much weaker.

527117 Solution Coating of Pharmaceutical Nanothin Films and Multilayer Nanocomposites with Controlled Morphology and Polymorphism

Abstract:

With the advent of combinatorial chemistry and high throughput screening, active pharmaceutical compounds (APIs) with low aqueous solubility have become increasingly pervasive in drug development pipeline. The formulation challenges of these drug candidates, that exhibit poor bioavailability, often encumber their route to pharmacy shelf from lab-bench. This issue is one of the major contributing factors for the skyrocketing cost of drug development today, which has reached over $1.3 billion per drug on average.[1]

In recent years, nanosizing is rapidly emerging as an effective and versatile method for enhancing solubility and thus the bioavailability of poorly aqueous soluble APIs. Previous studies have indicated that confining crystalline APIs to the nanoscale dimension exhibit increased solubility and enhanced dissolution, owing to large surface area and high surface curvature.[2] However, current industrially compatible nanonization processes mostly utilize top-down approaches where bulk crystals are broken down via mechanical forces. These methods of nanosizing are energy inefficient and require long processing time while other methods such as formation of polymeric micelles and nanoemulsions have poor stability and their processing mandates expensive instrumentation and high concentration of emulsifiers, respectively. In this scenario, industrially compatible techniques for large-scale manufacturing of API crystals with consistent crystal size distributions smaller than few hundred nanometers are urgently needed.

In this talk, we present meniscus-guided solution coating as a new technique of producing pharmaceutical films and multilayered composites at the nanoscale. This facile technique has previously been employed for depositing semi-conductor nanothin films for printed electronic devices and enables easy modulation of film morphology and polymorphism.[3-5] While API crystal morphology determines its dissolution rate, the packing modes of API molecules play crucial role in controlling the bioavailability, processability and stability of the APIs.[6] By changing simple process parameters such as API solution concentration and coating speed in order to modulate the driving forces involved in the film deposition, we demonstrate control of a) film thickness over more than two orders of magnitude (30 nm to 1.5 μm), b) thin film morphology and c) crystal polymorphism. In fact, solution coating technique also allowed us to systematically trap metastable polymorph in the film of a poorly soluble API. The controlled crystallization of metastable polymorph is advantageous in the case of poorly aqueous soluble APIs because metastable polymorphs usually exhibit higher solubility rate owing to their higher free energy.

Furthermore, we performed layer-by-layer solution coating of alternating nanometer-thick films of an API and a biocompatible polymer in a multilayered fashion, in a variety of substrates. This demonstrates the potential of solution coating technique in formulation of API doses consisting of multiple drugs which enables controlled and sequential release of each of the components as per patientÕs need, catering to the enormous enthusiasm of modern medical research towards personalized medicine.

References

1. DiMasi, J. A.; Grabowski, H. G.; Hansen, R. W., Innovation in the pharmaceutical industry: New estimates of R&D costs. *J Health Econ* **2016,** *47*, 20-33.

2. Kaptay, G., On the size and shape dependence of the solubility of nano-particles in solutions. *International Journal of Pharmaceutics* **2012,** *430* (1-2), 253-257.

3. Diao, Y.; Shaw, L.; Bao, Z. A.; Mannsfeld, S. C. B., Morphology control strategies for solution-processed organic semiconductor thin films. *Energ Environ Sci* **2014,** *7* (7), 2145-2159.

4. Diao, Y.; Tee, B. C. K.; Giri, G.; Xu, J.; Kim, D. H.; Becerril, H. A.; Stoltenberg, R. M.; Lee, T. H.; Xue, G.; Mannsfeld, S. C. B.; Bao, Z. N., Solution coating of large-area organic semiconductor thin films with aligned single-crystalline domains. *Nature Materials* **2013,** *12* (7), 665-671.

5. Diao, Y.; Lenn, K. M.; Lee, W. Y.; Blood-Forsythe, M. A.; Xu, J.; Mao, Y. S.; Kim, Y.; Reinspach, J. A.; Park, S.; Aspuru-Guzik, A.; Xue, G.; Clancy, P.; Bao, Z. N.; Mannsfeld, S. C. B., Understanding Polymorphism in Organic Semiconductor Thin Films through Nanoconfinement. *J Am Chem Soc* **2014,** *136* (49), 17046-17057.

6. Singhal, D.; Curatolo, W., Drug polymorphism and dosage form design: a practical perspective. *Advanced Drug Delivery Reviews* **2004,** *56* (3), 335-347.

538726 Optimization of Cooling Crystallization of an Active Pharmaceutical Ingredient Undergoing Degradation

Abstract:

Crystallization is one of the most important unit operations in pharmaceutical manufacturing. Crystallization is the â€œmissing linkâ€ between Drug substance and Drug product design and thereby is very important to the successful operation of downstream unit operations. The most important performance indicators of any crystallization process are the process yield and the end-of-batch mean particle size. Frequently Active Pharmaceutical Ingredients(APIs) degrade at higher temperatures due to various reasons viz. solvolysis, etc. This leads to loss of process yield since the crystallization process needs to be carried out from a lower starting temperature. Sometimes degradation of APIs can be so problematic that cooling crystallization has to be discarded as a crystallization strategy and alternate crystallization strategies viz. Anti-solvent crystallization, etc need to be considered.

Unseeded Cooling Crystallization processes typically follow a parabolic cooling profile where initially the system is nucleated very fast and then allowed to grow until the temperature is decreased sharply towards the end of the batch. In this work, the crystallization process has been modeled mathematically using population balance equations for an API undergoing degradation at high temperatures. The optimal cooling profile for the crystallization process has been obtained for different optimization objectives viz. Minimization of concentration of degraded product and Maximization of Mean Crystal Size at end of batch. It has been seen that the typical parabolic cooling profile is not always the optimal cooling profile when the API degrades and a completely counterintuitive optimal cooling profile results when the objective is to minimize the degraded product concentration. An online Ultra High Performance Liquid Chromatography was also used to monitor the concentrations of the API and the degraded impurities. It was seen that adsorption of the impurity on the crystal faces occurs for a particular class of cooling profiles. The process model was modified to account for the simultaneous degradation of API and adsorption of impurity and the parameters of the rate processes were estimated using the on-line measured concentrations.

530374 Antisolvent and Cooling Crystallization of Pharmaceuticals Using a Continuous-Flow Microfluidic Platform

Abstract:

Solid materials with different crystalline structure may be problematic when it comes to bioavailability, chemical stability or physical property. Controlling the crystal structures requires systematically study on the crystallization route and influential factors. Antisolvent and cooling crystallizations are two well-known routes for crystallization and are widely employed by pharmaceutical industries. Understanding the effect of experimental parameter such as supersaturation ratio, solvent type/composition, antisolvent and temperature on polymorphic solids is vital for synthesis of the desired polymorph. Moreover, identifying the growth rate as well as size distribution of crystals is an ultimate goal for any screening process. Here we are proposing a microfluidic crystallizer to study the polymorphism of crystals under Hydrodynamically controlled condition. The micro scale of this crystallizer is ideal to use significantly small amount of drugs particularly at early stage of development while screening numerous conditions. Our proposed microfluidic platform is fabricated using cheap and affordable material, eliminating the need of expensive lithography procedure. Using this platform antisolvent and cooling crystallizations have been studied for Acetaminophen and Glutamic acid have been selected for this study. In the antisolvent crystallization effects of different organic solvents on polymorphs of glutamic acid have been studied however for acetaminophen water is the only antisolvent. High throughput screening has also been performed within single device by increasing the number of stagnant zone and applying various ratios for entering flows. Finally, the synthesized crystals are compared with the reference materials to clarify the type of the polymorphs in different experimental condition.

522653 A Case of a Twisty Route to First Crystals

Abstract:

Nowadays, polymorph and salt screening are usually conducted in a high throughput (HT) fashion were hundreds of conditions are explored simultaneously. The HT studies are typically conducted by varying solvent systems, temperature programs, ways of creating supersaturation and counter ions. Quite often, HT polymorph and salt screening lead to several crystalline forms and the selection of the crystalline form to progress in development is based mostly on stability and purity attributes. However, in some cases HT polymorph and salt screening results in mostly oiling out and / or solids with poor crystallinity. In that situation, the search of crystalline solids is approached with a focused scientific rationale and conducted under narrower but better targeted conditions. In this presentation we present a case study where a focused scientific approach was utilized to identify and isolate a crystalline salt of a substance for which high throughput salt screening of hundreds of conditions only led to oiling out and/or amorphous solids in the. The presentation describes how a path to 1st crystals can be identified from learnings about the system behavior in terms of solubility and liquid-liquid phase separation (LLPS). The causes leading to oiling out and to the reluctance to crystal nucleation are discussed. In addition, an approach to perform a clarifying filtration when the solublities of both parent and salt do not permit a full dissolution is presented.

526549 Griseofulvin-Laden Extrudates Prepared Via Nanoextrusion: Impact of Dry-Milling on Dissolution Enhancement

Abstract:

Extrusion processes, including the traditional hot melt extrusion (HME) and a relatively novel nanoextrusion [1–4], have been used to enhance the dissolution rate and bioavailability of poorly water-soluble drugs. Nanoextrusion has been used as a continuous dryer to convert a wet-milled drug nanosuspension along with an extrusion polymer into nanocomposites [1,2]. During the process, the drug nanoparticles are dispersed in the polymeric matrix while water evaporates, which yields extrudates in the form of nanocomposites. Being a continuous process and having the capability to handle viscous drug suspensions, the nanoextrusion process is capable of preparing extrudates with good content uniformity even at very low drug loading [3]. In a recent investigation, Li et al. [4] used the same nanoextrusion process as a platform to produce extrudates either in the form of nanocomposites or amorphous solid dispersions (ASDs), thus enabling comparative assessment of nanocomposites vs. ASDs in a scientific manner.

The traditional HME process or the nanoextrusion process is usually followed by milling of the extrudate threads into powders for further downstream processing during the manufacture of final solid dosages forms. Modulating the particle (matrix) size via dry-milling of extrudates potentially allows for manipulating the drug dissolution performance because matrix size could affect drug release from the extrudates significantly [4]. In the majority of previous investigations where extrudates in the form of ASDs were produced via HME, the extrudates were dry milled and sieved into a *single size fraction*. For example, Fule et al. [5] prepared ASD of artesunate in the matrices of both Soluplus® and Kollidon® VA64, and the extrudates were milled and passed through a 200 µm sieve for various characterizations. Similarly, Juluri et al. [6] produced ASDs of griseofulvin and caffeine anhydrous in the matrix of kleptose linecaps DE17, and the milled extrudates between #40 and #50 sieves (297–420 µm) were collected and characterized. Pudlas et al. [7] produced the ASD of ibuprofen in the matrices of Soluplus® and copovidone, and the milled extrudates were sieved to exclude particles larger than 250 µm. Javeer and Amin [8] produced the ASD of carbamazepine in the matrix of hydroxypropyl methyl cellulose, and the extrudates were milled and passed through a #40 sieve with opening size of 420 µm. Also, Nagy et al. [9] produced the ASD of spironolactone in the matrix of Soluplus® and the milled extrudate was passed through a 300 µm sieve. Only in a limited number of studies, the milled extrudate particle size was actually measured. For example, Martinez-Marcos et al. [10] produced the ASD of albendazole in the matrix of polyvinylpyrrolidone. After dry milling and sieving the D_{10}, D_{50}, and D_{90} of the extrudate particles were reported in the range of 8.6 33.9 µm, 88.9 197.4 µm, and 425.1 463.6 µm, respectively. As exemplified above, in the majority of the published literature, the actual extrudate particle (matrix) sizes in the sieved fraction were not measured or reported. More importantly, the impact of matrix size from various sieved size fractions has not been examined and related to the specific surface area of the respective size fractions. Moreover, the impact of the extrudate matrix size on drug dissolution has not been studied. Similarly, when nanocomposites were produced by nanoextrusion [1–3], the impact of matrix size has not been thoroughly investigated. Hence, it is fair to assert that the impact of particle (matrix) size of the milled extrudates on drug dissolution has been studied thoroughly neither for nanocomposites nor for ASDs produced by an extrusion process, especially by the nanoextrusion process, which is the focus of this work.

The objective of current study is to examine the impact of matrix size, drug loading, and polymer–drug interactions, which modulate the solid state of the drug, on *in vitro* drug release from extrudate powders prepared via nanoextrusion. Soluplus® (Sol), Kolliphor P407 (Kol), and Hydroxypropyl cellulose (HPC) along with sodium dodecyl sulfate (SDS) were used to stabilize wet-milled griseofulvin (GF: drug) suspensions and form matrices of the extrudates. The wet-milled drug suspensions along with additional polymer (Sol/Kol/HPC) were fed to a co-rotating twin-screw extruder, which dried the suspensions and formed various extrudates. The extrudates with various polymer formulations and two target drug loadings (2% and 10% w/w) were dry-milled and sieved into three size fractions to examine the impact of matrix size. The drug nanosuspensions were characterized via laser diffraction, while the extrudates and their milled powders were characterized by laser diffraction, thermogravimetric analysis (TGA), scanning electron microscopy (SEM), and X-ray powder diffraction (XRPD). Digital microscopy was used to visualize the changes of different polymeric matrices when exposed to water. Drug wettability enhancement by Sol, Kol, and HPC solutions was studied using the modified Washburn method. Two drug doses, i.e., 8.9 mg and 100 mg, were used to investigate the dissolution response under non-supersaturating and supersaturating conditions in the bulk dissolution medium. It is hypothesized that extrudate matrices with smaller sizes can dissolve faster and that ASDs show a markedly and qualitatively different matrix size dependence of dissolution rate than nanocomposites under both dissolution conditions. Moreover, we assess the drug dissolution from ASDs vs. nanocomposites, which was modulated by various drug–polymer interactions. The impact of matrix size on GF dissolution from extrudate powders was significant especially for ASDs. We have found that the matrix size dependence of ASDs vs. nanocomposites is markedly and qualitatively different. Matrix size of ASDs can be modulated to control drug dissolution rate from ASDs; however, such modulation may not be very effective for nanocomposites that exhibit a weaker dependence of drug dissolution on matrix size. Overall, this study will provide significant insights to formulators about the impact of matrix size and differing dissolution response from nanocomposites vs. ASDs.

References:

1. R. Baumgartner, A. Eitzlmayr, N. Matsko, C. Tetyczka, J. Khinast, E. Roblegg, Nano-extrusion: A promising tool for continuous manufacturing of solid nano-formulations, Int. J. Pharm., 477 (2014) 1 11.
2. R. Baumgartner, J. Matić, S. Schrank, S. Laske, J. Khinast, E. Roblegg, NANEX: Process design and optimization, Int. J. Pharm., 506 (2016) 35 45.
3. X. Ye, H. Patil, X. Feng, R.V. Tiwari, J. Lu, A. Gryczke, K. Kolter, N. Langley, S. Majumdar, D. Neupane, Conjugation of hot-melt extrusion with high-pressure homogenization: a novel method of continuously preparing nanocrystal solid dispersions, AAPS PharmSciTech, 17 (2015) 78 88.
4. M. Li, N. Ioannidis, C. Gogos, E. Bilgili, A comparative assessment of nanocomposites vs. amorphous solid dispersions prepared via nanoextrusion for drug dissolution enhancement, Eur. J. Pharm. Biopharm., 119 (2017) 68 80.
5. R. Fule, V. Paithankar, P. Amin, Hot melt extrusion based solid solution approach: Exploring polymer comparison, physicochemical characterization and in-vivo evaluation, Int. J. Pharm., 499 (2016) 280 294.
6. A. Juluri, C. Popescu, L. Zhou, R.N. Murthy, V.K. Gowda, C.K. P, M.B. Pimparade, M.A. Repka, S.N. Murthy, Taste Masking of Griseofulvin and Caffeine Anhydrous Using Kleptose Linecaps DE17 by Hot Melt Extrusion, AAPS PharmSciTech, 17 (2016) 99 105.
7. M. Pudlas, S.O. Kyeremateng, L.A.M. Williams, J.A. Kimber, H. van Lishaut, S.G. Kazarian, G.H. Woehrle, Analyzing the impact of different excipients on drug release behavior in hot-melt extrusion formulations using FTIR spectroscopic imaging, Eur. J. Pharm. Sci., 67 (2015) 21 31.
8. S. Javeer, P. Amin, Solubility and dissolution enhancement of HPMC-based solid dispersions of carbamazepine by hot-melt extrusion technique, Asian Journal of Pharmaceutics, (2014) 119.
9. Z.K. Nagy, A. Balogh, B. Vajna, A. Farkas, G. Patyi, Á. Kramarics, G. Marosi, Comparison of electrospun and extruded soluplus®-based solid dosage forms of improved dissolution, J. Pharm. Sci., 101 (2012) 322 332.

10. L. Martinez-Marcos, D.A. Lamprou, R.T. McBurney, G.W. Halbert, A novel hot-melt extrusion formulation of albendazole for increasing dissolution properties, Int. J. Pharm., 499 (2016) 175 185.

530395 the Effect of Inorganic Salt on Disintegration of Tablets with High Loadings of Kollidon® VA64-Based Amorphous Solid Dispersion

Abstract:

Amorphous solid dispersions (ASDs) can improve dissolution of many poorly soluble drugs, and potentially enhance bioavailability in oral formulations. While the inclusion of ASDs in tablet formulations is increasingly common, tablets with high ASD loadings are often associated with slow disintegration, which hinders the development of immediate-release formulations with high drug loading. This slowdown of disintegration is believed to come from gel formation of the polymer, a major constituent of the ASD, upon exposure to aqueous media. In this study, we use a model ASD, composed of a hydrophobic drug with Kollidon® VA64 (a polyvinylpyrrolidone/ vinyl acetate copolymer) and a non-ionic surfactant, to explore formulation options that prevent such disintegration slowdown. We found that the pH, the ASD loading, and quite interestingly, the inclusion of inorganic salts in the tablet, all play important roles in the dissolution of VA64, and hence the tablet disintegration dynamics. We demonstrated that certain kosmotropic salts, when added in the formulation outside of the ASD, can significantly accelerate tablet disintegration. Contrary to expectation, we showed that the rank order of these salts in their effectiveness to help disintegration does not exactly follow the Hofmeister series; in addition to ionic effects, it also depends on other factors like particle size and dissolution rate of the salts. This highlights the dynamic nature of the tablet disintegration process, which is not to be fully predicted by steady-state properties. From our results, we provided a mechanistic explanation of the disintegration process: fast-dissolving kosmotropic salt builds high local concentration inside the restrained tablet matrix, thus lowering the cloud point of the polymer and preventing it from quickly dissolving and gelling, thereby helping the tablet break apart. This fundamental understanding has enabled design of a new platform formulation that accommodates higher ASD loadings for immediate release of poorly-soluble drug compounds.

527812 Miniaturized Formulation and Processability Screening for the Rational Design of Ethylene Vinyl Acetate Based Co-Extrudates

Abstract:

Intra-vaginal rings (IVRs) present an alternative strategy for the sustained release of estradiol (E2) in the vaginal tract. The currently marketed IVRs comprising E2 are reservoir-type silicon-based systems. One of the disadvantages of silicon elastomers is the requirement of additional curing and post-curing steps during the manufacturing process. Ethylene vinyl-acetate (EVA) copolymers represent a versatile polymeric carrier and they have been successfully used in a marketed IVR for contraception (NuvaRing®). EVA-based reservoir-type IVRs do not require additional curing steps and are manufactured via co-extrusion. A rapid, material-sparing and predictive screening methodology for IVR product development is, therefore, desirable for the rational (pre) formulation assessment and early feasibility studies.

The first aim of this study is to implement and evaluate a recently developed screening procedure (i.e., vacuum compression molding, VCM) to facilitate formulation development of EVA-based co-extrudates. The VCM screening procedure enables the manufacturing of small-scale reservoir systems (referred here as VCM systems) which can predict the rate and mechanism of release of full-scale IVRs. The second aim is to investigate the feasibility of EVA-based reservoir-type IVRs with broad daily in-vitro release rates of E2, considering the effect of the EVA copolymer type and drug loading. The equilibrium solubility of E2 in various EVA copolymers at low temperatures (i.e., 25 and 37 °C) was determined using polymer films immersed in an E2-saturated aqueous solution until equilibrium was reached. The drug-saturated polymer films were subsequently analyzed via HPLC. Additionally, the metastable solubility of E2 in the same copolymers was estimated via hot-stage microscopy (HSM) in samples produced with the VCM tool. Small-scale reservoir systems (VCM systems) with different drug loadings and different core-skin EVA copolymer combinations were produced. To evaluate the predictability of the screening procedure, full-scale IVRs were prepared with hot-melt co-extrusion to elucidate the impact of process parameters (e.g., shear forces) on the E2 drug release. The average daily E2 amount release (per cm^2) from the IVR was higher by a factor of 1.154 ± 0.160 compared to the corresponding VCM systems, dependent on the drug core loading. This fact can be explained by the enhanced E2 solubility in the core polymer, due to the high shear forces during co-extrusion. Still, the average release rates per cm^2 were in the same order of magnitude. All formulations showed near zero-order release ($R^2 > 0.99$) over 28 days. It is well known that drug release from a reservoir-type system is a function of the core and skin polymer type, the drug loading and the solubility of the drug in the core polymer. Adaptation of these parameters, based on the outcome of the rational screening approach, resulted in average E2 release rates between 1.87 ± 0.65 and 21.01 ± 3.84 $\mu g/day/cm^2$.

The generated data show that the applied screening method can be used as a time- and cost saving procedure to predict and facilitate formulation design of IVRs prior to co-extrusion. Moreover, it was shown that a broad and constant E2 release from EVA-based reservoir-type systems can be achieved by optimizing formulation parameters, such as the polymer type and drug loading. IVR formulations with broad release rates of E2 could present a versatile strategy for sustained vaginal drug delivery expanding from topical and systematic menopausal symptoms treatment to contraception purposes.

535109 *Advances in Dissolution Modeling for Oral Dosage Forms with Amorphous Solid Dispersions*

Abstract:

We present further developments in a mathematical model of disintegration and dissolution of amorphous solid dispersion-based oral dosage forms. Amorphous Solid Dispersions (ASDs) of poorly water soluble active pharmaceutical ingredients (APIs) in hydrophilic polymeric matrices are a commonly used strategy to increase the solubility and dissolution rate of oral-dosage forms [1]. The dissolution kinetics of ASD-based dosage forms is substantially different from crystalline based dosage forms due to API supersaturation capability, the gelling and swelling of the polymer matrices during dissolution and the nucleation, precipitation and micellization kinetics.

The proposed models allows the streamlining of oral dosage formulation and process development in achieving an optimal tradeoff between mechanical strength, disintegration time and dissolution profile. This is achieved by an early assessment of dissolution profile with a first principles mathematical model which takes into account intrinsic material properties of the formulation such as drug diffusivity in the polymer matrix [2], the polymer swelling/erosion rate [3] and whether the API is crystalline or is in an amorphous solid dispersion. The disintegration dynamics are handled with a Population Balance Equation [4] and are coupled to the dissolution model. Whenever direct assessment of intrinsic material properties are not possible, these parameters are obtained via curve-fitting the model to a reference experimental dissolution data using a particle swarm optimization algorithm.

We present a case-study on the application of the methodology to the downstream formulation and process development of an amorphous solid dispersion which illustrates the advantages of guiding the development with a computational tool in terms of process and drug performance.

Figure 1: Dissolution profile of a tablet formulated from an amorphous solid dispersion intermediate obtained both experimentally and from the presented mathematical model.

Figure 2: Snapshot of the predicted drug and polymer distribution during dissolution as a function of the radial position within a dissolving particle showing the polymer-rich/drug-poor particle surface caused by the flux of drug dissolved. The abscissae is normalized by the instantaneous radius of the particle which evolves over time as the polymer swells and erodes

References

[1] Perrie Y., Rades T. "Pharmaceutics - Drug Delivery and Targeting", Pharmaceutical Press, 2010.

[2] Wang, Y., Abrahamsson, B., Lindfors, L. & Brasseur, J.G. "Comparison and Analysis of Theoretical Models for Diffusion-Controlled Dissolution", Mol. Pharmaceutics, 9, 1052-1066, 2012.

[3] Siepmann, J., Kranz, H., Peppas, N. A., & Bodmeier, R. "Calculation of the required size and shape of hydroxypropyl methylcellulose matrices to achieve desired drug release profiles", Int. J. of Pharm., 201(2), 151–164, 2000.

[4] Wilson, D., Wren, S. & Reynolds G., "Linking Dissolution to Disintegration in Immediate Release

Tablets Using Image Analysis and a Population Balance Modelling Approach", Pharm. Res., 29:198-208, 2012.

531867 Modeling the Bead-Dissolution Kinetics for Composite Melt-Spray Congealed (MSC) Multiparticulates

Abstract:

Melt-spray congealed (MSC) beads are a promising drug-delivery platform that offers tunable dissolution profiles and drug-release characteristics, taste masking, and ease of dosing for pediatric and geriatric populations. These small, composite, multiparticulate beads (~50-500 mm) are comprised of a mixture of wax and pore former. This wax and pore-former matrix is embedded with active-pharmaceutical ingredient (API) particles, i.e., the drug powder. Once submerged, the pore former rapidly dissolves away leaving behind a porous network. The API then dissolves into the fluid-filled pores, diffuses through the pores, and is finally released at the bead surface into the dissolution media.

Theoretically, the pore-former level, API particle size, bead size, and other adjustable parameters can be engineered to obtain the desired dissolution profile. In reality, this presents a challenge as the exact quantitative dependence of the dissolution rates on the adjustable formulation parameters (pore-former level, drug level, API size, bead size) and unchangeable API material properties (solubility, diffusion rate) are not well understood. To address this issue, this present work describes a model, entitled BeaDiPUB (an acronym for Bead Dissolution Predictions Using BeaDiPUB), that is developed to predict the dissolution of MSC beads as a function of these formulation and material parameters. The current model version is focused on an immediate-release-type formulation without any protective coating.

A calibration/validation procedure is developed for BeaDiPUB which allows application of the model to real compounds. Subsequently, parametric studies are performed to analyze the sensitivity of the dissolution kinetics to the formulation and material properties. Finally, the various regimes for drug release with different rate-limiting steps are examined. The intent of the model is to allow for virtual exploration of the formulation-design space, and thereby reduce the number of experimental iterations needed to manufacture multiparticulates with the desired dissolution characteristics.

535336 Linking Process, Product and Performance By Raman Imaging Analysis

Abstract:

Linking process, product and performance by Raman Imaging analysis

We employ Raman imaging to assess the spatial distribution of the components of a pharmaceutical tablet and its impact on performance measures such as *in vitro* dissolution and mechanical strength. We focus on tablets containing amorphous solid dispersion powders produced by spray drying and we vary the degree of heterogeneity by altering process parameters during sieving and blending steps.

Tablets containing Amorphous Solid Dispersions (ASDs) of poorly water soluble active pharmaceutical ingredients (APIs) in hydrophilic polymeric matrices are a commonly used strategy to increase the solubility and dissolution rate of oral-dosage forms [1]. Furthermore, many of the constituents of the amorphous dispersion/tablet are functional excipients selected to further enhance the bioavailability of the API [2].

Raman Imaging is a versatile analytical tool to provide chemical and spatially resolved information on dosage forms. Mapping the chemical composition throughout a section of a tablet provides a quantitative measure of the heterogeneity of each of the components within that section as shown in Figure 1. Additionally, it allows the detection of clusters which may arise from agglomerates not broken down during the blending step.

Figure 1: Raman imaging of the centre section of a pharmaceutical tablet indicating the spatial distribution of one of the components.

Herein a case-study is presented for a model drug where the same ASDs spray dried powder is formulated into a tablet with different process parameters following a design of experiments (DoE). In the DoE we vary the sieve mesh size and blending shear rates, known to have impact in individual component distribution within the tablet. Advanced analytical characterization by Raman imaging will be employed to evaluate tablet homogeneity. The influence of those parameters on the tablet tensile strength, friability, disintegration time and dissolution will also be studied, allowing an integration between process and product performance.

[1] Jermain, S. V., Brough, C., & Williams, R. O. / Amorphous solid dispersions and nanocrystal technologies for poorly water-soluble drug delivery – An update. International Journal of Pharmaceutics, 535 (1–2) (2018) 379–392

[2] S. Poozesh et al. / Understanding the process-product-performance interplay of spray dried drug-polymer systems through complete structural and chemical characterization of single spray dried particles.

Powder Technology 320 (2017) 685–695

537663 Repurposing Pollen Grains for Oral Delivery of Biologics

Abstract:

Background:

In the recent years, use of biologics as therapeutic agents has significantly increased. However, oral delivery of biologics is challenging, especially due to the need to protect them from the harsh environment of the stomach. Other challenges include limited permeability of biologics through the intestinal epithelium. To overcome some of these challenges we have repurposed pollen grains (PGs). PGs are naturally occurring microcapsules with a tough outer shell. Their natural function is to safely transport the male gamete contained within the pollens to the flower for fertilization. To perform this function, the PG wall is naturally tough and can withstand the assault of low pH and enzymes in the stomach. Furthermore, due to microscopic ornamental features on their surface PGs are mucoadhesive, which can increase their residence time over the use of polymeric particles. Thus, we postulated that PGs could be used as â€˜Trojan Horsesâ€™ to safely carry biologics through the stomach. However, before PGs can be used for oral administration they need to be cleaned to remove intrinsic biomolecules and allergens. Herein, we have developed a new treatment method that uses sequential steps of an organic solvent, acid and an alkali to treat PGs. To augment the properties of PGs we have also created a matrix of an enteric polymer in the PGs to offer protection to the drug and a controlled release property.

Methods: PGs were treated with different protocols to help remove biomolecules from the inner cavity of the PGs, and to obtain clean empty shells. Elemental analysis, scanning electron microscopy (SEM), and Fourier-transform infrared (FTIR) spectroscopy techniques were used to characterize the PGs after treatment. Ragweed (*Ambrosia elatior*) pollen were then chosen as the model pollen species to evaluate their ability to formulate bovine serum albumin (BSA) as a model biologic. Eudragit L100-55 was used as the enteric polymer and glycofurol was used as the solvent. Formulation was optimized with respect to the amounts of Eudragit L100-55 and BSA that were required to achieve high encapsulation efficiency without causing clumping of PGs (to have them free flowing). The formulation was tested for its ability to protect BSA by incubation in simulated gastric fluid. The release profiles of the formulation were also evaluated in the simulated gastric and intestinal fluids.

Results: SEM and elemental analysis showed that our new chemical treatment method can produce clean PGs from multiple species. FTIR spectra confirmed loss of lipid peaks. Video analysis provided insight into the mechanism of the cleaning steps. After cleaning, apertures, which are naturally weak spots in the PG wall got opened producing orifices in the pollen wall measuring about 1 Î¼m in diameter. To ensure that encapsulated proteins do not leak out from these orifices we successfully produced a matrix of enteric Eudragit L100-55 in the PGs. BSA was encapsulated at about 12% efficiency in this matrix. When PGs with Eudragit+BSA were incubated in simulated gastric fluid, the matrix offered protection against denaturation as determined through measurement of intrinsic fluorescence of tyrosine and tryptophan residues. Lastly, we measured the release kinetics of the PGs with Eudragit+BSA in simulated gastric and intestinal fluids. As expected, the presence of PGs provided controlled release, and this effect was accentuated with the use of Eudragit L100-55. The formulation had a slower release as compared to just the Eudragit polymer. We also confirmed the mucoadhesive nature of ragweed PGs by their ability to stick to ex vivo pig intestine, offering PGs a significant edge over smooth polymeric particles.

Conclusions and comparison of PGs to synthetic polymeric particles: These results demonstrate the ability of our new treatment protocol to clean a variety of PGs, and pave the way to investigate various applications of these naturally-occurring pollen particles. Furthermore, the protection of proteins and peptides in the PGs using the enteric matrix creates a new oral delivery system that could substitute parenterally administered biologics and vaccines. The natural mucoadhesive property of PGs can significantly increase their residence time, which makes them very attractive over synthetic polymeric particles. Synthetic polymeric particles also often require the use of organic solvents for encapsulation, and this can harm the proteins to be encapsulated. In contrast glycofurol is protein-friendly. From a cost and manufacturing perspective, PGs are already commercially harvested in large amounts to cater to the need of the allergy industry. In allergen field, PGs are incubated with buffers to extract allergens for immunotherapy and allergy testing. The PG shells are a waste product of this industry. Cost wise PGs are relatively inexpensive. Overall, use of PGs for oral delivery of biologics is an interesting and a viable option that should receive more attention.

533674 Strategies for Crystallization Development of Process Intermediates in Early and Late Stage Manufacturing Processes

Abstract:

Several tools are available for modeling crystallization attributes such as in silico solubility prediction using discreet sets of solubility screening data and regression modeling of temperature dependent solubility to aid in design of crystallization. These techniques are sometimes underutilized during intermediate isolations but provide a framework for making better judgements in initial crystallization design for process intermediates. In one case study, this approach was used to discover a suitable replacement crystallization for isolation of crude API from a high salt load process stream and was further developed to engineer Particle size control in the API isolation. In a second case study, use of distillation models in Dynochem led to the discovery of a suitable solvate form of a process intermediate that enabled complete purge of impurities originating from vendor lots of raw material in the chemical process, providing a "purity gate" which simplified the Quality attribute landscape in a fast moving development project. Prior to this discovery the material was an amorphous oil. A similar approach was used in later intermediate steps to control impurities and improve crystallization predictability and performance.

534999 Polymorphism Control Via Combined Cooling and Antisolvent Crystallization in Continuous Mixed Suspension Mixed Product Removal Crystallizers

Abstract:

This work proposes a model based polymorphism control in 2-Aminobenzoic acid system via combined cooling and antisolvent crystallization (CCAC) in continuous mixed suspension mixed product removal (MSMPR) crystallizers. A systematic approach is devised to achieve a specific polymorph in the three dimensional space of the operating variables – temperature, solvent composition and residence time. The solubility of the two polymorphs is determined at different temperatures[a] and solvent compositions by thermogravimetric method. The induction and polymorph transformation times are studied to identify the operational window of residence times. The polymorphic crystallization process is modelled using population balance equations (PBEs) in which nucleation, growth and dissolution kinetics are considered to account for solvent mediated polymorphic transformation. A systematic experimental design is applied to estimate the kinetic parameters. The growth rate of the stable polymorph (form I) at a particular supersaturation was found to increase with increase in solvent content (isopropyl alcohol) in the solution. The estimated kinetic parameters are applied to the population balance model for the polymorphic system for the calculation of optimal temperature, solvent composition and residence time in continuous CCAC for achieving a specific polymorphic form. The presentation will show the detailed systematic approach to control polymorphism via CCAC in continuous MSMPR crystallizers and experimental validation of the model.

[a]Simone, E., Saleemi, A.N., Nagy, Z.K., 2014. Active polymorphic feedback control of crystallization processes using a combined Raman and ATR-UV/Vis spectroscopy approach. *Cryst. Growth Des.* 14 (4), 1839–1850

521869 Anticipate and Avoid Oiling out in Crystallization Using Molecular Dynamics Simulations

Abstract:

"Oiling out" is a phenomenon observed commonly in anti-solvent crystallization. The solute, instead of precipitating out of solution as a solid, forms a separate liquid phase with the anti-solvent. A significant effort is needed to recover the product from the oiled out solution. It can often lead to equipment damage if the mass solidifies into hard balls, etc. It is highly desirable to prevent such conditions that lead into oiling out phenomenon, especially in the regular production batches. Therefore, it is important to understand the molecular dynamics of the molecule from the clustering perspective which will help us to prevent crashing out of the compound as a solid or oil in the solution.

We have used molecular dynamics simulations for this purpose. We ran a series of simulations with the supersaturated solute in the presence of varying quantities of anti-solvent to investigate the differences in nucleated structures between onset of crystallization and oil-out phases. The simulations demonstrate that system can be triggered to nucleate into a separate liquid phase rather than forming a solid phase depending on the interfacial energy of the nucleated or seeded crystals. If the energy is favorable for adsorption of the "oil" molecules rather than API molecules, the system oils out.

We have identified conditions, such as solvent composition, mode of addition, degree of supersaturation, etc., that favor adsorption of one species over another to predict "oiling out" following a simple thermodynamic-and-kinetic model.

In our study, we used this approach to verify robustness of the crystallization protocol. In addition, the method also illustrates an example of the quality-by-design (QbD) approach where a range of crystals can be produced with varied particle size distribution by making small modifications to the procedure thereby avoiding oiling out.

534266 Controlling Crystal Size Via Fine Particle Dissolution in a Closed Loop Wet Milling Crystallization

Abstract:

Controlling Crystal Size via Fine Particle Dissolution in A Closed Loop Wet Milling Crystallization

Kiran Ramisetty[1], Denise Crocker[1], Aaron Cote[2], Tom O'Ceallaigh[3], Ake Rasmuson[1]

1. *Department of Chemical and Environmental Science, Synthesis and Solid State Pharmaceutical Centre, Materials and Surface* Science *Institute, University of Limerick, Limerick, Ireland.*

2. *Merck & Co., Inc. (2000 Galloping Hill Road, Kenilworth, New Jersey, USA 07033.*

3. *Meck & Co., Inc., Ballydine, Kilsheelan, Clonmel Co Tipperary, Ireland.*

Controlling crystal size in pharmaceutical industry is a challenging task; several approaches has been used to produce crystals with narrow crystal size distribution. One of the novel technique is integrating a wet mill to the crystallizer in order to break the crystals in the suspension[1]. However, this method often produces particles with more number of fines. In the presentation, we will show how the crystal size with a narrow distribution can be obtained via wet milling of the suspension from a crystallizer (CR) followed by fines dissolution in a separate crystallizer (CD - crystal dissolver) which is maintained at a higher temperature than CR and followed by slow cooling to grow the crystals for better yield. We will also show how the particle size produced from milling can be altered or controlled through optimization of the number of milling cycles, milling speed and the initial properties of the crystals (shape and size). To study this, different size crystals produced from CR by generating different supersaturation regimes via varying parameters such as cooling rate, antisolvent addition rate and initial amount of the solute. Additionally, different shape of the crystal produced with different solvent. Both the reactors were equipped with process analytical techniques like FBRM, ATR-FTIR and PVM. Change in crystal size and number continuously monitored with FBRM in crystallizer and dissolver to understand the crystallization and dissolution kinetics. ATR-FTIR used to monitor the change in concentration during dissolution and growth phases in the dissolver. This work will also show the possibility to implement batch process to continuous crystallization with integrated milling and dissolution to produce narrow crystal size distribution.

References:

1. Yang Yang, Liangcheng Song, Tianyue Gao, and Zoltan K. Nagy Integrated Upstream and Downstream Application of Wet Milling with Continuous Mixed Suspension Mixed Product Removal Crystallization, *Cryst. Growth Des. 2015, 15, 5879−5885*

This work has been supported by Science Foundation Ireland, Grant number: 12/RC/2275.

534468 Sonoseeding: An Alternative Approach for Scale-up of Batch Sonocrystallization

Abstract:

Sonoseeding: An alternative approach for scale-up of batch sonocrystallization

Kiran A. Ramisetty, V Kannuchamy, Åke C. Rasmuson*

Department of Chemical and Environmental Science, Synthesis and Solid State Pharmaceutical

Centre (SSPC), Bernal Institute, University of Limerick, Limerick, Ireland.

This work demonstrates the strategic way of controlling the product crystal size distribution using two-reactor system of generating seed crystals with ultrasound followed by growth. A saturated solution of Form III Piracetam at 50°C in isopropanol was prepared in a 50ml crystallizer and preheated to 55 °C to ensure complete dissolution. This prepared solution was cooled down to its MSZW (Meta stable zone width) at 35°C and subjected to ultrasonication induce nucleation. Wet seed suspension loaded into crystallizer with 10times larger volume than seed generator to grow crystal for better yield. Initially growth experiments conducted isothermally for concentration to reach equilibrium and followed by a slow cooling profile. A constant FBRM count number confirmed the negligible amount of secondary nucleation during the growth phase. Temperature the crystallizer maintained same as the temperature in the seed generator. Sonication mode, time and operating temperature are effecting parameters in producing uniform seed size distribution. Scale-up studies proved the capability of this technique in delivering narrower product crystal size distribution.

526495 Crystal Form and Morphology Control of Obeticholic Acid by Crystallization in Selected Solvents

Abstract:

Crystal Form and Morphology Control of Obeticholic Acid by Crystallization in Selected Solvents

Shichao Du[1, 2], Yan Wang[1, 2], Junbo Gong[1, 2*], Xiaoyue Tan[3]

[1] National Engineering Research Center of Industry Crystallization Technology, School of Chemical Engineering and Technology, Tianjin University, Tianjin, China;

[2] The Co-Innovation Center of Chemistry and Chemical Engineering of Tianjin, Tianjin University, Tianjin, China;

[3] Department of Pathology, Medical School of Nankai University, Tianjin China

*E-mail address: junbo_gong@tju.edu.cn

Abstract

Obeticholic Acid (OCA, Fig. 1), a novel effective drug to cure primary biliary cirrhosis, non-alcoholic fatty liver disease, etc., has an extensive clinical application prospect. By now, most literatures about OCA have focused on the medical aspects, but research on the crystal form and the particle property of OCA is poor. The product of OCA is easy to agglomerate, which results in poor dispersibility and fluidity, thus making the formulation more difficult.

In this work, we aim to develop novel crystal form and better morphology of OCA by controlling the crystallization process. First, solution crystallization and solvent assisted grinding were applied to screen new crystal forms. Then the formation mechanism of the developed crystal form and crystal morphology were investigated by PAT analysis and molecular simulation method. Moreover, the crystallization thermodynamics and kinetics of OCA were studied and provided important instructions for the crystallization process control.

Two novel crystal forms of OCA (Fig. 2) were successfully developed and analyzed by PXRD, SXRD, DSC, TGA, FTIR, Raman, etc. Form I was crystallized by anti-solvent method with two selected organic solvents (ester + methylbenzene), which was needle-like shape and difficult to be optimized by altering the crystallization parameters. Interestingly, form II could be obtained when adding some water into (ester + methylbenzene) mixed solvents though water can cause liquid-liquid phase separation. Moreover, form II was block-shape and improved the dispersibility and fluidity, indicating that water played a vital role of controlling both crystal form and morphology in the crystallization of OCA.

It was also found that OCA had a wide metastable zone and led to extensive nucleation, making the crystallization uncontrollable. PAT analysis indicated that temperature, seeding, mixing and solvent composition were the main factors influencing the crystallization. We optimized these parameters based on the thermodynamic and kinetic studies of OCA. Finally, we developed a controllable crystallization process using selected solvents to obtain the pure crystal form II with good morphology which showed a promising prospect.

Fig. 1 The chemical structure of OCA

Fig.2 PXRD patterns of OCA

541565 25 by 25: Chemical Engineering in the Next 25 Years

Abstract:

In 2008, the centennial year for AIChE, we asked 25 chemical engineers for their visions of the profession from around the US and the world: Nine faculty, eight from industry, and eight post-docs and graduate students. We posed four questions related to each individual's industrial sector or area of research, asking for extrapolation, new impacts on existing sectors, new sectors, and an optional, more open-ended comment on the future of the profession as a whole.

In this 110th year of AIChE, we have revisited these questions with many of the same people. Ten years has seen many developments in our profession, in society, and in our thought-leaders' experiences. Here we report on their perspectives, comparing them to the 2008 visions.

541542 The Future of Chemical Engineering Itself

Abstract:

Chemical engineeringâ€™s future is bright because it is vital to so much of society and industry -- and its future is complicated by its rapidly expanding universe of activities. It must represent depth in core skills centered on the chemical sciences including biology and materials. At the same time, it must be transdisciplinary; it must exemplify the need and value of reaching across disciplines in borderless ways.

One way is to create and disseminate a compelling vision and strategy for chemical engineering, research, and education for the future. AIChE and the National Academies are working together to that end. In 1988, the National Academies published "Frontiers in Chemical Engineering: Research Needs and Opportunities.â€ Its committee of authors outlined a wide-ranging roadmap to turn promising research opportunities into reality, while calling for researchers to embrace new frontiers.

Since 1988, that has happened -- along with many other transformative changes like the Web, fracking, personalized medicine, and social media. A new nationwide effort, aided by international contributions, will address both the identity of chemical engineering and its future directions. Its goal is to ensure that the people in chemical engineering are primed to meet society's needs and rapid changes.

544708 Accelerating Innovation through Academic-Industrial Partnerships

Abstract:

In 2011, The Dow Chemical Company launched the University Partnership Initiative (UPI) to sponsor collaborative research at leading U.S. Universities. This 10-year commitment aims to address fundamental questions in industrial chemistry and chemical engineering with alignment to Dowâ€™s next-generation process technology and product development efforts. To date, this portfolio of projects has generated an impressive array of intellectual property, publications, and numerous examples of the collaborative developments implemented at Dow. This presentation will introduce UPI program and highlight several contributions from our collaborators and leaders in the chemical engineering profession.

540559 Maximizing Uptime, Efficiency, and Safety of Industrial Operations through Early Risk Detection

Abstract:

Achieving zero incidents while maximizing process uptime and efficiency is a key vision for operating companies. An important building block towards this vision is establishment of a proactive risk mitigation culture, supported by effective systems and a strong management. Studies have shown that most of the incidents and unexpected process failures can be avoided if the operating teams get timely information about developing risks and take preventive actions early on. While advances have been made in process monitoring and automation in the past decades, there still remains significant technological and behavioral gaps that prevent plant operations from proactively mitigating process risks.

A modern industrial plant monitors thousands of parameters, generating upwards of 50-100 million data points every day. Thanks to recent breakthroughs in machine learning and artificial intelligence approaches, today there are autonomous systems that can sift through this data and point out meaningful and timely insight. This can help operating teams ascertain process issues that are hidden in the data, long before process variables reach critical levels.

In this presentation, a new autonomous system, Dynamic Risk Analyzer™ (DRA), will be introduced that points out early indicators of process issues using proprietary machine learning. The risk indicators enable operations team to respond and make required process changes, days and sometimes weeks ahead of any alarm, avoiding late stage (expensive) fixes. To realize these objectives fully, a proactive management workflow must also be in place. The presentation will discuss new initiatives and workflows developed by plant operations to implement a proactive culture. Real-life case studies from plants using DRA technology will be presented on how this culture change resulted in increased uptime, efficiency and safety of plant operations.

531309 Gaussian Processes for Hybridizing Analytical & Data-Driven Decision-Making

Abstract:

Surrogate models are widely appreciated in process systems engineering [1]. The typical setting focuses on expensive-to-evaluate, possibly uncertain functions. Examples include: modular process simulators [2], integrated gasification combined cycle processes [3], a carbon capture absorber [4], and many other applications, e.g. [5-8]. Resources are typically limited, so effective decision making requires data-efficient learning.

The data science and statistical machine learning communities typically focus on models learned solely from observed data. But chemical engineering applications may also require explicit, parametric models, e.g. modeling known process constraints, operations constraints, and cost objectives [9]. So, work has integrated semi-algebraic functions with those learned from data [10] or developed semi-physical modeling techniques [11, 12].

This presentation surveys the state-of-the-art in hybridizing analytical and data-driven decision making [13]. We consider three probabilistic modeling applications to these hybrid situations:

Design of experiments for model discrimination [14]. We bridge the gap between classical, analytical methods [15] and Monte Carlo-based approaches [16]. Classical methods may have difficulty managing non-analytical model functions and data-driven Monte Carlo approaches come at a high computational cost. We replace the original, parametric models with probabilistic, non-parametric Gaussian process surrogates learned from model evaluations. The surrogates are flexible regression tools that extend classical analytical results to non-analytical models, while providing us with model prediction confidence bounds and avoiding the computational complexity of Monte-Carlo approaches.

Multi-objective optimization [17, 18]. We make novel extensions to Bayesian multi-objective optimization in the case of one analytical objective function and one black-box, i.e. simulation-based, objective function. The resulting method has been applied to a bone neotissue application [19] and a more general test suite.

Scheduling plant operations under uncertainty. For processes with equipment degradation, we use Gaussian processes to approximate large-scale, mixed-integer optimization problems.

We close by offering a broad outlook on applying probabilistic surrogate models to chemical engineering. Statistical machine learning has recently attracted significant interest in process systems engineering [20]. Here we show that state-of-the-art research in Gaussian processes [21, 22] and probabilistic modeling more generally [23] can have a big impact on chemical engineering.

References

[1] Bhosekar A, Ierapetritou M. Advances in surrogate based modeling, feasibility analysis and and optimization: A

review. *Comput Chem Eng*. 108: 250-267, **2018**.

[2] Caballero JA, Grossmann IE. An algorithm for the use of surrogate models in modular flowsheet optimization. *AIChE J*. 54(10):2633-2650, **2008**.

[3] Lang Y, Zitney SE, Biegler LT. Optimization of IGCC processes with reduced order CFD models. *Comput Chem Eng*. 35(9):1705-1717, **2011**.

[4] Cozad A, Sahinidis NV, Miller DC. Learning surrogate models for simulation-based optim-ization. *AIChE J*. 60(6):2211-2227, **2017**.

[5] Boukouvala F, Ierapetritou M. Surrogate-based optimization of expensive flowsheet modeling for continuous pharmaceutical manufacturing. *J Pharm Innov*. 8: 131, **2013**.

[6] Soepyan FB, Cremaschi S, Sarica C, and others. Estimation of percentiles using the Kriging method for uncertainty propagation. *Comput Chem Eng*. 93:143-59, **2016**.

[7] Shokry A, Ardakani MH, Escudero G, Graells M, Espuña A. Dynamic Kriging-based fault detection and diagnosis approach for nonlinear noisy dynamic processes. *Comput Chem Eng*. 106:758-76, **2017**.

[8] Tran AP, Georgakis C. On the estimation of high-dimensional surrogate models of steady-state of plant-wide processes characteristics. *Comput Chem Eng*. DOI 10.1016/j.comp-chemeng.2018.02.014, **2018**.

[9] Boukouvala F, Hasan MF, Floudas CA. Global optimization of general constrained grey-box models: new method and its application to constrained PDEs for pressure swing adsorption. *J Glob Optim*. 67(1-2): 3-42, **2017**.

[10] Boukouvala F, Floudas CA. ARGONAUT: AlgoRithms for Global Optimization of coNstrAined grey-box compUTational problems. *Optim Lett*. 11(5): 895-913, **2017**.

[11] Pearson RK, Pottmann M. Gray-box identification of block-oriented nonlinear models. *J Process Contr*. 10(4):301-315, **2000**.

[12] Cozad A, Sahinidis NV, Miller DC. A combined first-principles and data-driven approach to model building. *Comput Chem Eng*. 73: 116-27, **2015**.

[13] Boukouvala F, Misener R, Floudas CA. Global optimization advances in mixed-integer nonlinear programming, MINLP, and constrained derivative-free optimization, CDFO. *Eur J Oper Res*. 252: 701-727, **2016**.

[14] Olofsson S, Deisenroth MP, Misener R. Design of Experiments for Model Discrimination Hybridising Analytical and Data-Driven Approaches. *arXiv:1802.04170*. **2018**.

[15] Buzzi-Ferraris G, Forzatti P, Emig G, Hofmann H. Sequential experimental design for model discrimination in the case of multiple responses. *Chem Eng Sci*. 39(1): 81-85, **1984**.

[16] Vanlier J, Tiemann CA, Hilbers PAJ, van Riel NAW. Optimal experiment design for model selection in biochemical networks. *BMC Syst Biol*, 8(20), **2014**.

[17] Olofsson S, Mehrian M, Geris L, Calandra R, Deisenroth MP, Misener R. Bayesian Multi-Objective Optimisation of Neotissue Growth in a Perfusion Bioreactor Set-Up. *Computer Aided Chemical Engineering*. 40: 2155-2160, **2017**.

[18] Beykal B, Boukouvala F, Floudas CA, Pistikopoulos EN. Optimal design of energy systems using constrained grey-box multi-objective optimization. *Comput Chem Eng*. DOI 10.1016/j.compchemeng.2018.02.017, **2018**.

[19] Mehrian M, Guyot Y, Papantoniou I, Olofsson S, Sonnaert M, Misener R, Geris L. Maximizing neotissue growth kinetics in a perfusion bioreactor: An *in silico* strategy using model reduction and Bayesian optimization. *Biotechnol Bioeng*. 115(3):617-29, **2018**.

[20] Lee JH, Shin J, Realff MJ. Machine learning: Overview of the recent progresses and implications for the process systems engineering field. *Comput Chem Eng*. DOI 10.1016/j.compchemeng.2017.10.008, **2017**.

[21] Damianou A, Lawrence N. Deep Gaussian processes. In *Artificial Intelligence and Statistics*. 207-215, **2013**.

[22] Deisenroth MP, Ng JW. Distributed Gaussian processes. In *32nd International Conference on International Conference on Machine Learning*. 37: 1481-1490, **2015**.

[23] Ghahramani Z. Bayesian non-parametrics and the probabilistic approach to modelling. *Phil Trans R Soc A*. 371(1984):20110553, **2013**.

531876 A Comparative Study of Residence Time Distribution and Cooling Crystallization in a Continuous Dynamic/Oscillatory Baffle Crystallizer Versus a Stirred Tank

Abstract:

Crystallization has become a predominant technique in particle design technology and it is widely applied in the pharmaceutical industry. Traditionally crystallization has been carried out in batch mode where operating conditions vary with time, often resulting in inconsistent products.[1] Continuous crystallization, on the other hand, has been identified to improve reproducibility, reduce operating cost and simplify scale-up.[2] A novel commercial dynamic baffle crystallizer(DBC) operated as a mixed-suspension-mixed-product-removal(MSMPR) system has been studied. It consists of a reactor tank and dynamic â€˜donutâ€™ shaped baffles to provide oscillatory flow resulting in more uniform mixing.[3] Oscillatory systems have been well studied for batch reactions and synthesis but few studies have been carried out to study the DBC for continuous two-phase operations such as crystallization.

In the work proposed here, liquid and solid residence time distributions were studied systematically in the DBC in comparison with a traditional continuous stirred tank crystallizer(STC). In-situ process analytical technology(PAT) tools such as UV-vis, Infrared(IR), and focused beam reflectance measurement(FBRM) were used to measure system response to step or pulse changes in order to construct RTDs. It is observed that the uniformity of both liquid and solid RTDs was significantly improved in the DBC than the STC indicating potential improvement in size uniformity during crystallization processes. A preliminary continuous cooling crystallization of paracetamol was carried out for two different oscillating conditions in the DBC as well as the STC at the corresponding rotational speeds yielding (approximately) equal power density. Online PAT tools were used to monitor the processes in addition to sampling for offline characterization at every residence time (45min). Nucleation event took place earlier in the DBC than the STC possibly due to the improved mixing mechanism. At early stages of operation, DBC produced crystalline particles of comparable or improved size uniformity. At later stages of operation (>7 residence times), particle sedimentation or severe particle breakage occurred in the STC causing process deviation from steady state while such events did not take place in the DBC and steady state was maintained. In the RTD studies as well as the preliminary crystallization study, the DBC showed improved system uniformity and consistency demonstrating potentials for more in-depth crystallization studies in the DBC.

1. Nagy ZK, Aamir E. Systematic design of supersaturation controlled crystallization processes for shaping the crystal size distribution using an analytical estimator. *Chem Eng Sci*. 2012;84:656-670. doi:10.1016/j.ces.2012.08.048.
2. Alvarez AJ, Singh A, Myerson AS. Crystallization of Cyclosporine in a Multistage Continuous MSMPR Crystallizer. *Cryst Growth Des*. 2011;11(10):4392-4400. doi:10.1021/cg200546g.
3. Hewgill MR, Mackley MR, Pandit AB, Pannu SS. Enhancement of gas-liquid mass transfer using oscillatory flow in a baffled tube. *Chem Eng Sci*. 1993;48(4):799-809. doi:10.1016/0009-2509(93)80145-G.

528279 Design of Fluoropolymer Membrane Systems for the Dehydration of Organic Solvents for Use in Continuous Flow Chemistry Pharmaceutical Processes

Abstract:

Dehydration of solvents is an important step in many pharmaceutical and specialty chemical manufacturing processes. Currently, the most commonly used technology for this purpose is distillation which is both energy-intensive and limited by azeotropes between certain solvents of interest (i.e., isopropyl alcohol and water). Alternatively, membranes can be used to separate water from these solvents in an energy-efficient and scalable process. However, it has been difficult to find membranes with both sufficient separation capabilities as well as the ability to withstand the conditions in which the separation must occur (i.e., chemical and thermal stability).

We will discuss a new class of fluoropolymer membranes which are designed specifically to withstand harsh chemical environments while permeating water through the membranes to reach very low water content within a variety of solvents including THF, peroxides, ketones, alcohols, etc. The talk will cover the separation characteristics of the membrane materials as well as how to design a full pervaporation system for operation within a larger process. This will include how to use membranes in batch mode as well as in a flow chemistry process. Issues covered will include process parameters, heat control, compatibility, etc. Specifically, we will share several case examples where pervaporation was used as part of a larger process.

523621 Advanced, Material-Aware Model Predictive Control Strategies for Evaporation Processes in the Pharmaceutical Industries

Abstract:

Pharmaceutical plants are required to operate with very strict product quality specifications and deal with complex and highly integrated processes, varying production targets and incoming material variability. The process design step, in particular, is an important and major factor in determining how well controlled the process can be. This step is time consuming, expensive and resource intensive. Model Based Control (MbC) is an important advanced control technique for complex multivariable control problems. It is able to predict the future behavior of a system, deal with complex dynamics, and to incorporate process constraints (safety, business, equipment, etc.) within a single optimization based problem formulation.

This work illustrates how using model based control approaches greatly improves the time and resource utilization for the development process. Moreover, certain pharmaceutical process design problems are transformed into process control problems. Consider a pharmaceutical process for the manufacture of a newly active pharmaceutical ingredient (API). The traditional approach requires a series of Design of Experiments (DoE) to be performed for the new process to be designed. An alternative MPC approach, by using the process model and being aware of material properties can make the necessary adjustment to adapt to the new API/solvent mixtures where only information regarding the material properties of the new APIs/solvent mixtures are required. Open loop experiments can still be performed and used for parameter estimation. This will eliminate the need of DoEâ€™s as well as the need for the definition of design space since this is an open loop concept that loses relevance in the presence of a controller for the quality attributes.

We develop an advanced multi-parametric model predictive controller (mp-MPC) for a semi-continuous evaporation process designed to work with different APIs/solvent mixtures. For the general design, a high fidelity model of the process was developed that is further used for testing and validation purposes [1]. Possible thermodynamic scenarios and their corresponding effects on the control strategies are then identified and classified to determine what type of controller design works for each scenario. Therefore, for every new API/solvent mixture, the user only needs to feed the corresponding thermodynamic properties, production targets and concentration setpoints to the controller, which will adjust the design and tuning parameters accordingly. To test the performance and limitations of the developed control strategy, it is applied to different mixtures for varying operating targets and process disturbances.

References

1. N. Pistikopoulos, E., et al., *PAROCâ€”An integrated framework and software platform for the optimisation and advanced model-based control of process systems.* Vol. 136. 2015.

527650 Development of a Novel Drying Technology for Drying of Wet APIâ€™s

Abstract:

There has been a growing interest of the pharmaceutical industry to link primary and secondary manufacturing. Especially with regard to cost effectiveness and cost reduction in the drug product development process.

In continuous manufacturing, not only the products quality but also its processability have to be considered. Active pharmaceutical ingredients (APIs) have a generally poor flowability since their particles are mostly small, making the powder very cohesive. Moreover, while the API synthesis may be a continuous process, the subsequent steps of filtration, washing and drying are still performed using batch process. This is primarily due to the lack of appropriate equipment.

In terms of linking primary and secondary production, drying is one of the most challenging process steps. Not only does it consume a vast amount of energy, it can significantly change the particle properties, which may often be problematic. In a crystallizer, for example, particle properties are tailored with high precision and should not be modified during the downstream processes. During drying, particle collision or shearing may lead to attrition or agglomeration of single particles, which may affect the final properties of the primary API particles.

In our work, ibuprofen was used as a test substance. The needle-shaped particles make processing particularly challenging and are thus perfectly suited for a processability study. In addition, ibuprofenâ€™s poor flowability may lead to agglomeration and restricted flow in the dryer, affecting the residence time distribution (RTD). Ideally, the RTD should be narrow and no material should be found on the walls and in dead spaces. This novel drying prototype was developed for the challenges occurring when drying cohesive powders. The new drying mechanism was developed for a mass flow of 0.5- 2 kg/h of dry substance, and showed strong improvements compared to available drying equipment. It was tested with material (ibuprofen) having various moisture levels and with different process parameters. Neither the particle size of ibuprofen nor the needle-shaped structure of the primary particles changed due to the drying process, suggesting that the contact-convective drying technology is suitable for drying cohesive materials, such as ibuprofen, at low temperatures. This study indicates that the contact-convective drying technology is a gentle and effective technology, which allows to achieve a final moisture below 1%.

527677 Possibilities and Limitations of Static Mixers in Precipitating Environments

Abstract:

Static mixers are extensively applied in industrial environments to improve mixing and heat transfer. Being a continuous processing system, they offer higher productivity, safer operating conditions, reduced waste production and lower energy consumption. Static mixers have a small footprint, enable sharp residence time distributions, intensified mixing and isothermal processes and minimize byproducts in a broad range of applications in the pharmaceutical, food processing, chemical, pulp and paper, polymer synthesis, water treatment and petrochemical industries.

Fouling is a major engineering challenge in industrial chemical processes. Various types of fouling have been identified based on its diverse mechanisms: precipitation fouling, particulate fouling, corrosion fouling, chemical reaction fouling, solidification, biofouling and composite fouling. It is a complex process affected by a large number of variables. In general, fouling on a surface contains the following consecutive stages: Initiation, transport, attachment, removal and aging.

Static mixers used in harsh industrial environments (e.g., for purification of crystals via solvent exchange washing) undergo combined composite fouling, mainly determined by particulate fouling (i.e., sedimentation) and precipitation fouling (i.e., precipitation of solids out of solutions or suspensions due to the addition of an anti-solvent). The precipitation takes place either directly on a surface (i.e., static mixer) resulting in deposits that are difficult to remove or in the bulk liquid, which involves rather loose agglomerations that are easier to remove. As more depositions are created over time, the cross-sectional area is reduced, increasing the flow velocity. The depositions may be periodically removed, altering the fouling resistance progression from linear to asymptotic.

This work investigates the application of static mixers in harsh, precipitating environments. Technical opportunities and limitations of static mixer designs were evaluated for various equipment sizes and environmental conditions. It was found that, co-axial mixing of feed and anti-solvent in the static mixer by application of ultrasound resulted in the most stable process.

530227 Coupling Flow Synthesis and Formulation By Electrospinning

Abstract:

INTRODUCTION:

In the recent decades continuous manufacturing spread in almost every industrial sector while pharmaceutical technology still relies on traditional batch methods. Albeit batch manufacturing results in slower and more expensive production and deviations in product quality over time, strict regulations restrain any technological improvements. In contrast, real time analysis of the steady state flow coupled with immediate feedback in a continuous system ensures constant product quality. Moreover, the investment and operating costs of continuous pharmaceutical devices are lower and these systems require less inventories for the storage of intermediates while the entire process is accelerated.

There have been many studies of drug syntheses using continuous flow reactors, however, the direct formulation of the drug coming from the synthesis was found to be rather challenging. The only exemplary research on such an integrated system producing a solid dosage form was conducted in the laboratory of the Massachusetts Institute of Technology (MIT) [1]. Nevertheless, the connection of synthesis and formulation was resolved by a melting process unfavorable for thermosensitive drug compounds.

Electrospinning provides a unique opportunity to convert the final liquid flow of the synthesis into solid product immediately. During the process micro- and nanofibers are formed by applying high voltage and even non-volatile solvents can be removed in a gentle way under ordinary circumstances. The release rate of the drug can be adjusted by selecting the appropriate polymeric matrix for the fibers. Furthermore, ES can also be a suitable continuous formulation method of choice when coupling it with flow operations due to its scalable productivity in the range of 0.1-100 g h^{-1}. Despite all these considerations, ES has not been applied as a part of a continuous-flow system to process liquid API streams into dry fibers serving as a base to produce solid dosage forms.

Thus, in this work we developed an end-to-end continuous model system (CMS) using ES as the key technology to turn the synthesized API solution into a fibrous solid product with the incorporation of a polymer [2]. Both synthesis and formulation were monitored by PAT tools with chemometric analyses combined with overall process control. The synthesis of acetylsalicylic acid (ASA) was chosen as model reaction using excess amount of acetic anhydride as reagent, and the side product acetic acid with high boiling point as well as the solvents of the reaction mixture had to be removed during the simultaneous formulation of the product.

RESULTS AND DISCUSSION:

Optimization of ASA synthesis:

ASA was synthesized using salicylic acid (SA) as starting material, acetic anhydride as reagent and phosphoric acid as catalyst. Following acetylation a quenching step was also required to get rid of the excess acetic anhydride and the quenchable byproducts. The optimal composition of the reaction mixture was chosen based on batch pre-experiments. Applying flow conditions, design of experiment studies were carried out for the discovery of the design space and for the identification of optimal reaction temperature and residence time in the acetylation and in the quenching step as well. At last, >99% conversion and >95% purity was achieved meeting the 3% regulatory limit for SA.

Connecting electrospinning to the flow reactors:

Polyvynilpyrrolidone K30 (PVPK30), a water-soluble polymer was introduced to the liquid stream of ASA synthesis after acetylation. Thus, after quenching the reaction mixture could be readily electrospun into fast-dissolving solid nanofibers with high purity applying high voltage. The fibers were collected on a water-soluble carrier film made of pullulan, forming a double-layered solid dosage form for peroral administration. The carrier film was continuously moved in front of the electrospinning needle, and the created double-layered product was conveyed further after the deposition of the fibers. An automated cutting mechanism cut the strip into single dosage units suitable for peroral administration.

Product characterization and CMS testing:

The physical state of ASA was investigated in the nanofibrous product with DSC and XRPD. Both analytical techniques showed that ASA turned into an amorphous form during ES.

Dissolution tests were carried out with different doses of layered PVPK30-ASA nanofibers on pullulan carrier. The ODWs were dissolved in 10 mL purified water without additional stirring at 25°C. Between the range of 1 and 25 mg ASA doses the dissolution of the fibers was instantaneous, the nanofibers disappeared in less than 2-3 seconds.

Residual solvent content measurements were conducted in steady state of the CMS for 4 hours. The results revealed satisfactorily low values in the fibrous product. The quantities of EtOH, EtOAc and AcOH were below the regulatory limit (5000 ppm) in all cases.

Repeated content uniformity measurements were conducted for 8 hours after the synthesis unit had reached steady state conditions. The settings of the formulation unit (i.e., film speed and cutting frequency) were determined earlier for 5 mg ASA dose. The collected layered ODWs showed mean contents close to the original target dose of 5 mg with satisfactorily low fluctuations.

The CMS was successfully operated for 24 hours, i.e., approximately 7 times of the residence time of the whole system including formulation. The purity of ASA in the nanofibers produced from the reaction mixture reached the >95% level not long after passing the nominal residence time. The SA content converged below the regulatory limit of 3% complementary with ASA purity. The purity of the ODWs produced by the CMS was comparable to a marketed ASA tablet formulation and both met the regulatory requirements for SA content.

Integration of PAT-based control strategies:

Real-time monitoring of quality was integrated into the CMS with spectroscopic PAT tools. The CMS was also equipped with PAT-based control loops to ensure quality. A Bruker Alpha FTIR spectrometer was used with an ATR flow cell to analyze the purity of ASA in the synthetized reaction stream. The flow cell was placed after the BPR and directly before the ES unit. Purity was calculated from the spectra by a quantitative model based on partial least squares (PLS) regression built with different ASA-SA ratios.

A Raman probe was applied for the inspection of the fibers on the pullulan strip placed before the cutting mechanism. The probe was motorized providing transversal movement of the laser beam patrolling on the surface of the nanofibrous film.

CONCLUSIONS:

A novel approach is presented based on ES for the continuous production of a solid dosage form from the synthesis of the API to the final formulation of layered ODWs with integrated PAT quality assurance. As for productivity, at 5 mg dosage strength ~1016 doses per day can be reached. Hence, a continuous benchtop device such as the one presented is able to produce considerable amounts of dosage units with ES especially when formulating high potency APIs.

Besides the optimization of ASA synthesis together with ES for purity and fiber morphology, solid phase analyses were conducted on the nanofibers verifying the amorphous form of the drug. Ultrafast dissolution could be observed with the pullulan-based nanofibrous composites at 1-25 mg ASA dose levels. The CMS was operated for 24 hours covering several cycles of the mean residence time; detailed measurements showed excellent stability of drug purity over time in addition to content uniformity and residual solvent content. Real time control of both synthesis and formulation was developed using ATR-FTIR and Raman tools combined with chemometric analyses.

REFERENCES:

[1] S. Mascia et al., Angew. Chem. Int. Ed., 52, 12359â€“12363 (2013)

[2] A. Balogh et al., Chem. Eng. J., 350, 290-299 (2018)

543729 Dynamic Modulation of Protein Functions by Strand Displacement

Abstract:

Inspired by the remarkable ability of natural protein switches to sense and respond to a wide range of environmental queues, here we report a strategy to engineer synthetic protein switches by using a generalizable strand displacement format to dynamically organize proteins with highly diverse and complex architectures. We showed that both DNA and coiled-coil peptide strand displacements can be used to dynamically control the spatial proximity and the corresponding functions of different proteins for both in vitro and in vivo applications. The general utility of the strategy was used to performing Boolean logic operations to provide explicit control of protein function using multi-input, reversible, and amplification architectures suitable for a wide range of applications.

543730 Fabrication of Cellulosic Fibers for use as Functional Materials

Abstract:

Current growing demands in renewable materials have driven a new wave of technology development for processing of cellulosic fibers, targeting a broader range of applications far beyond the traditional paper and packaging material industries. In particular, surface affinity fabrication with engineered biomolecules and nanostructures have been showing promising in developing functional materials for selective absorbents, composites and electrochemical active materials. This talk will review recent advances in fabrication of cellulosic fibers for uses as functional materials.

543731 Nanomaterials for Combination Therapies and Immunomodulation

Abstract:

This talk will focus on how nanomaterials can be designed and developed to deliver combination therapies for cancer treatment, as well as activate the immune system and serve as adjuvants for sub-unit vaccines.

The first part of the talk will focus on how nanomaterials developed in our laboratory can be used to deliver combination therapies for the treatment of pancreatic cancer. Pancreatic cancer has a very high fatality rate, and the current treatment using gemcitabine (GEM) does not work effectively, partly due to desmoplasia. We have developed dual delivery nanoscale devices to deliver miRNA and GEM together, resulting in downregulation of the Sonic hedgehog signaling pathway and leading to inhibition of desmoplasia, pancreatic stellate cells and cancer stem cells. This, in turn, improved therapeutic outcomes of GEM in PC in mice through improving its perfusion in the tumor, and also led to significant reduction of metastasis.

The second part of the talk will focus on nanovaccines that can deliver pathogenic proteins, activate the immune system, and act as vaccine adjuvants. Respiratory pathogens such as influenza affect large segments of the population. Universal influenza vaccines that can provide effective protection against various strains of the pathogen, and in elderly populations as well, are sorely needed. We have developed virus-mimicking nanomaterials or nanovaccines that provide a depot for the antigen and enhance and modulate the immune response by targeting immune cells. We have shown long-term sustained neutralizing antibody titers, reduction of viral loads and effective protection against viral challenge in both young and aged mice as a result of administration of the nanovaccines.

543732 A Sense of Balance: Exploring the Role of Metabolic Pathway Modularization in the Microbial Production of Chemicals

Abstract:

Efforts in designing synthetic cells for metabolic engineering applications have been almost exclusively based on redirecting fluxes towards product synthesis by enzyme overexpression. However, such strategies result in draining large amounts of carbon precursors and energy resources (such as ATP and NAD(P)H) from the host. The imposed metabolic burden can cause unexpected physiological changes and dramatically reduces cell performance. For example, our simulations show that cell catabolism is abundant in wild type strains so that they can support a certain amount of metabolic burden without having apparent biosynthesis deficiency (e.g., without showing a slower growth after mutations). However, cell burden usually increases when genetic modifications are introduced or fermentative products are accumulated. When cell catabolism and respiration is unable to support the increasing ATP expenditure, the biosynthesis yield will suddenly drop, forming a “cliff” (“the straw that broke the camel’s back”).

In order to address the issue of metabolic burden, our laboratory has focused on developing novel approaches that allow metabolic pathway balancing and division of labor (DoL) in order to mitigate and if possible eliminate the negative effects of metabolic burden on cellular physiology. In one example, we have modularized the fatty acid biosynthetic pathway in E.coli, and improved the final titer as well as the production rate by balancing the different modules using different plasmid copy numbers and ribosome-binding sites of different strengths. A similar approach was used to balance a synthetic metabolic pathway that allows conversion of single-carbon (C1) substrates to multi-carbon natural products in a recombinant E.coli strain. In more recent research, we have used co-culture and poly-culture approaches where multistep pathways for producing phytochemicals have been split into different production strains in order to alleviate metabolic burden, allowing the de novo production of high-value chemicals from recombinant microorganisms for the first time. In this talk, I will highlight our recent work on engineering such strategies and its implication for industrial biomanufacturing.

543733 Repairing the Brain After Stroke: aÂ Biomaterials Strategy

Abstract:

Stroke is the leading cause of disability due to the brainâ€™s limited capacity to regenerate damaged tissue. After stroke, an increased inflammatory and immune response coupled with severely limited angiogenesis and neuronal growth results in a stroke cavity devoid of normal brain tissue. However, stroke also induces the formation of a pro-repair/plastic region in the area adjacent to the stroke core that commences in the weeks to days after stroke and subsides months after stroke. We propose that the delivery of pro-repair biomaterial scaffolds in synchronization with the endogenous pro-repair window will result in repair of the stroke cavity. This talk will focus on our current efforts to promote brain repair after stroke utilizing biomaterials.

533985 Mathematical Models for the Stability of Atropisomer-Forming Drug Substance in a Spray-Dried Dispersion Formulation

Abstract:

Spray dried dispersions are commonly implemented to enhanced the delivery properties of many oral dosage drug products. Despite their favorable absorption profiles, adoption of this formulation may result in a drug product of decreased physical or chemical stability compared to formulations incorporating a crystalline form of the API. Predicting the long term stability of SDDâ€™s under ICH recommended storage conditions is generally required in order to establish appropriate control strategies, determine target specification, and toxicological qualification limits to enable a viable commercial formulation. We describe the development of a mathematical model to quantify the rate of atropisomer interconversion in the SDD drug product of a chiral compound with four potential atropisomers. The interconversion kinetics in the liquid state were adequately described by the Wei-Pratter formalism, however in the amorphous dispersion the kinetics followed a Kohlrausch-William-Watts stretched exponent function. The KWW model was used to establish expectations on the observable levels following prolonged storage. Bayesian analysis was incorporated to quantify the risk of Out of Specification instances following commercialization of the product. Stretched exponent functions are recurrent in the kinetic phenomena in the amorphous state therefore this formalism can be extended to address the risk of SDD formulations from early accelerated stability data.

537716 Study of Nimesulide Recrystallization Kinetics By Particle Sizing, Microscopy, and Raman

Abstract:

The stability of a common preclinical dosage form comprised of amorphous spray dried (ASD) material was studied using a new analytical probe developed by Blaze Metrics, that combines particle size, microscopy, and Particle Focused Raman spectroscopy. An amorphous dispersion of Nimesulide, a class of nonsteroidal anti-inflammatory drugs, was prepared by spray drying with HPMCAS-HG polymer. At a 10ml working scale, the stability of the ASD in differing concentrations of methyl cellulose was followed after exposure to serial dilutions in simulated gastric and intestinal fluids. Recrystallization kinetics were captured in the dual oral compartment model by simultaneously monitoring microscopy, particle size, and Particle Focused Raman using the single 3 in 1 probe. Samples were also taken, quenched, and analyzed by x-ray diffraction throughout the process as an independent monitor of crystallinity. It was found that increasing concentrations of methyl cellulose slowed the rate of recrystallization. While the microscopy, particle size, Raman and XRD results were all in agreement, Blaze in-process microscopy was able to detect recrystallization before the other techniques. Interestingly, the Particle Focused Raman spectroscopy revealed a short lived intermediate during the recrystallization process indicating a perhaps previously undiscovered form of Nimesulide. The insights from this data confirm the suitability and usefulness of this analytical probe technology for preformulations work. Additional (potential) applications in formulations will also be presented.

536923 Revealing Polymorphic Phase Transformations in Polymer-Based Hot Melt Extrusion Processes

Abstract:

The pharmaceutical industry is exploring new manufacturing technologies such as hot melt extrusion (HME) that enables continuous formulation of complex products for peroral administration. To date, aside from thermal stability, the polymorphic stability of an active pharmaceutical ingredient (API) is thought to be a prerequisite when HME is employed to produce solid dosage pharmaceutical formulations. Polymorphism, a phenomenon that enables molecules to exhibit multiple crystalline phases, is one of the most scrutinized critical quality attributes during the manufacturing of solid dosage formulations. Polymorphism is estimated to occur in 80% of molecules with pharmaceutical applications, affecting properties of the solid state (stability, solubility, dissolution rate, and bioavailability), and therefore, the quality and efficacy of the final drug product. The unsubstantiated notion that undesired polymorphic phase transformations might occur during the HME process limits the application of this technique to produce crystalline solid dispersions for a narrow set of APIs that are both thermally stable and monomorphic (âˆ¼20%). The present study aims to understand polymorphic phase transformations occurring during the production of crystalline solid dispersions by decoupling the effect of critical process parameters (CPPs) that accompany a HME process (temperature, drug-polymer composition, polymer properties such as molecular weight, and residence time) to relate them to the polymorphic outcome of the formulated product. Other relevant CPPs such as pressure, shear stress, and their combination with temperature, composition, and residence time will be explored in future investigations. Since conventional laboratory hot melt extruders (1) do not allow the independent study of CPPs, (2) require considerable quantities (10âˆ'800 g/h) of materials per experiment, and (3) do not permit *in situ*monitoring along the process, this work probed the effect of CPPs on the polymorphic phase transformations of physical mixtures using a combination of polarized optical hot-stage microscopy (HSM), offline powder X-ray diffraction (PXRD), and *in situ*time-resolved PXRD. Such an approach served to identify correlations among the selected CPPs and, therefore, will help to minimize the number of experiments needed to be tested in future laboratory-scale HME processes. This work demonstrates that flufenamic acid (FFA), one of the most polymorphic APIs known, thus far, can be processed with polyethylene glycol (PEG) as polymeric carrier using temperature-simulated HME. At temperatures above the transition point of FFA forms III and I (42 °C), the induction time of the polymorphic phase transformation is longer than the average reported residence time in conventional HME processes (5 min). Moreover, it was demonstrated that thorough understanding of the thermodynamic and kinetic design space for the PEG-FFA system leads to polymorphic control in the produced crystalline solid dispersions. Ultimately, this investigation helps to gain fundamental understanding of the processing needs of crystalline solid dispersions, which will lead to the broader application of HME as a continuous manufacturing strategy for drug products containing APIs prone to polymorphism, representing about 80% of all APIs.

529925 Development of a Tablet Manufacturing Line Via Hot-Melt Extrusion and Strand Pelletization

Abstract:

Development of a tablet manufacturing line via hot-melt extrusion and strand pelletization

T.R. Hörmann[1,4], O. Scheibelhofer[2], J. Rehrl[2,4], A. Funke[3], A. Paudel[2,4], J.G. Khinast[1,2,4] [1] Institute for Process and Particle Engineering, Graz University of Technology, Graz, Austria

[2] Research Center Pharmaceutical Engineering, Graz, Austria

[3] Bayer AG, Chemical and Pharmaceutical Development, Germany

[4] European Consortium on Continuous Pharmaceutical Manufacturing (ECCPM), 8010 Graz, Austria

ABSTRACT

Hot-melt extrusion (HME) is a versatile, inherently continuous process for making amorphous solid dispersions (ASD) with improved bioavailability of poorly soluble active pharmaceutical ingredients (APIs)[1]. Pellets can be made via cold or hot-strand cutting of the extrudate and offer excellent flowability and handling properties. However, the production of tablets from hot-melt extruded pellets is challenging, since the pellets are dense, non-porous particles with lower surface-to-volume ratio, thus not facilitating inter-particle bonding in comparison to powders. Moreover, many polymers exhibit elastic deformation upon compression. In our work we addressed the corresponding challenges to develop an effective formulation and a robust process.

The investigated line is shown in Figure 1. The aim of the HME formulation development was the generation of a stable ASD from a poorly soluble API yielding immediate API release from HME pellets. Five extrudable polymers were screened for ASD stability, miscibility and processability. The screening resulted in a ternary formulation with 10 % (w/w) in two polymers, i.e., Eudragit®E and HPMC E5 (Methocel®E5). The composition ratio of these polymers was optimized using small scale techniques to identify the best trade-off between processabiliy in HME and pelletization, and ASD performance (stability and dissolution profile). Molecular interactions and miscibility were characterized extensively in order to gain a detailed product understanding. Process development for the HME and pelletization process was done in two steps: i) a process parameter screening and ii) a robustness study of process and product performance upon small deviations from the optimal settings. Subsequently, a design space was established for the process, which was experimentally validated.

Figure 1. Semi-continuous processing line for the production of tablets from HME pellets. Dashed lines indicate manual material transfer during the study.

Tableting formulation development consisted of an initial small-scale screening step using several binders and disintegrants on a compaction simulator, followed by formulation optimization with Avicel PH102 and Kollidon CL on a rotary tablet press. It was possible to identify a formulation with 40% (w/w) of HME pellets, yielding mechanically stable tablets with desired API release profile. Next, downstream process development was performed with the selected formulation.

Powder blending was optimized in terms of mass flow rates aligned to the HME-pelletization step and the residence time distribution (RTD) was characterized. Since RTD was not equal for pellet and powder components in the process unit, the RTDs of the two materials were determined separately. This facilitates accurate material tracking through the continuous process line setup. This approach was also used to evaluate pellet and powder RTDs in semi-continuous tableting runs. The combination of findings from tablet formulation development and downstream process characteristics allows a performance prediction of the complete continuous processing line. This can be deployed for example, to develop advanced process control strategies.

Process development studies for the continuous HME-pelletization process could be performed in a very efficient way by investigating a number of process parameter (DoE) set points in short sequence without process interruptions. A comparison of the effort needed showed a potential to reduce process development time by approx. 50% compared to traditional batch-wise fluidized bed granulation. The API demand during the process development phase could be reduced by approx. 30%. This demonstrates the ability to develop continuous processes more efficiently.

REFERENCES

1. J. Breitenbach, Melt extrusion: from process to drug delivery technology, Eur. J. Pharm. Biopharm. 54 (2002) 107–117

528691 Generation of Stable Nanobubbles Following Reconstitution of Lyophilized Protein Formulations: Effects of Excipient Structure on Nanobubble Formation

Abstract:

Recently, reconstitution of lyophilized formulations has been shown to result in the formation of large numbers of stable bubbles approximately 150 nm in size. These nanobubbles exhibit remarkable stability, but slowly coalesce during storage over the course of days to weeks. We hypothesize that nanobubbles are formed during reconstitution as excipients dissolve around nano-sized voids present within the cake structure. Nano-sized voids may result from dehydration of regions of locally higher water concentration created during freeze concentration of solutes. These regions can be explained based on the previously reported clustering of water in polyhydroxy compounds resulting from hydrogen bonding interactions which create segregated regions of water molecules within an amorphous matrix. However, we have recently observed that lyophilization of mannitol, an excipient well known to crystallize during lyophilization, significantly increases nanobubble generation following reconstitution. Analysis of mannitol formulations using X-ray diffraction exhibited a correlation between the extent of mannitol crystallization and nanobubble generation. We believe nanobubble generation can result from dehydration of a hemihydrate crystal polymorph of mannitol to the anhydrous polymorph during drying. This dehydration process may release water back into the cake potentially generating nano-sized voids. The formation of nanobubbles following reconstitution is concerning as nanobubbles have been shown to promote aggregation of a therapeutic protein, recombinant human interleukin receptor antagonist-1. We will show that protein adsorbs to the nanobubble air-water interface resulting in significant change in nanobubble surface charge, dependent on protein net charge. Reductions in the magnitude of nanobubble surface charge following protein adsorption could be used to decrease colloidal stability of nanobubbles in protein formulation and potentially the extent of protein aggregation. Finally, we will describe the effect of metal ions leached from borosilicate glass vials on protein aggregation in nanobubble suspension. We have observed that nanobubbles stored for one week in borosilicate glass induce significantly increased monomer loss compared to suspensions stored in cyclic olefin polymer vials. Addition of a chelating agent inhibits aggregation in nanobubble suspensions further supporting the role of metal ions. The observed aggregation and particle formation in protein formulations containing nanobubbles could significantly impact both the safety and efficacy of lyophilized protein therapeutic products.

527050 Hydrodynamics Characteristics of a Stirred Tank Provided with Angle-Mounted Impellers Using Computational and Experimental Approaches

Abstract:

Most industrial stirred tanks and reactors are provided with a centrally mounted impeller to ensure adequate mixing and with wall baffles to prevent the swirling effect that would otherwise occur in an unbaffled vessel. However, an alternative approach to reduce swirling and promote mixing in an unbaffled tank is to place the impeller asymmetrically and mounted on an angled shaft. This type of configuration is common in a number of industrial operations and especially in the pharmaceutical industry, since the absence of baffles reduces the potential for contamination and ensures thorough cleaning of the vessel between batches while introducing adequate top-to-bottom liquid recirculation of the vessel content and effectively eliminating swirling effects. Although industrially relevant and relatively common in the industrial practice, only limited information is available on the operation of tanks provided with angle-mounted impellers and even less is known of their hydrodynamics. Therefore, the objective of this work was to study the hydrodynamics of a scaled-down version (diameter: 316 mm) of a tall industrial vessel with an elliptical bottom and provided with two angle-mounted (by 5Â° off the vertical) A310 Lightnin impellers under different operating conditions in order to determine the mixing characteristics of the industrial system. Both Particle Image Velocimetry (PIV) and Computational Fluid Dynamics (CFD) were used to quantify the hydrodynamics and provide guidance on the optimal operation of these industrial system. The PIV apparatus was used to experimentally determine the velocity flow field in water at different agitation speeds so as to validate the CFD predictions. A multiple reference frame (MRF) computational approach and turbulence models based on the k-Îµ method were applied for all cases in the CFD simulations. The results revealed substantial agreement between the CFD predictions and the PIV experimental results. A complex flow in the vessel was observed. Due to the axial pumping action of the impellers, a downward flow impinging the vessel bottom was created resulting in the splitting of the flow. A portion of the zone near the bottom of the vessel, which is the preferential location for the sedimentation of settling solids in the liquid, was dominated a less well-mixed zone formation. Additionally, increasing agitation speeds had no significant effect on the presence of that zone.

530926 Experimental and CFD Study of Mixing Two Fluids of Different Properties and Its Application in Biological Drug Product Manufacture

Abstract:

During biological drug product manufacture (solution in syringe or vial), concentrated drug substance in solution, typically with high viscosity, is either used as is or diluted with buffer to desired concentration. The high viscosity drug substance is amenable for subcutaneous administration but is difficult to administrator through intravenous injection (IV) without dilution. While subcutaneous administration can be conducted at home by a non-healthcare professional, intravenous injection must be administrated by a healthcare professional. Therefore, highly concentrated drug substance with high viscosity is desired at the starting point of drug product process to allow flexibility between subcutaneous administration (used as is) and intravenous injection (used after dilution). Efficient dilution of highly concentrated drug substance can be achieved at an appropriate mixing speed in a given mixer. However, the max mixing speed is set by the sensitivity of the biological molecules to shear and interfacial stress. A speed too fast may raise stability concerns. Hence, an optimal mixing condition balances between efficiency and product stability.

In this study, sucrose solution of various concentrations is mixed with water or water like liquid to mimic the two-fluid mixing system in biological drug product formulation process(Sucrose solution: drug substance, water: buffer). The impacts of initial concentration of sucrose and impeller speed on the mixing time are investigated using both experiments and computational fluid dynamic (CFD) simulation at three different scale (impeller bottom driven) mixing tanks, 250ml, 10L and 400L respectively. The experimental results in 250ml and 10L tank show that there exists a minimum mixing speed at which two fluids homogenous mixing occurs. It is also found that the impact of the initial concentration of sucrose solution on mixing time decreases at the mixing speed increases. To scale up two fluids mixing system using the traditional dimensionless number like Reynold number or tip speed is challenging due to the complexity of two-fluid mixing. A CFD two-fluid mixing model recognizing the difference of viscosity and density between the two fluids is developed for scale up. The model captures the characteristics of two-fluid mixing as described in experiments. After validated by the experimental at 250ml and 10L tanks, the model is applied at a 400L mixing tank. The model suggests key mixing operation parameters based on the liquid volume in the tank. The suggested parameters work as expected from the model.

* Presenter.

533597 Mini Continuous Stirred Tank Reactors (mini-CSTR) for Cell and Tissue Culture Applications

Abstract:

An ideal cell culture system will provide a well-controlled, homogeneous, and steady environment for cells and tissues. For instance, well-controlled steady states would greatly benefit organ-on-chip experiments, stem cell culture, and tissue propagation (among other relevant biomedical applications). However, at present, no continuously stirred mini-reactors are commercially available for lab-scale culture applications.

We are developing simple, low-cost, and user-friendly miniaturized continuously stirred tank reactors (CSTRs) for biomedical and biotechnological agitations. These well-mixed mini-CSTRs will enable cost-efficient continuous culture at small scales. We use high resolution 3D-printing to fabricate these CSTRs and an Arduino platform to measure and control key parameters, such as temperature and pressure, in small portable incubators. Nutrients are fed by syringe pumps, and well-controlled low-speed (benign) agitation is provided by a custom-made magnetic system. Since the reactor behaves as a well-mixed reservoir, all bulk-liquid concentrations can be measured at the outlet stream, thereby greatly reducing the need for intrusive instrumentation. Here we present an exhaustive validation of the mixing performance of this system, including results from computational fluid dynamic experiments (CFD) at different agitation speeds. In addition, we present preliminary results of the use of this culture platform in two model applications: (a) the extended culture of breast cancer spheroids, and (b) the culture of Chinese Hamster Ovary Cells (the warhorse for biopharmaceutical production) for continuous production of biopharmaceutical compounds.

533915 Power Dissipation and Power Numbers for a Retreat-Blade Impeller in Pharmaceutical Mixing Tanks and Reactors Using an Experimentally Validated Computational Approach

Abstract:

Mixing in stirred vessels is a very common operation in many pharmaceutical processes. Glass-lined tanks and reactors are especially commonly used in the in pharmaceutical industry for the synthesis of Active Pharmaceutical Ingredients (APIs) because of their corrosion resistance, cleanability, and reduced contamination levels. In such cases, a retreat-blade impeller with a low impeller clearance off the tank bottom is commonly used in and the reactors are cylindrical and provided with a torispherical vessel bottom and equipped with a single baffle and a retreat-blade impeller (RBI). The power, *P*, dissipated by the impeller is critical to any mixing process since mechanical energy dissipation is needed to homogenize the vessel content, disperse immiscible phases, suspend solids, increase mass transfer, and, in general, produce the desired mixing level. *P* depends not only on the type of impeller used, agitation speed, and physical properties of fluid but also on the geometry of the system, including the location of the impeller in the mixing vessel. Despite their common use, little information is available on the hydrodynamics of RBIâ€™s is such systems and their mixing performance, including their power dissipation. In this work the power, *P*, dissipated by a RBI, and the non-dimensional Power Number, Po, in different fluids were obtained both computationally via Computational Fluid Dynamics (CFD) and experimentally, using a strain-gage based torque measuring device, for different Reynoldâ€™s Number values in a 61-Liter, scaled-down version of a typical glass-lined reactor for the synthesis of pharmaceutical APIs. Additionally, P and Po were obtained for a fully baffled system (four rectangular baffles) and for an unbaffled system. *P* and Po were obtained for a very large range of the Reynolds Number, Re (1<Re<1,000,000). Po was found to vary significantly with baffling type and Re. Excellent agreement was obtained between the CFD-predicted and the experimentally obtained values for Po. Correlation equations relating Po with Re using baffling type as the parameter were additionally obtained. These equations can be used by industrial practitioners to obtain Po and hence P for their system and thus help quantify and optimize pharmaceutical mixing processes typically involving the synthesis of APIs in this type of reactors.

534745 Identifying Nucleation and Growth Kernels of Crystallization

Abstract:

Crystallization, a separation process, is simply attachment and detachment of molecules across time scales. Minor changes in the operating conditions have proven to change the outcome of crystallization and controlling the process of crystallization remains a challenge. Nucleation and crystal growth are the two processes of the crystallization and are modeled using the thermodynamic approaches. Mechanism of both processes is the same, but the time scales of the two processes are different. The challenge in modeling the process of crystallization is to derive nucleation rate of crystal structure and growth rates of energetically stable faces on the crystal. The theories used to find these rates rely on calculating interfacial free energies which is often difficult and lacks generality. With the help of a novel approach based molecular simulations, we were able to find the nucleation and growth rates. Once the rates are known, it is necessary to connect the operating conditions to outcome of crystallization. A set of operating conditions which lead to a specific outcome of crystallization will be called as kernel of crystallization. To effectively capture the time scales and the mechanism of crystallization without lack of generality, we propose the use of Population Balance Modeling (PBM). PBM finds the rate of change of internal coordinate and for crystallization, the internal coordinates are nucleation and growth rates. The rates obtained from different operating conditions can be connected to the outcomes of crystallization and hence PBM successfully finds the kernel of crystallization. Knowledge of these kernels can successfully control the process of crystallization and also shows potential to solve the problem of polymorphism.

538036 A Novel Mode of Supersaturation Feedback Control: Semi-Batch Cooling Crystallization By Feeding Flow Rate Profiles

Abstract:

Concentration Feedback Control (CFC) approaches, based on ATR-FTIR, ATR-UV/Vis or Raman spectroscopy coupled with chemometrics, have been widely applied in cooling and antisolvent crystallization systems at laboratory as well as industrial scales, by which the improved product quality have been extensively demonstrated [1]. In the feedback loop, traditionally the supersaturation is controlled to an optimal set point by manipulating the solution or cooling jacket temperature. However, the traditional batch cooling crystallization is limited by the temperature dependency and has limited applicability in some specific, but often encountering situations like crystallization of heat sensitive materials or polymorphic crystallization systems.

In this work, we propose a novel CFC implementation of semi-batch mode (feed hot saturated solution into lower temperature saturated solution), which can achieve the feedback control of supersaturation by manipulating the feeding flow rate rather than the temperature. In such, the crystallization temperature can be fixed throughout the process and thus eliminate the temperature dependence of the system. The semi-batch mode complements the classical mode in CFC implementation to some crystallization systems, such as enantiotropic polymorphic system, heat sensitive system, temperature dependent reversible reaction-crystallization and also could be designed as a start-up strategy for continuous MSMPR crystallizers [2].

Using the proposed operational method for enantiotropic polymorphic system, in the feeding step the process temperature is fixed or varied in a narrow range. This is advantageous over the cooling crystallization of much more stable polymorphic form, because the polymorph stability order is a function of temperature for enantiotropic polymorphic system. For instance, the transition temperature of α and β form of para-amino benzoic acid (pABA) in pure ethanol [3], is 13.8   (β form is stable under 13.8   ). Therefore, it is difficult to obtain pure β form with CFC in batch cooling mode because the operating temperature range was limited by this transition point. Whereas the crystallization temperature is fixed in the proposed CFC with semi-batch mode, which can be flexible for crystallization of easily design for each polymorphic form. In this context, the semi-batch operation to generate different polymorphic forms combines the advantages of batch and continuous crystallization: applicable to small scale production and stably produces one or the other pure polymorphic forms.

We modelled, simulated and evaluated three supersaturation control procedures for cooling crystallization processes: batch crystallization (T-SSC), semi-batch crystallization (F-SSC) and combined cooling and semi-batch crystallization (TF-SSC). For this, the monovariate population balance equation was used, involving nucleation and growth, solved by the method of moments and extended with the solute mass balance equation [4]. In this work, the comparison of these three scenarios will be presented through numerical simulations, which enables the identification of optimal operation modes of cooling crystallizers for various types of crystallization systems. Proof of concept experimental demonstrations for cooling and semi-bath crystallization will be also presented to support the numerical results followed by the implementation of T-SSC and F-SSC for enantiotropic polymorphic system (o-ABA) to achieve the efficient crystallization of both polymorphic forms.

References:

[1] Nagy Z K, Braatz R D. Advances and new directions in crystallization control[J]. *Annual Review of Chemical & Biomolecular Engineering, 2012, 3(5):55.*

[2] Su Q, Nagy Z K, Rielly C D. Pharmaceutical crystallisation processes from batch to continuous operation using MSMPR stages: Modelling, design, and control[J]. *Chemical Engineering & Processing Process Intensification, 2015, 89:41-53.*

[3] Hao H, Barrett M, Hu Y, et al. The Use of in Situ Tools To Monitor the Enantiotropic Transformation of p-Aminobenzoic Acid Polymorphs[J]. *Organic Process Research & Development, 2012, 16(16):35-41*

[4] Poehlein G W, Wenzel L A. Theory of particulate processes-analysis and techniques of continuous crystallization: By Alan D. Randolph and Maurice A. Larson, Academic Press, New York, *Journal of Colloid & Interface Science, 1972, 40(1):130-130.*

524824 Parameter Range Optimization and Modeling for a Reactive Crystallization Process

Abstract:

Chemical purity control is an essential component of quality risk management in active pharmaceutical ingredient (API) manufacturing. Impurity control typically relies on optimization of reaction parameters to minimize impurity formation, with downstream unit operations such as extractions and crystallizations serving as additional means of purity upgrade. In cases where crystallization occurs during the reaction, process performance can be more difficult to predict as the nucleation and uncontrolled crystallization of the product introduces stochastic variability with respect to impurity entrapment in the crystal lattice. This presentation details an industrial case study focused on process analysis and design of a robust impurity control strategy for an API step. In this step, due to a large solubility difference between API and penultimate starting material, high supersaturation conditions are achieved in the reaction, resulting in API self-nucleation and crystallization. Variability in nucleation and crystallization dynamics observed during process parameter space development resulted in variable levels of starting material entrapped in the isolated product. Extensive experimental studies were conducted to determine factors that influenced residual starting material impurity levels in isolated API, including reagent charge rate, solvent composition, and mixing conditions. A model was produced to assess process robustness over wide parameter space and quantify risks associated with operation on edges of or outside of the recommended ranges.

524971 *Enhanced Mass Transfer Process Control Via Hollow Fiber Membrane Assisted Antisolvent Crystallization*

Abstract:

Enhanced Mass Transfer Process Control via Hollow Fiber Membrane Assisted Antisolvent Crystallization

Xiaobin Jiang[a*], Linghan Tuo[a], Xuehua Ruan[b], Wu Xiao[a], Gaohong He[a,b*]

a State Key Laboratory of Fine Chemicals, Engineering Laboratory for Petrochemical Energy-efficient Separation Technology of Liaoning Province, School of Chemical Engineering, Dalian University of Technology, Dalian, Liaoning 116024, China

b School of Petroleum and Chemical Engineering, State Key Laboratory of Fine Chemicals, Dalian University of Technology at Panjin, Panjin 124221, China

Corresponding author: xbjiang@dlut.edu.cn; hgaohong@dlut.edu.cn

Abstract

Antisolvent crystallization plays an important role in various fields as an effective separation[1-4] and crystal manufacture technology[5-10]. Due to the operation under ambient temperature and atmospheric pressure, antisolvent crystallization is environment friendly and low energy consumption [11,12].

In this work, a novel hollow fiber membrane assisted antisolvent crystallization (MAAC) was proposed to enhance the antisolvent crystallization process control. PES membrane module was introduced as the key device for antisolvent mass transfer and solution mixing. An antisolvent liquid film formed on the membrane surface was renewed by the crystallization solution. The liquid film can additionally prevent the membrane from contacting with crystallization solution. By controlling the shell side flow velocity and the antisolvent mass transfer process, the antisolvent permeate rate can achieve sensitive, stable and accurate control during long term utilization.

Benefited from the nano and sub-nano size pores on the membrane surface and high packing density of the membrane module, MAAC greatly improved the micromixing between antisolvent and crystallization solution and reducing the local supersaturation degree. The calculated interfacial mass transfer rate of MAAC was approximately 1/50 of the one of the conventional antisolvent crystallization. On the basis of this effective and accurate antisolvent mass transfer control, MAAC provided the erythritol crystal products with better morphology, narrower CSD and smaller C.V. compared to the products manufactured by conventional antisolvent crystallization under the same antisolvent adding rate and operation duration.

Additionally, due to the the intrinsic feature of antifouling of the proposed MAAC operation, the permeate rate of MAAC could keep a stable value in a long running and the permeate performance of the membrane module was repeatable after more than twenty four times utilization. With well-controlled nucleation and crystal growth in the membrane module reducing the required retention time and the size of the tank for suspension liquid in MAAC, the production capacity *P.C.* (kg products¡¤m^{-3}¡¤h^{-1}) can be also enhanced via MAAC.

Figure. 1 Comparison of size distributions of aspect ratio (A), crystal morphology (B) and interfacial mass transfer rate (C) between erythritol produced by MAAC and conventional antisolvent crystallization after 1 hour operation.

Acknowledgements

We acknowledge financial contribution from National Natural Science Foundation of China (Grant No. 21527812, 21676043, U1663223, 21606035), Changjiang Scholars Program (T2012049), the Fundamental Research Funds for the Central Universities (DUT16TD19, DUT17ZD203) and Education Department of the Liaoning Province of China (No. LT2015007).

References

1. Biradha K, Su CY, Vittal JJ. Recent Developments in Crystal Engineering. *Crystal Growth & Design.* 2017;11(11):875-886.

2. Cisternas LA, V¨¢squez CM, Swaney RE. On the design of crystallization-based separation processes: Review and extension. *Aiche Journal.* 2010;52(5):1754-1769.

3. Simon L, Allan MS. Continuous antisolvent plug-flow crystallization of a fast growing API. Paper presented at: International Symposium on Industrial Crystallization - Isic2011.

4. Ulrich J, Frohberg P. Problems, potentials and future of industrial crystallization. *Frontiers of Chemical Science and Engineering.* 2013;7(1):1-8.

5. Lu H, Wang J, Wang T, Wang N, Bao Y, Hao H. Crystallization techniques in wastewater treatment: An overview of applications. *Chemosphere.* 4// 2017;173:474-484.

6. Park M-W, Yeo S-D. Antisolvent crystallization of carbamazepine from organic solutions. *Chemical Engineering Research and Design.* 2012;90(12):2202-2208.

7. Zhang C-t, Wang H-r, Wang Y-l. Internally generated seeding policy in anti-solvent crystallization of ceftriaxone sodium. *Chemical Engineering and Processing: Process Intensification.* 2010;49(4):396-401.

8. Aditya NP, Yang H, Kim S, Ko S. Fabrication of amorphous curcumin nanosuspensions using ¦Â-lactoglobulin to enhance solubility, stability, and bioavailability. *Colloids and Surfaces B: Biointerfaces.* 2015;127:114-121.

9. Zijlema TG, Geertman RM, Witkamp GJ, Rosmalen GMV, Graauw JD. Antisolvent Crystallization as an Alternative to Evaporative Crystallization for the Production of Sodium Chloride. *Industrial & Engineering Chemistry Research.* 2000;39(5):1330-1337.

10. † GXZ, Mitsuko Fujiwara, Yi Woo, ¡ì Xing, et al. Direct Design of Pharmaceutical Antisolvent Crystallization through Concentration Control. *Crystal Growth & Design.* 2006;6(4):892-898.

11. Lindenberg C, Krättli M, Cornel J, Mazzotti M, Brozio J. Design and Optimization of a Combined Cooling/Antisolvent Crystallization Process. *Crystal Growth & Design.* 2009;9(2):1124-1136.

12. Nowee SM, Abbas A, Romagnoli JA. Antisolvent crystallization: Model identification, experimental validation and dynamic simulation. *Chemical Engineering Science.* 2008;63(22):5457-5467.

525202 Experimental Implementation of a Model-Free Feedback Controller for the Size and Shape of Needle-like Crystals Growing in Suspension

Abstract:

Particle size and shape have a major impact on downstream processes in a crystallization process. The size and shape influences the physical characteristics of the final product obtained at the end of a crystallization step. Feedback control-based approaches have been proposed to control the size and shape of a population of crystals. Application of feedback control in a process can be particularly powerful as they have the ability to guide a population towards a desired target despite the presence of un-modeled phenomena and process disturbances. The prerequisite for the development and implementation of such feedback controllers are the availability of online monitoring and characterization tools for both the solid and liquid phase.

Studies aimed at experimental application of feedback control for size and shape control are scarce in literature [1]. This could be partly attributed to the unavailability of tools that could characterize the size and shape of a population of crystals. Recent advancements in imaging based tools have nevertheless addressed the above-mentioned issue of solid phase characterization [2]. Independent simulation studies performed using model-based and model-free feedback controllers in a growth-dominated seeded batch cooling crystallization process have led us to conclude that a model-free controller can robustly reach final target dimensions with a comparatively low level of complexity. The model-free approach requires only knowledge of the thermodynamics eliminating the need to develop multi-dimensional kinetic models for the system-of-interest [3].

In this work, a feedback controller was implemented in a lab scale batch crystallizer to guide two different seed populations of β L-glutamic acid in water towards desired target dimensions under growth-dominated conditions. The size and shape of the population of crystals was characterized using the online monitoring tool, μ-DISCO [2]. The experimental campaign was a two-step process, involving namely constant supersaturation controller (CSC) and model-free path following controller (PFC) experiments. The CSC experiments were performed at four different constant supersaturation levels to serve two purposes. First, to verify the monotonic behavior of the ratio of the rate of change of the average particle dimensions in the population with temperature, which is essential for the PFC. Second, to estimate an attainable region in the particle size and shape space where the controller could act on the population of crystals.

Upon the completion of the first step, PFC experiments were performed on both population of crystals, to reach two different target dimensions that falls within the attainable region. Repeatability experiments for all the PFC experiments were also performed. The controller was able to successfully reach target dimensions with only the thermodynamic knowledge of the system despite the controller acting on different batches of seed crystals, and operating conditions. The robustness provided by the controller used and experimental verification of the same leads us to conclude that model-free approach towards size and shape control could be a promising alternative to model-based control strategies.

References

[1] Eisenschmidt, H.; Bajcinca, N.; Sundmacher, K. Optimal Control of Crystal Shapes in Batch Crystallization Experiments by Growth-Dissolution Cycles. *Cryst. Growth Des.* **2016**, *16* (6), 3297–3306.

[2] Rajagopalan, A. K.; Schneeberger, J.; Salvatori, F.; Bötschi, S.; Ochsenbein, D. R.; Oswald, M. R.; Pollefeys, M.; Mazzotti, M. A Comprehensive Shape Analysis Pipeline for Stereoscopic Measurements of Particulate Populations in Suspension. *Powder Technol.* **2017**, *321*, 479–493.

[3] Ochsenbein, D. R.; Schorsch, S.; Vetter, T.; Mazzotti, M.; Morari, M. Growth Rate Estimation of β L-Glutamic Acid from Online Measurements of Multidimensional Particle Size Distributions and Concentration. *Ind. Eng. Chem. Res.* **2014**, *53* (22), 9136–9148.

537109 Morphology Based Adsorption Kinetics of a Selectively Modulated Metal-Organic Framework

Abstract:

Metal organic frameworks (MOFs) have gained interest in recent decades due to their high porosity, tunable pore aperture, and easily synthesized crystalline structures for applications such as catalysis, separations, sensing, gas storage, medicine, and energy. MOFs are composed of metal ions or oxo-metallic secondary building units (SBUs) coordinated with organic linkers to form a porous, open framework. The rational selection of these components in addition to modulating species and structure directing agents allows for the control of pore geometry and size at the surface of MOF crystals. By tuning the crystal size, morphology, and the dominant surface pore as determined by the crystal orientation, it is possible to tune the selectivity of a MOF to enhance the adsorption and separation of a given species of gas molecule. Previous results have shown that a copper based MOF with anisotropic pore distribution in the unit cell, $[Cu_2(ndc)_2(dabco)]_n$, can show controlled crystalline morphology through the addition of modulators during the synthesis. In this work, we leverage the modulation of crystallization growth processes to synthesize MOF surfaces with selective pore sizes and orientations. We demonstrate how changing the morphology of MOFs can affect the selectivity and adsorption kinetics of select species. The relationship between particle morphology, surface pore size and the adsorbed species is discussed. Additionally, tuning MOF morphology and orientation for higher selectivity and separations is discussed.

525827 Effects of Scale-up on the Mechanism and Kinetics of Crystal Nucleation

Abstract:

Crystal nucleation is an important step in solution crystallization of pharmaceutical compounds as it determines the crystal shape, polymorphic form, chiral form and the crystal size distribution of the product. Understanding this fundamental phase separation of molecules from the liquid to the solid phase is essential to exhibit process control. Nucleation is a stochastic process for unseeded crystallization processes involving low supersaturation levels or for small crystallizer volumes. Stirred small volume solution methods can be utilized to measure the stochastic nature of nucleation and to estimate nucleation kinetics. This way, probability distributions of induction times can be obtained from isothermal induction time experiments which can be used in conjunction with the classical nucleation theory (CNT) to estimate the nucleation rate and other important nucleation parameters. However, it remains unclear how scale-up affects the nucleation mechanism and nucleation kinetics in solution crystallization processes as the isothermal induction time method is currently limited to small scale experiments.

In this work, the effects of scale-up on the mechanism and kinetics of crystal nucleation will be presented. Four different crystallization setups were investigated, ranging from small magnetically-stirred 10 mL solutions to overhead-stirred solutions of 680 mL. The onset of crystallization was characterized using a Mettler Toledo Focussed Beam Reflectance Measurement (FBRM) probe. The crystallizer was programmed to start a new experiment after the FBRM probe detected crystallization. This way, many induction times were recorded in an automated fashion. The nucleation kinetics were estimated from the probability distributions of the induction times.

Nucleation kinetics were estimated for paracetamol in 2-propanol in the different crystallization platforms and at different supersaturation ratios. The nucleation rate was an order of magnitude faster in the magnetically-stirred crystallizer as compared to the crystallizers involving overhead stirring. The thermodynamic part of nucleation did not significantly change the nucleation rate whereas the kinetic nucleation parameter was found to be the rate-determining process when the crystallization process was scaled-up. In particular, the shear rate was rationalized to be the part of the kinetic parameter that changes most significantly when the crystallization process was scaled-up. The shear rate increases with volume and this effect becomes stable after the experiment has sufficiently increased in volume. In addition to these kinetic results, experiments involving the chiral sodium chlorate showed that the single nucleus mechanism is the underlying nucleation mechanism in all four tested crystallization setups. On the other hand it was found that cooling crystallization experiments were driven by a multi-nucleus mechanism. The presence of the single nucleus mechanism provides support for the use of probability distributions of induction times to estimate nucleation kinetics across all four tested crystallization setups.

In conclusion, a robust methodology was used to automatically record induction times using an FBRM probe. This approach is easily transferable and allows to capture induction times in different crystallizer types and volumes. The results presented herein on paracetamol and sodium chlorate contribute to a better understanding of the scale-up effects on the mechanism and kinetics of crystal nucleation of fine chemicals.

530421 Accelerating High-Throughput Experimentation with Templated Statistical Analysis

Abstract:

Automated high throughput experimentation for process development allows researchers to rapidly evaluate the effects of process conditions on the performance of synthetic steps. At BMS, an established workflow is used to conduct highly automated experimental designs. The final step of the workflow involves merging the automated reaction conditions with sample chromatograms (HPLC, GC, or LC-MS reports) and analyzing the data using statistical software. Often this is a bespoke, manual process that may be challenging to reproduce when the need arises in the future.

This presentation reports the development of a statistical analysis template that automates much of the repetitive, manual work of data analysis. Through a web interface, statistical analysis reports are reproducibly created from the raw sample data and shared with the project team. The work of merging data files, empirical model building, plotting trends in the data, response surfaces, and suggested future experiments are performed automatically with a minimal amount of user input. In addition to standardized data products, project teams have the flexibility to customize the analysis to address specific hypotheses. This template has been successfully applied at BMS to both automated and manually executed experimental designs.

536403 Process Fingerprinting Tools in the Development of an Alkylation Reaction

Abstract:

The final chemical transformation in the synthesis of doravirine (an NNRTI drug intended for the treatment of HIV) is an alkylation reaction with several key process variables and interactions that had historically produced high sensitivities in reaction performance. In this work, high-throughput experimentation and data-intensive PAT tools are applied to characterize and optimize this reaction for greatly improved performance. With the use of automated reactors and sampling devices, large data sets are produced to measure reaction kinetics, build models, and define ideal operating parameters to reduce impurity generation and process variability. In-situ FTIR, FBRM, and PVM are also applied to characterize the multi-phase reaction mixture by measuring solubility and addressing scale-up effects. This case study highlights the ability of these tools to enable rapid, data-driven decision making in various phases of chemical process development.

537623 Reactor Design for Continuous Synthesis of Pharmaceutical Intermediates: Correlation of Computational Fluid Dynamics and Design of Experiment Approaches

Abstract:

Reactor design principals as applied to the continuous production of pharmaceutical intermediates are important due to the increasing need to control processes and reduce costs. The Bill and Melinda Gates Foundation is dedicated to the reduction of costs of critical AIDS treatments for the developing world. Investigations into the cost reduction of the integrase inhibitor, dolutegravir, have encouraged the exploration of continuous reactor design and optimization in the conversion of keto-ester 1 into intermediate enamine-ester **3**. Thermodynamic and kinetic studies of the reaction have been used as inputs into comprehensive computational fluid dynamic studies of the reaction in order to model mixing efficiency and extent of reaction. In a parallel approach, design of experiment was used in order to study reactor design factors using conversion as an output. The result of both approaches will be discussed along with attempts to develop a predictive correlation for the production of this important intermediate in the dolutegravir synthesis.

528173 Optimal Design of Experiments for Building Fundamental Models of Pharmaceutical Production Processes

Abstract:

In many chemical and pharmacological systems, fundamental models play a significant role in design and optimization of industrial processes. When building a fundamental model, knowledge about chemical, physical and biological laws is often used to develop a model or propose rival models that can describe the underlying process behaviour. Usually, these models contain adjustable parameters that need to be estimated using available experimental data. Often available experimental data are not sufficiently informative for model building and validation, so modeler developers must perform additional experiments to obtain data to achieve improved parameter estimates and model predictions. Experimental runs may be expensive and time-consuming, especially if the new operating conditions lie outside of the routine operating range of the system. Therefore, experiments should be designed carefully to save time and effort while obtaining sufficient additional information.

Different experimental design strategies can be applied to obtain improved parameter estimates and model predictions Factorial and fractional-factorial experimental designs are commonly used because they are easy to implement and interpret. However, these designs are somewhat inflexible because they do not account for prior experimental data, nor for any structural information contained in the fundamental model equations. To overcome these difficulties, sequential model-based optimal design techniques can be used to select appropriate experimental settings that will lead to accurate parameter estimates and model predictions

Objective functions for sequential design may be A-optimal (aimed at reducing variance of parameter estimates), D-optimal (aimed at reducing the size of the joint confidence region for the parameters) or V-optimal (aimed at reducing uncertainty in model predictions at key operating conditions). An important computational step when performing traditional A-,D- and V-optimal design calculations is inversion of the Fisher information matrix (FIM). Unfortunately, in practical sequential design situations for fundamental models with numerous parameters, the FIM may be non-invertible. The main objective of the current study is to investigate the effectiveness of the several approaches for addressing the problem of noninvertible FIM when designing new sequential model-based experiments for pharmaceutical production processes. Proposed approaches for dealing with this problem include: i) a â€œleave outâ€ approach where problematic parameters are identified and left out of the design problem (LO approach), ii) a pseudo-inverse approach where the Moore Penrose pseudo inverse is used (PI approach) and iii) an approach where modelers select new experiments based on their prior knowledge without using a formal experimental design technique (MS approach). For the MS approach, the new experimental settings are chosen randomly to be either the lower or the upper bounds for the corresponding decision variables.

This study focuses on sequential A-optimal design because the A-optimality criterion is relatively easy for modelers to interpret and is the most readily usable design criterion for situations where the FIM is noninvertible. Two case studies are considered. The first case study is a simple linear example model with 7 parameters. In this example, old experimental results are available that result in perfectly correlated parameters, which makes the FIM noninvertible. Additional sequential experiments are designed to obtain parameter estimates with the lowest total variance. The second case study involves a Michael addition reaction used by a pharmaceutical company. Old experimental data are available at settings that result in a noninvertible FIM. New A-optimal sequential experiments are designed to select the recipes for one or two additional batch experiments that will lead to the best possible parameter estimates. The investigations are carried out using Monte Carlo simulations. Simulated datasets with random noise are generated for both case studies. The effectiveness of the LO, PI and MS approaches are compared using an A-optimal criterion.

The results for the two case studies show that the LO approach is better on average for achieving parameter estimates and model prediction that are closer to their true values. In addition, the results illustrate that, even with a badly conditioned or non-invertible FIM, it is advantageous to design sequential experiments using the LO approach rather than arbitrarily choosing new experimental settings at the bounds of the operating range. Designing more than one new experiment improves the quality of the parameter estimates. Also, if the number of sequential experiments is large enough, the FIM can become invertible so that the LO and PI approaches give the same results.

525884 Model-Based Design of Experiments for Pharmaceutical Reaction Development

Abstract:

Optimal experimental design has long been a strategy for maximizing knowledge capture while minimizing experimental cost in many modeling applications for diverse fields such as geology, physics, biology, and economics. Where physical models are available to describe data, model-based design of experiments provides a framework for efficient experimentation to develop and refine model systems and their parameters. Pharmaceutical process development is no exception to the need for efficient experiments with limited material, resources, and time. In this work, model-based design of optimal experiments has been applied to systems of synthetic organic chemical reactions relevant to pharmaceutical process chemistry. Model discrimination techniques are used to discern potential mechanisms with targeted experimentation. Parameter identifiability and estimability analysis is applied to justify model structure and suggest optimal analytical methods. Parameter precision is maximized by predictions of sensitivity within traditional and novel objective functions applied to parallel experimental design. These model-based optimization strategies are combined in a workflow that provides resource-efficient reaction model building in short development timelines.

534959 Defining the Optimal Operating Window for Pharmaceutical Reactions Using the Dynamic Response Surface Methodology for All Measured Species

Abstract:

For pharmaceutical reactions, it is of interest to identify the end-of-reaction conditions that lead to a final mixture in which the product is above a certain benchmark concentration and the concentrations of the by-products or intermediates are below target values. For robust manufacturing processes, the end of reaction conditions must be stable to ensure that impurity growth does not occur during the in-process control sample procedure or during other events that may require extended cycle times. Numerically, this robustness can be defined as the conditions that maximize the time window where the batch concentrations remain within the product quality constraints. This is achieved here by utilizing the Dynamic Response Surface Methodology (DRSM) Model[1,2] a recently proposed data-driven method that models time-resolved outputs.

Here, we apply a constrained variation of the DRSM-2[2] model in a simulated pharmaceutical reaction network of eight reactions involving ten species, as well as experimental data from the two collaborating pharmaceutical companies. The DRSM models for the time evolution of the measured species concentrations under different operating conditions play a crucial role in the solution of the above optimization problem. The optimal operating widows is initially estimated from the normal predictions of the DRSM model. This can be updated by accounting for the prediction interval of the DRSM. We examine an additional robustness issue related to the sensitivity of the calculated operating window with respect to a possible variation of the operating conditions from the calculated optimal ones.

References:

1. Klebanov N, Georgakis C. Dynamic Response Surface Models: A Data-Driven Approach for the Analysis of Time-Varying Process Outputs. *Industrial & Engineering Chemistry Research.* 2016;55(14):4022-4034.
2. Wang ZY, Georgakis C. New Dynamic Response Surface Methodology for Modeling Nonlinear Processes over Semi-infinite Time Horizons. *Industrial & Engineering Chemistry Research.* 2017;56(38):10770-10782.

534317 Sensitivity Analysis and Identification of Feasible Region of a Wet Granulation Continuous Pharmaceutical Manufacturing Process

Abstract:

Driven by the advances in continuous pharmaceutical manufacturing when compared to batch processes, extensive research is focused on developing predictive models for continuous unit operations. While this is leading to detailed and more accurate models, it is accompanied by an increase in model complexity and computational expense. This is further amplified when several such models are integrated to develop a flowsheet model that simulates the dynamic behavior of a plant scale manufacturing process. The limitations are particularly exacerbated for integrated models simulating the wet granulation solid oral dosage manufacturing route as different approaches are used to model unit operations in this route. For example, feeders and blenders are modeled using semi-empirical approaches where as units involving particle size changes such as granulation and milling are simulated using population balance models[1]. Discrete element method models are developed to understand and capture particle scale interactions that would render more accuracy to the model. In order to fully realize the potential of the developed models through applications such as sensitivity analysis and feasibility analysis, there is a need to develop and implement strategies to overcome the computational limitations.

Sensitivity analysis is the study of how uncertainty in a model output can be assigned to different sources of uncertainty in the model input[2]. Sources of uncertainty exist through potential variability in raw material quality, operating conditions and production rate demands. As flowsheet models have several input and output variables, it is important to identify the variables that have a significant effect on the properties of interest. This allows prioritization of the input factors and simplification of the high dimensional problem that aids in further applications such as feasibility analysis. Feasibility analysis is used to identify the feasible region of a process that is characterized by the range of input variables within which the process meets equipment, quality and production constraints[3]. Traditional methods to identify feasible regions incur large sampling costs which is prohibitive for a wet granulation flowsheet model. In addition, some constraints are not available in closed form expression. Alternatively, they may be available but expensive to evaluate[4].

Current work utilizes surrogate based feasibility analysis method[5] that is based on building surrogate models used to approximate the function of interest. In this work, a reduced order model is built for feasibility function, which characterizes the maximum constraint violation. The surrogate model accuracy is improved through selection of new samples that are identified using an adaptive sampling strategy. The strategy uses a modified Expected improvement function to direct search towards feasible region boundaries and unexplored regions[6]. The developed approach is utilized to identify feasible regions for an integrated wet granulation line that simulates a plant scale operation.

References

1. Boukouvala F, Chaudhury A, Sen M, et al. Computer-Aided Flowsheet Simulation of a Pharmaceutical Tablet Manufacturing Process Incorporating Wet Granulation. *J Pharm Innov.* 2013;8(1):11-27.
2. Saltelli A, Annoni P, Azzini I, Campolongo F, Ratto M, Tarantola S. Variance based sensitivity analysis of model output. Design and estimator for the total sensitivity index. *Computer Physics Communications.* 2010;181(2):259-270.
3. Wang Z, Escotet-Espinoza MS, Ierapetritou M. Process analysis and optimization of continuous pharmaceutical manufacturing using flowsheet models. *Computers & Chemical Engineering.*
4. Rogers A, Ierapetritou M. Feasibility and flexibility analysis of black-box processes Part 1: Surrogate-based feasibility analysis. *Chem Eng Sci.* 2015;137:986-1004.
5. Wang Z, Ierapetritou M. A novel feasibility analysis method for blackâ€box processes using a radial basis function adaptive sampling approach. *Aiche J.* 2017;63(2):532-550.
6. Boukouvala F, Ierapetritou MG. Derivativeâ€free optimization for expensive constrained problems using a novel expected improvement objective function. *Aiche J.* 2014;60(7):2462-2474.

542243 Process Intensification Strategies in Chemical Synthesis: From Batch to Continuous Operation

Abstract:

As the worldâ€™s population rises and new economies emerge, society requires novel solutions to meet its most basic needs, including energy, water, housing, food, health, and transportation. However, limited natural resources entail their efficient use while simultaneously enhancing the quality of life of current and future generations. In this respect, process intensification is becoming an immensely important tool to achieve overall sustainability of the chemical process and for the reduction of energy footprint through controlled plant size. Energy efficiency can be achieved through fundamental understanding of process thermodynamics, kinetics, and optimization of reactor designs. Further, the benefits of continuous processing for process intensification cannot be over-emphasized.

This talk will focus on two separate examples of energy efficiency through continuous processing of epoxy resins for adhesives and dimensionally confined nanomaterials for optoelectronics. Specifically I will discuss development of a tubular microreactor system for cost effective processing of hazardous divinyl benzene dioxide resins and Indium phosphide based dimensionally confined nanomaterials for electronic displays. Influence of processing conditions on product properties and conversion efficiency will also be presented.

542244 Process Fundamentals in a Counter-Current Moving Bed

Abstract:

The counter-current moving bed has been widely applied in industry, e.g., solvent removal for polyolefins and food products, and moisture reduction for biomass or plastics. ExxonMobil applied the purge bin for their polyolefin technology. After being generated in a reactor, these polyolefin materials are known to have high levels of dissolved hydrocarbon content, and adequate purging is required for safe operations and regulatory compliance. On the other hand, excessive purge gas flow rate could introduce unwanted N2, and increase flaring. This work focuses on understanding the multiscale, fundamentals of a moving bed purge bin.

A phenomenological model based on process fundamentals is developed which incorporated the hydrodynamics, kinetics, and thermodynamics to understand the performance of a purge bin. These components are modeled in a transient manner (not just simulated at steady state) in the gPROMS software platform. After comparing with the literature data, a sensitivity study is presented to understand the effects of operating parameters such as purging dimensions, temperature, pressure, purge gas flow rate and composition, and purge gas injection location, on the purging performance. The mal-distributions of gas and solids mass flux in a purge bin were shown to have a large effect on purging performance.

542245 Energy Technology for a Carbon Constrained World Can It be Relevant for the Smaller Scale Chemical Processes

Abstract:

The need to reduce carbon emissions has motivated the development of a number of reaction-separation processes that may translate to other areas of reactor engineering. It would of interest to determine if the methods of analysis used for this energy processes, such as free energy of reaction and separation could be applied as a a tool for process selection. Other methods include sorption enhanced reaction, where a reaction product is selectively remove to carry out a reversible reaction to completion or the use of reverse flow chromatographic reactor to substitute the very complex simulated moving bed reactors

542246 Automated Continuous Crystallization and Mechanism Assessment of Zeolites

Abstract:

Downstream pharmaceutical crystallization is typically performed in well-controlled batch crystallizers. This process, however, convolutes several complex phenomena: 1) gradients in transport (heat transfer and mass transfer/mixing effects), 2) multiphase mechanics/thermodynamics, including metastable crystal formation, 3) stochastic and competitive nucleation and growth kinetics. These processes are further complicated by dynamic transients in the liquid phase composition with time. Achieving a desired crystalline phase, monodisperse particles, and efficient/rapid crystallizations can then become rather challenging.

Similar analogs are observed with inorganic crystallization. Zeolite synthesis has conventionally progressed through classical batch processing with high pressure, unstirred autoclaves and is notoriously difficult to predict structures/formulations a prior. Furthermore, many fundamental questions remain about the non-classical crystallization mechanism/pathways.

In this presentation, techniques for using continuous flow chemistry to systematically control the transport and transients in the crystallization process will be explored. An automated reactor system used to access crystallization mechanisms will be discussed and applied to zeolite synthesis.

539173 Development and Qualification of a Custom Continuous Blending System for Clinical Manufacturing

Abstract:

Merck has used Design Thinking to develop a custom continuous blending system for our Solid Dosage Pilot Plant that is capable of meeting the needs of many users across the Product Development functions, including capabilities and flexibility for both small-scale development and clinical manufacturing. This presentation will step through the design process on through qualification and implementation of the system including early data and experience from the system. The discussion will focus on sharing lessons learned throughout the process of designing and implementing this new technology.

527695 Feeder Discharge Prediction Using Discrete Element Method (DEM) Simulations

Abstract:

Feeder discharge prediction using Discrete Element Method (DEM) simulations

Martina Trogrlic[1], Peter Toson[1], Eva Siegmann[1], Dalibor Jajcevic[1], Johannes Khinast[1]

&

Pankaj Doshi[2], Daniel Blackwood[2], Mary T. am Ende[2]

[1] Research Center Pharmaceutical Engineering, Inffeldgasse 13, 8010 Graz, Austria

[2] Pfizer, Inc., Worldwide Research and Development, Groton, CT, USA

Keywords: Pharmaceutical Manufacturing; Solids Flow, Handling, and Processing; Equipment

The use of powder feeder is one of the essential processes enabling continuous manufacturing in the pharmaceutical, chemical, food and other industries. In the drug production process, the ability to feed the correct amount of active pharmaceutical ingredient (API) and excipients can make the difference between good and poor product quality. Challenges with a constant feed rate emerge from the material properties of the powder, where optimal feeder parameter settings for one powder often fails for a different one with e.g. different material properties like inter-particle cohesion level and/or particle size distribution (PSD). The feed rate is also influenced by the packing density of the powder in the hopper, which depends on the powder state of consolidation, environmental factors, and events such as refilling[1]. Significant effort has investigated ways of achieving a consistent feed factor (mass fed per screw rotation), since the variance in the feed rate is often transmitted downstream to the next unit operation[2].

As shown in Fig.1, a twin-screw feeder consists of a hopper, impeller, barrel, two screws and a screen plate at the end. The role of the impeller is to break clumps of the powder and to prevent bridging. The screen at the end is intended to break up any agglomerates formed as the powder progresses along the length of the screw.

DEM method provides insight into feeder discharge time, feed rate, screw torque and solid fraction in the screw. Another advantage of using DEM simulations is the ability to predict the feeding behavior for a range of powders that are available in scarce amounts or are too costly to run full hopper experiments.

The number and the size of particles are key factors that influence duration of the simulation[3]. Particle number in the case of full hopper reaches on the order of several million, and in combination with the rotating geometries, the computation time can become significantly long. For simulating the complete discharge of the feeder in Fig.1 with calibrated cohesive material, it would take over 5 months of calculation time, for approximately 20 minutes of real process time. In order to reduce the computational time, instead of running one discharge simulation of full hopper, multiple piecewise simulations were run where a part of the powder bed was replaced with a newly developed "pressure boundary condition". Since solid fraction in the screw and consequently the feed rate changes directly with the hopper fill level, different fill levels were modeled by changing the force acting on the upper boundary surface. The resulting feed rate of the piecewise simulations was compared with the feed rate of the full hopper discharge, each modeled with linear spring-dashpot (LSDP) contact model, and very good agreement was demonstrated. In this way the computational time for feeder discharge simulation was reduced to approximately 1 month, resulting in computational savings of 4 months.

Virtual design of experiments (vDOEs) with different screw and impeller speed, screen presence, hopper fill level and calibrated materials have been performed. The results reveal very good agreement for the feed rate. The difference between cases with and without an end screen shows that the screen ensures a constant feed rate during discharge process. The particle contact model and material parameter affect the powder solid fraction inside the screw, and thus also the feed rate. DEM simulations with a cohesive force model are in better agreement with experimental results than the classic non-cohesive linear spring-dashpot model. This shows that for capturing the flow behavior of the powder, material calibration is of essential importance.

Fig. 1 Feeder with cut-away powder bed inside hopper to reveal the impeller feeding the screw region.

[1] W. E. Engisch and F. J. Muzzio, "Feedrate deviations caused by hopper refill of loss-in-weight feeders," *Powder Technol.*, vol. 283, pp. 389–400, 2015.

[2] R. Weinekötter and L. Reh, "Continuous Mixing of Fine Particles," *Part. Part. Syst. Charact.*, vol. 12, no. 1, pp. 46–53, 1995.

[3] D. S. Nasato, C. Goniva, S. Pirker, and C. Kloss, "Coarse graining for large-scale DEM simulations of particle flow - An investigation on contact and cohesion models," *Procedia Eng.*, vol. 102, pp. 1484–1490, 2015.

534190 Towards Continuous Manufacturing (CM) of Drug Loaded Strip Films: Continuous Mixing of Dry Micronized Drug Particles with Film Precursor Via Twin Screw Extruder (TSE)

Abstract:

Continuous manufacturing (CM) is an emerging technology in pharmaceutical industry. In this timely work, CM is being incorporated for strip film production where it is a natural addition for the current laborious mixing process. This strip films serve as a flexible dosage regimen for patient populations such as pediatrics and geriatrics, and those with dysphagia. Recent studies have explored the critical quality & material attributes of strip films produced via batch mixing. The novelty of the present study is continuous mixing via twin screw extruder at the room temperature, for the manufacturing of strip films containing BCS Class II drug particles. Fenofibrate and Hydroxyropyl Methyl cellulose (HPMC-E15LV) were used as the model BCS Class II drug and the polymeric matrix, respectively. Three different mixing speeds were used as critical processing parameters in TSE. Impeller mixer was also used as a batch mixer for the comparison purposes. Very fine micronized particles posed significant feeding challenge due to their highly cohesive nature. Current and future efforts involve improving powder feeding into TSE. Except for the lowest mixing speed, drug recovery upon dried film sample redispersion was excellent. In all cases, final products with very good content uniformity were successfully manufactured at the room temperature via TSE. However, uniformity of the final product reduced (express as RSD%) when such drug particles were used in the batch mixing process. The residence time in continuous mixing was remarkably low (~30 sec) compared to that of the batch mixing (3h). Overall, these results explore the introduction of micronized drug particles for TSE mixing, paving way to establishing continuous manufacturing for the strip films. Current efforts are directed towards integrating continuous film drying and overcoming cohesive drug powder feeding challenges.

537720 Material Properties Characterization and Ingredient Agglomerate Behavior in Continuous Direct Compaction Process

Abstract:

In the modern pharmaceutical process, the appropriate material characterization and process setting are required to achieve robustness of product. we are proposing a decision tree methodology for rapid screening of the direct compression formulations, which enable us based on the API characteristics to define the appropriate formulation and target process setting for direct compression continuous process. The first step is to identify the failure modes. Based on our experience, direct compression continuous processes my present issues, the most important failure modes can be captured by several questions, such as Can we feed each ingredient at the required flow rate? Do material properties change during process? Do the ingredients stick and agglomerate? Can we achieve blend homogeneity? In this study, we develop characterization method focusing on the questions of material properties, ingredient sticking on equipment and agglomerate.

This work consists of two sessions. In the first session, we investigate the properties of post-feeder material. We have recently observed some material sticks on the feeder surface and some material properties might have change during feeding process. Our process cases are high and low feeding speed. More than twelve materials were studied and characterized before and post feeding on compressibility, shear cell flowability, density and electrostatic charge density. Our experimental results supported that materials with both very small particle size and high electrostatic charge potential represent increasing of compressibility and electrostatic net charge density and decreasing on density. This phenomenon is driven by the electrostatic charge accumulating when material is discharged by feeder. In the second session, we study the ingredient agglomerate potential in continuous process. A method was developed that can quantify the post-feeder agglomerate size based on the image analysis. When analyzing agglomeration of more than 25 material, we would map the material agglomerate potential based on the particle size and compressibility characterization. Moreover, a case study of semi-fine APAP was performed to determine the agglomerate characterization on the blends and tablets, and the agglomerate elimination/mitigation will also be discussed.

533346 Comparing a Semi-Continuous Tablet Coating Process at Different Scales Using CFD-DEM

Abstract:

Tablet coating is a widely used unit operation in the pharmaceutical industry. Traditionally, large-scale drum coater systems are loaded with hundreds of kilos of tablets and rotate at relatively low rotation rates (Fr<<1). This leads to several hours of batch processing in order to obtain sufficient coating uniformity. The ConsiGma^{TM} coater from GEA is a tablet coater which can be integrated into the GEA ConsiGma^{TM} continuous process line. The ConsiGma^{TM} coater operates with low drum loads (5-7 kg) but very high rotation rates (Fr~1) leading to much shorter cycle times in the range of a few minutes. To achieve better coating results the tablets are jetted away from the rotating drum wall by so-called air knives in order to form a stable, gravity-free cascade. This air flow needs to be considered in a simulation model to accurately predict the coating process in the ConsiGma^{TM} coater. Depending on the desired throughput rates there are three different types of the GEA ConsiGma^{TM} coaters, differing in the depth of the coating drum from 80 to 320 mm.

We developed a computational CFD-DEM model that allows the simulation of the coating process inside the GEA Consigma^{TM} coaters including tablet fluid interaction. We will show a model to simulate accurately biconvex tablets including a drag model that takes into account the non-sphericity of the tablets. At the end we will compare two different drum scales of the GEA Consigma^{TM} coater (160 mm width and 320 mm width) and compare the scalability of the drum design. The comparison will include the mixing inside the Consigma^{TM} coater, as well as the coating performance and the heat transfer during the coating process. Guidance on how to scale between the two tablet coaters will be developed with the help of the simulation results.

528027 Continuous Fluidized Bed Drying of Pharmaceutical Granulations: Prediction of the Moisture Content

Abstract:

Continuous Fluidized Bed Drying of Pharmaceutical Granulations: Prediction of the Moisture Content

Hao Chen [a], Subham Rustagi [a], Emily Diep [a], Timothy A. G. Langrish[b], Benjamin J. Glasser [a]

[a]Department of Chemical and Biochemical Engineering, Rutgers University, Piscataway, NJ, 08854

[b]School of Chemical and Biomolecular Engineering, The University of Sydney, NSW 2006, Australia

In recent years, continuous manufacturing has been adopted in a number of pharmaceutical manufacturing processes. Traditional batch manufacturing requires long processing times and consumes a tremendous amount of materials during process development. There is the potential for continuous processing to reduce development times and processing times. Although there are no regulatory hurdles for implementing continuous manufacturing, the lack of understanding of continuous pharmaceutical unit operations has limited the integration of unit operations into continuous lines. For drying of pharmaceutical granulations, batch fluidized bed drying has been the traditional way of removing the moisture due to good mixing and gentle drying behavior. In continuous drying, there are many unanswered questions in terms of the effect of process conditions on the effluent moisture content.

In this study, experiments were conducted in a modified Glatt GPCG-2 continuous fluidized bed dryer. A twin-screw feeder was used to feed wet powder into the continuous dryer. A model for predicting the effluent moisture content was developed by combining the drying kinetics and residence time distribution (RTD) of powder in the dryer. The drying kinetics were obtained from a batch fluidized bed drying process. The RTD was characterized by carrying out a tracer response test in the continuous fluidized bed. A tank-in-series model was used to describe the RTD curves. A maximum mixedness model (MMM) was adopted for characterizing the micromixing in the continuous fluidized bed. All of the resulting information from the batch experiments (drying kinetics) and information from the flow of powder through the continuous dryer (RTD) was combined into a model to predict the effluent moisture content. The results of the model were compared to experimental results and we discuss how the operating conditions affect the resulting effluent moisture content.

520565 SEMI-Continuous Manufacturing Process for Generic Drug Products

Abstract:

Claims are made regarding the semi-continuous manufacture for a drug that has extremely limited demand and can only be economically produced using batch process. Generics has an opportunity to capitalize on economies of scale to create better and more robust continuous or batch processes. Since generic companies know the molecule while the drug is under patent, they have ample time to create a process that is efficient, superior and can be commercialized after patent expiration. Continuous manufacturing does not fit every hand. It has to be custom fitted. It is not a fashion that is economical and suits everyoneâ€™s needs. Principles of chemistry and chemical engineering can be applied for the manufacture of chemicals and their blends. A distinctive operations management and regulatory strategy is needed to stay competitive amidst the radical changes occurring in the small molecule drug product manufacturing industry.

Organizations must retain a large product portfolio to remain visible. They will also need to deliver products from their portfolio with a high supply chain velocity to meet customer expectations. This trend has become more prominent with the consolidation of wholesalers and pharmacy benefit managers in the U.S. market. Consequently, supply chain flexibility is a key element to success in the generic business. Manufacturing operations are expected to increase throughput while simultaneously minimizing operational disruptions resulting from deviations or failures. Scaling products with variable batch sizes is an option; however, this incurs substantial R&D costs and increased CAPEX spending due to additional infrastructure and time requirements. Capacity optimization across global operations is typically conducted by larger establishments. The best solution would be manufacturing operations that provide ample supply chain flexibility.

This paper/presentation discusses the fit for purpose semi-continuous manufacturing operations for generics. The proposal of converting existing batch processes to semi-continuous processes for solid dose manufacturing sites also involves application of integrated PAT, statistical process control (SPC), and predictive modeling software technologies. Implementing fully continuous manufacturing lines would result in engaging multiple fixed assets until the end of the manufacturing process, and hence may not be a desired option for manufacturers just yet. The semi-continuous strategy will improve production planning flexibility, enabling just-in-time manufacturing and injection of orders based on market demands, while avoiding backorder scenarios. Manufacturing of drug products only as required would allow for minimal overstock and prevent short expiry situations that would have otherwise resulted in further erosion of margins. The semi-continuous model enables manufacturers to maintain larger product portfolios with flexible supply chain prospects. The batch size flexibility, minimal footprint, and prospective decrease in development costs provides a distinct advantage. The improved ability for capacity optimization and enhanced production controls will ensure supply chain reliability, resulting in customer satisfaction.

537421 Application of Mechanistic Models for the Digital Design and Online Control of Pharmaceutical Processes

Abstract:

Mechanistic models are becoming more commonly applied for Research and Development in the pharmaceutical sector. Traditionally, the output from this activity, namely a validated mechanistic model, which is capable of quantitatively predicting the behaviour of the various Critical Quality Attributes (CQA) for typical batch or continuous pharmaceutical processes for a wide range of Critical Process Parameters (CPP). However, these tools are almost exclusively employed in an offline manner currently, primary aimed at assessing process robustness and variability, with very little subsequent online application of the model for control or soft sensing purposes.

Traditionally, online control companies utilize statistical models that require a significant level of tuning and verification, with the online full-scale plant, using Pseudo-Random Binary Sequences (PRBS) to vary the process parameters and observed the process responses. This requirement for the application of rigorous Model Predictive Control (MPC) using statistical model approaches commonly leads to reduced appetite for uptake of the technology in the Pharmaceutical sector. These online trials for tuning the MPC may result in off-specification productions, potentially leading to significant losses in productivity, as the controllers pushes the process to observe responses at extreme points. However, most of these drawbacks can be overcome by the integration of mechanistic models, developed using laboratory scale data with MPC system, such as that offered by Perceptive Engineering, namely PharmaMV.

In this work we outline, the application of an advanced process modelling tool, namely gPROMS FormulatedProducts, to describe a number of pharmaceutical processes. The process model and the mechanistic model kinetic parameters were validated using process data gathered from the literature and from lab-based experiments. The lab-based process was subsequently used to predict the behaviour of the full scale production scale. In order to make the model predictive, some refinement of the kinetic parameters for secondary nucleation was required using minimal experimental data from the typical plant runs.

The mechanistic model was integrated with PharmaMV to develop and tune the MPC against the mechanistic simulation of the process. The PharmaMV platform was subsequently transferred to the physical process. With this approach the MPC derived from the mechanistic model was utilized to accurately control the defined CQAs, such as final particle attributes, potency or moisture.

520705 Impact of Physico-Mechanical Properties of Co-Processed Excipients on the Tableting Performance By DM3 Approach

Abstract:

The aim of present investigation was to evaluate tableting performance of co-processed excipients (CPEs) MicroceLac® 100, StarLac® and CombiLac® at different humidity conditions using our previously published DM3 approach. The DM3 approach involves the design of experiments (D) and multivariate analysis techniques (M3) to determine the effect of material properties (M1), and manufacturing factors (M2) on product attributes like compressibility, compactibility, and tabletability. The powders were exposed at different relative humidities of 33.0 % ± 1.4 % RH, 55.0 % ± 4.8 % RH, 75.0 % ± 4.3 % RH. These exposed powders were compressed at 5mm/minute compression speed and at 100 MPa compression pressure using bench-top compaction simulator. The evaluated material properties were sorption-desorption behavior, equilibrium moisture content (%), enthalpy of dehydration and crystallization, and heat of fusion. The compression properties were modeled with 'in-die' Heckel descriptors. The material and manufacturing factors together were used to study impact on the compressibility, compactibility and tabletability of these materials. Multivariate Analysis of Variance (MANOVA), Principal Component Analysis (PCA), and Principal Component Regression (PCR) were the statistical tools used to analyze the data. Increased moisture content gave a significant reduction in TMS. On the other hand, the reduced tablet porosity indicated increased compressibility. The Heckel analysis showed a reduction in YPpl (MPa) and YPel (MPa) values indicating increased plastic deformation and elastic recovery. Reduction in WoC and WoE was also observed on increasing humidity levels. It seems that the spray dried co-processed excipients sorbed moisture on exposing at increasing humidities. The lubricating and plasticizing effect of moisture caused an increase in compressibility. However, the moisture caused weakening of interparticular bonds and decreased particle-particle interaction thereby decreasing the TMS. Thus, present investigation can be utilized for a careful selection of CPEs in the formulation.

529371 Pharmacy on Demand: On-Demand Drug Product Manufacturing in a Miniaturized and Portable System

Abstract:

Typically, in the pharmaceutical industry, drug substances (active pharmaceutical ingredients, APIs) are synthesized in one facility and formulated to finished dosage forms so-called drug products in another facility, which could be located anywhere else worldwide. Due to the complex nature of the pharmaceutical supply chain, the industry faces several major challenges when it comes to ensuring an adequate supply of quality drug products. These challenges are not only the causes of supply chain disruptions and financial loss, but can also prevent underserved and remote areas from receiving life-saving drugs. Another concern for orphan drugs that pharmaceutical company generates small sales in comparison to the development and manufacturing cost in a large footprint facility. To address all the above issues, at MIT we have developed a continuous manufacturing platform that combines both drug substance and final drug product development in a custom-built miniature system[1-3]. In this work specifically, drug product as oral tablet manufacturing from drug crystals in the miniaturized, portable, re-configurable and automated system will be presented[1,2].

The compact, portable system, roughly the size of a household oven, [72.4 cm (length) ×53.3 cm (width) ×134.6 cm (height)] is built to manufacture on-demand pharmaceutical tablets on a scale of hundreds to thousands per day. Its design combines off-the-shelf devices with custom-designed automated mechanisms. A direct compression approach is considered for making tablets. The whole unit is divided into upper process streams and lower process streams. The upper process streams start with powder feeding and end with blending. The lower process streams include powder dispensing thru to making tablets. These two process streams operate independently under high-level software control, allowing for smooth integration between the two. As this is for on-demand pharmaceutical drug manufacturing, a simplified formulation approach is adopted by eliminating excipients required for stability. Hence, only filler, glidant, lubricant, and additional functional excipients, if necessary are considered for formulation.

Several APIs having different dosage strength were investigated. Here ibuprofen (200/340 mg) and diazepam (10/250 mg) were shown as the example. Assay, content uniformity, hardness and in-vitro testing of tablets were conducted to confirm it meets U.S. Pharmacopeia standards. Formulated blended powder is freely flowing (flow function coefficient (ffc)>10) and has a good bulk density (0.65-0.72 g/cm^3) that are suitable for making tablets via direct compression. Assay of ibuprofen and diazepam tablets are found 98.8% and 103.7 % of the labeled amount, respectively. The acceptance value in content uniformity/weight variation test is 6.12 and 9.36, respectively for ibuprofen and diazepam tablets. The tensile strength of ibuprofen and diazepam tablets are 1.14 and 0.52 MPa, respectively. Tablets dissolution profiles for both of drugs meet USP standard criteria (85% dissolved in 30 minutes).

This custom-built miniature system has the potential to (1) help tackle regional drug shortages (a threatening global problem), (2) be advantageous for drugs with a short shelf life, (3) be put into immediate production of drugs based on demand, bypassing the need to stockpile drugs (e.g., for humanitarian needs), and (4) reduce formulation complexity relative to products needing yearlong stability. We envision that this system will enable the creation of small-scale, portable, fully integrated and closed loop pharmaceutical solids dosage manufacturing that can advance the treatment of human diseases at the point-of-care.

References:

1. Azad M, Osorio Caicedo J, Brancazio D, Hammersmith G, Klee D, Rapp K, Myerson A 2018. A compact, portable, re-configurable, and automated system for on-demand pharmaceutical tablet manufacturing. Int. J. Pharm. 539 (1-2):157-164.
2. Myerson A, Azad M, Hammersmith G, Brancazio D, Osorio Caicedo J. Systems and methods for the fabrication of Ttablets, including pharmaceutical tablets. U.S. Provisional Patent Application 62/621,429, Filed on January 24, 2018.
3. Adamo A, Beingessner RL, Behnam M, Chen J, Jamison TF, Jensen KF, Monbaliu J-CM, Myerson AS, Revalor EM, Snead DR 2016. On-demand continuous-flow production of pharmaceuticals in a compact, reconfigurable system. Science 352(6281):61-67.

538133 Applying Dynamic Similarity Principles to a Narrow Therapeutic Index Drug Powder Blending Process Scale-up: Laboratory Case Studies and Relevant Regulatory Experience

Abstract:

Powder blending process scale-up is one of the critical aspects to be evaluated during scientific and regulatory assessment of oral solid dosage form manufacturing. It directly impacts content uniformity (CU), especially for potent drug applications such as narrow therapeutic index (NTI) drug applications and low dose drug applications. Theoretically, geometric, kinetic, and dynamic similarity should be collectively considered during manufacturing process scale up. However, quantitative process scale-up relationships are rarely used to support powder blending processes in new and generic drug applications. In the past few years, certain powder blend process scale-up deficiencies have been identified during the scientific and regulatory reviews of the drug applications. In this work, we report our feasibility evaluation of applying dynamic similarity principles to blending process scale-up for an NTI drug powder. Data were obtained using a laboratory-scale high-shear mixer with PAT real time process monitoring. Two types of dynamic similarity rules, i.e., maintaining either Froude number (Fr) or impeller tip speed, were applied for guiding the model formulation for blending process scale. This presentation will also cover relevant regulatory review experience on powder blending process scale up. This work provides case studies which demonstrate the utility of process engineering scale-up principles in drug development and manufacturing for rational process design space selection and better product quality.

525692 Implementation of Advanced Process Control System into Continuous Pharmaceutical Manufacturing Pilot-Plant

Abstract:

Advanced process control (APC) such as model predictive control (MPC) has several advantages in compare to conventional control algorithms [1-3]. Therefore, the pharmaceutical industries are moving toward continuous manufacturing (CM) integrated with advanced process control (APC) system. This modernized integrated system open the possibilities of achieving Quality by Control (QbC) and real time release (RTR) paradigm. However, there are different levels of challenges associated with design and implementation of an efficient control system into pharmaceutical manufacturing plant. Therefore, this is still an open area of research.

The different control strategies for continuous tablet manufacturing process has been developed and implemented into our pilot-plant facility using industry standard control platform. The critical control variables are drug concertation, powder level, main and pre compression forces, tablet weight and hardness. The different control architectures under three control algorithms (PID, MPC, hybrid MPC-PID) have been implemented and compared for control of CPP's and CQA's. The flexible control system is designed in such a way that any control scheme can be easily selected. The control system is also modular in nature that can be easily transferred to any other control platform and manufacturing plant. The assurance of CQA's have been achieved through two levels strategy. In first level, the CPP's and CQA's are controlled in real time using advanced process control system while in second level, the none-confirming products are diverted in real time in waste to assure the final CQA's of qualified tablet lots. This two level strategy assures the maximum production efficiency while assuring the final product quality. The drug concentration of blend is controlled before its compaction using PAT and MPC. An efficient control strategy for powder level control in a chute placed in between blender and tablet press unit operation of continuous tablet manufacturing process has been developed, implemented and evaluated [1]. A novel noninvasive technique based on change in electric field concept has been used for real time monitoring of powder level in continuous manufacturing pilot-plant. The sensor has been integrated with a control panel (DeltaV (Emerson)) through relays and charms. The control panel then send the signal (4-20 mA) to control platform where the control strategy, a program to operate the sensor and a program to convert the sensor signal into powder level have been implemented. Subsequently, an advanced model predictive controller as well as PID controller for powder level have been implemented into our continuous pharmaceutical manufacturing pilot-plant facility. The performance of both control strategies have been practically evaluated for set point tracking and disturbance rejection. The APC has been used to control tablet press as well [2]. The pre and main compression forces have been controlled using a novel decoupled strategy so that the desired tablet weight and hardness can be achieved simultaneously. Similarly, the other CPP's and CQA's have been controlled. It has been observed that the MPC performs better than PID. A systematic methodology for material traceability has been also developed and coupled with integrated control system [3]. The methodology starts with material and process specifications followed by scanning the powder drums and transmitting the material and time information to the control platform. The control platform then utilizes the information stored in the database, generated through residence time distribution model and information collected through real time monitoring to track the material during continuous manufacturing. All the relevant data generated during continuous manufacturing has been systematically collected, stored and organized in a data hub (OSI PI) and cloud system.

The objective of this presentation is to demonstrate the performance of integrated advanced process control (APC) system implemented into continuous pharmaceutical tablet manufacturing pilot-plant.

References

1. Singh, R. (2017). A novel continuous pharmaceutical manufacturing pilot-plant: Advanced model predictive control. Pharma, Issue 28, PP 58-62.
2. Bhaskar, A., Barros, F. N., Singh, R. (2017). Development and implementation of an advanced model predictive control system into continuous pharmaceutical tablet compaction process. International Journal of Pharmaceutics, 534 (1-2), 159-178.
3. Billups, M., Singh, R. (2018). Material Traceability in Continuous Pharmaceutical Tablet Manufacturing. Pharmaceutical Technology 42 (2), 32-35, 59.

536425 The Role of Software Tools in Quality By Design: A Case Study on Monoclonal Antibody Production

Abstract:

The role of software tools in Quality by Design

A case study on monoclonal antibody production

Maria M. Papathanasiou[a], Nilay Shah[a], Efstratios N. Pistikopoulos[b]

[a]Dept. of Chemical Engineering, Centre for Process Systems Engineering (CPSE), Imperial College London SW7 2AZ, London, U.K

[b]Artie McFerrin Department of Chemical Engineering, Texas A&M University, College Station TX, 77843

Over the last few years, the U.S Food and Drug Administration (FDA) has highlighted issues with drug shortages arising mostly from deficiencies in the pharmaceutical quality and manufacturing. Moreover, regulatory bodies raise concerns that the process currently in use are not state-of-the-art, while little emphasis is given to the manufacturing process which can constitute 25% of the total product cost. Along those lines, FDA urges the pharmaceutical industry to move towards novel process setups that promise: (i) risk-free scale up, (ii) seamless monitoring and (iii) high product quality [1]. In the case of monoclonal antibodies, the aforementioned challenges come along with competition arising due to patent expiration. Based on recent reports, the global pharmaceutical industry is threatened with $194 billion loss on sales, as biosimilars are expected to take 4-10% of the market by 2020 [2]. Therefore, the necessity for the use of novel, eco-efficient processes with higher product titers becomes eminent.

In this work we present an in-silico software platform to assist computational experimentation in monoclonal antibody manufacturing. The platform comprises two main parts, namely the upstream and the downstream processing. The latter is looking into both the initial capture step, as well as the chromatographic separation unit used for the purification of the antibody. The platform is designed and executed following the systematic PAROC framework and software platform that comprises four main steps: (i) model development and validation, (ii) model reduction/approximation, (iii) controller development and (iv) controller validation under in an in-silico, 'closed-loop' fashion [3]. For the controller design we employ multi-parametric Model Predictive Control (mp-MPC) strategies that allow control laws to be easily tested and implemented on embedded device. The optimal control laws are retrievable immediately through simple function evaluations, enabling fast response without the computational requirements of a computer in the background [4].

For the upstream system, we consider a GS-NS0 cell culture system operated in fed-batch mode [5], while the downstream configuration is focusing on the simulation of the capture step as presented in [6] and the Multicolumn Countercurrent Solvent Gradient Purification (MCSGP) process [7]. Following their development, the models are used in order to evaluate the performance of the studied units under continuous operation and their performance is assessed [8]. The latter falls under the umbrella of process intensification that promises shorter production times, decreased costs, tighter product quality and significantly lower energy consumption. The presented, in-silico results indicate that continuous operation could lead to prolonged culture times and cell viability in the upstream. In the case of the downstream, the MCSGP unit is simulated under the operation of the mp-MPC controller that returns periodic input profiles and successfully handles disturbances resulting from variations in the composition of the upstream mixture. Lastly, this work focuses on the key challenges faced during the design of the aforementioned tools and how they can be addressed. Moreover, the in-silico integration of the aforementioned processes is discussed, along with suggestions to overcome issues with unavailable measurements during their experimental application.

References

[1] U.S. Food and Drug Administration, "Guidance for Industry Quality Systems Approach to Pharmaceutical CGMP Regulations," 2006.

[2] EvaluatePharma, "World Preview 2016, Outlook to 2022," 2016.

[3] E. N. Pistikopoulos, N. A. Diangelakis, R. Oberdieck, M. M. Papathanasiou, I. Nascu, and M. Sun, "PAROC - An integrated framework and software platform for the optimisation and advanced model-based control of process systems," *Chem. Eng. Sci.*, vol. 136, 2015.

[4] P. Dua, K. Kouramas, V. Dua, and E. N. Pistikopoulos, "MPC on a chip—Recent advances on the application of multi-parametric model-based control," *Comput. Chem. Eng.*, vol. 32, no. 4–5, pp. 754–765, Apr. 2008.

[5] M. M. Papathanasiou, A. L. Quiroga-Campano, F. Steinebach, M. Elviro, A. Mantalaris, and E. N. Pistikopoulos, "Advanced model-based control strategies for the intensification of upstream and downstream processing in mAb production," *Biotechnol. Prog.*, vol. 33, no. 4, 2017.

[6] F. Steinebach, M. Angarita, D. J. Karst, T. Müller-Späth, and M. Morbidelli, "Model based adaptive control of a continuous capture process for monoclonal antibodies production," *J. Chromatogr. A*, vol. 1444, pp. 50–56, Apr. 2016.

[7] T. Müller-Späth *et al.*, "Two step capture and purification of IgG2 using multicolumn countercurrent solvent gradient purification (MCSGP)," *Biotechnol. Bioeng.*, vol. 107, no. 6, pp. 974–984, Dec. 2010.

[8] K. B. Konstantinov and C. L. Cooney, "White paper on continuous bioprocessing May 20-21, 2014 continuous manufacturing symposium," *J. Pharm. Sci.*, vol. 104, no. 3, pp. 813–820, 2015.

520125 Continuous Mixing Technology: Process Design with Discrete Element Method (DEM) Simulations

Abstract:

Continuous Mixing Technology: Process Design with Discrete Element Method (DEM) Simulations

Peter Toson[1], Eva Siegmann[1], Martina Trogrlic[1], Dalibor Jajcevic[1], Johannes Khinast[1]

&

Pankaj Doshi[2], Daniel Blackwood[2], Mary T. am Ende[2]

[1] Research Center Pharmaceutical Engineering, Inffeldgasse 13, 8010 Graz, Austria

[2] Worldwide Research and Development, Pfizer Inc. Groton CT, USA

Keywords: Pharmaceutical Manufacturing; Mixing; Process Design & Development

The pharmaceutical industry is currently transforming from batch to continuous production processes. Advantages of continuous manufacturing include better control of process parameters and product quality as well as lower down-times compared to batch processing. One of the unit operations in a continuous process is powder mixing.

An overview of the Continuous Mixing Technology (CMT) design is given shown in Figure 1a. Feeders provide a constant inlet material flow to the CMT device. The top region is a de-lumping zone that breaks up agglomerates that may have formed upstream. The mixing happens in the bottom conical region of the CMT – the mixing zone. The mass throughput is actively controlled by upstream feeders and hold-up mass in the CMT is controlled by changing the exit valve opening. Therefore, the mean residence time of particles in the mixer, which is defined by ratio of hold up mass to mass throughput, is also constant. The main advantage of this vertical CMT design is that the shear rate can be controlled independently by changing the impeller speed without changing the residence time. This presents more opportunities for process design.

In this work, the CMT device has been modeled with discrete element method (DEM) simulations. The cohesive contact model of the DEM particles has been calibrated with compression, spring back, and shear cell tests. The calibrated contact model is then used in simulations of the CMT device.

The DEM simulations allow the analysis of particle trajectories, velocity profiles, powder bed shape, mixing behavior, as well as residence time distribution and travel distance distribution (Figure 1b). To understand the influence of process parameters, such as impeller speed, hold-up mass, and mean residence time on these quantities, a full factorial virtual design of experiments (DoE) with 4 impeller speeds and 5 hold-up masses has been performed. The mass throughput has been kept constant at 10kg/h. The highest examined hold-up mass (500g) has been modeled with 4.1 million DEM particles.

DEM simulation results show that the powder bed shape changes significantly with impeller speed. At impeller speeds of up to 450rpm the mixing zone is filled with powder and narrow valve openings are required to keep the hold-up mass constant. However, high impeller speeds >500rpm exert high centrifugal forces to the particles, lifting the powder bed at the bottom away from the valve. High valve openings are required in this case to ensure a constant material outflow.

One of the most important parameter which defines the extent of mixing is the residence time distribution (RTD) of particles in CMT. It has been evaluated directly from DEM particle data and mimics a tracer impulse experiment (Figure 1c). A set of particles created over the course of one second of simulation are defined as tracer particles (tracer impulse input). For each particle, the time between entering and exiting the CMT is stored. A histogram of these residence times yields then the RTD.

Based on the calculation of residence time distributions for all operating conditions, process maps have been developed which display regions with near-perfect exponential distribution as described by the ideal continuous stirred tank reactor (CSTR) model, and regions with less-than-ideal mixing.

The data – particle trajectories, powder bed shape, and residence time distribution – are available at each of the 20 operating points defined by the virtual DoE, making DEM simulations a valuable tool to design the continuous mixing process.

Figure 1. (a) CMT overview. (b) A cut through the CMT to reveal the powder bed shape in the mixing zone at 350 rpm. (c) Tracer particles (red) used to determine the residence time distribution. Snapshot taken 2s after insertion of the tracer impulse.

527712 Continuous Mixing Technology: Residence Time Distribution Modeling

Abstract:

Continuous Mixing Technology: Residence Time Distribution Modeling

Peter Toson[1], Eva Siegmann[1], Martina Trogrlic[1], Dalibor Jajcevic[1], Johannes Khinast[1]

&

Pankaj Doshi[2], Daniel Blackwood[2], Alexandre Bonnassieux[2], Mary T. am Ende[2]

[1] Research Center Pharmaceutical Engineering, Inffeldgasse 13, 8010 Graz, Austria

[2] Worldwide Research and Development, Pfizer Inc. Groton CT, USA

Keywords: Pharmaceutical Manufacturing; Mixing; Process Design & Development

The pharmaceutical industry is in the midst of a transformation from batch to continuous production processes. There are several advantages of continuous manufacturing which include better control of process parameters and product quality as well as lower down-times compared to batch processing. One of the most important unit operations in a continuous process is powder mixing.

An overview of the Continuous Mixing Technology (CMT) design is given in Figure 1 (a). Multiple feeders provide a constant inlet material flow to the CMT device. The top region is a de-lumping zone that breaks up agglomerates that may have formed upstream. The mixing happens in the bottom conical region of the CMT – the mixing zone. The mass throughput is actively controlled by upstream feeders and hold-up mass in the CMT is controlled by changing the exit valve opening. Therefore, the mean residence time of particles in the mixer, which is defined by ratio of hold up mass to mass throughput, is also constant. However, discrete element method (DEM) simulations and tracer experiments have revealed that even if the mean residence time (MRT) is actively controlled, the shape of the residence time distribution (RTD) changes with different operating conditions, e.g. impeller speeds and hold-up mass.

The RTDs are described by a generalized model of *n* continuous stirred tank reactors in series (*n*-CSTR). Whereas the standard tanks-in-series (TIS) model only allows integer values of *n*, the residence time distribution formula can be generalized to any positive, fractional value of *n*. DEM simulations and tracer experiments showed that a typical value of *n* for the CMT device is between 0.5 and 1.1. . A value of $n=1$ corresponds to the ideal CSTR model with an

exponential residence time distribution. Values $n>1$ describe over-mixing (just as the TIS model, but with finer granularity), values $n<1$ describe under-mixing. A value of 0.5 is an indication of short-circuiting in the mixing process indicated by an early peak in the RTD. Thus, the value of *n* quantitatively describes the mixing quality. An example fit with n=0.95 is shown in Figure 1 (b).

The analytical RTD described by the generalized n-CSTR model is then used to describe the response at the CMT outlet to upstream fluctuations in the feeder. Three kinds of fluctuations in the input mass flow are considered:

1. general low-amplitude noise

2. short, high peaks in the mass flow, e.g. a cohesive agglomerate falls off the feeder screw (feeder bearding)

3. long drifts in the feeder signal, e.g. if the feeder controller switches to volumetric flow with an incorrect feed-factor and slowly corrects the mass flow. This case is demonstrated in Figure 1 (c).

The mass flow at the CMT outlet is calculated as a convolution integral of the input mass flow and the residence time distribution of the CMT. From the outlet mass flows of the individual components, it is possible to calculate whether or not the resulting blend composition is still within content uniformity specifications. Because different operating points result in different RTDs, it is possible to link the results back to the operating conditions. Thus, RTD modeling helps to define process operating space to ensure blend composition meet the specification.

Figure 1. (a) CMT overview. (b) Comparison of the RTD obtained from discrete element method (DEM) simulations (histogram), a fit with n=0.95 (black dotted line), and a tracer experiment (green solid line). (c) Response of the CMT outlet to a feeder drift under different process conditions. Although the feeder starts at 150% of the target and takes 90s to correct the mass flow, the CMT is able to damp this peak to 112%-115% of the target value, depending on the process conditions.

537097 Residence Time Distribution Determination in Pneumatic Conveying Processes for the Continuous Manufacturing of Pharmaceutical Products

Abstract:

Conveying powder form one unit operation to the next is a critical, and frequently problematic, aspect of all oral solid dose (OSD) manufacturing paradigms. Pneumatic conveying of materials offers several advantages over gravitational methods. For one, the height of the manufacturing suite can be limited. When gravitation conveying is utilized, the manufacturing space must be tall enough to allow unit operations to be stacked on top of each other. This can require expensive modification to existing facilities. In addition, pneumatic conveyors can transport poorly flowing powders that would pose considerable challenges in gravitational systems. When used in conjunction with continuous manufacturing system the pneumatic conveying of powder form one unit operation to the next can represent a large impact to the residence time distribution (RTD) and holdup of the system. Understanding the effects of this is important for the design a continuous systemâ€™s equipment as well as its operation. This talk will discuss the determination of RTD and holdup in pneumatic conveying systems.

525683 RTD Based Control System for Continuous Pharmaceutical Manufacturing Process

Abstract:

Currently, the pharmaceutical companies are going through a paradigm shift from conventional batch to continuous manufacturing (CM) process integrated with advanced automation and control system [1-2]. In CM, Real Time Release Testing (RTRT) can be facilitated by the development of a highly efficient control system that can monitor and correct process variables in real time when a fault is detected. But, there may be some out of specification tablets that need to be diverted in real time, in order to ensure the quality of final products. Specifically, the drug concertation of each tablet must be guaranteed before it can be released to the market. However, currently no methods and tools are available that can assure the drug concentration and divert the non-confirming products in real time. Therefore, the systematic methods and tools are needed for real time assurance of tablet drug concentration and diversion of non-confirming products.

In this work, a systematic framework including the methods and tools have been developed for design and implementation of control system to assure the drug concertation of final pharmaceutical tablet. The framework is based on the residence time distribution (RTD) model and a novel tablet diversion system. The work has been performed in two phases. In first phase, a mathematical model based Insilco study has been performed to identify the best strategy that can be employed to divert the non-confirming tablets in real time. In CM, the drug concentration is measured in real time before the tablet compaction (chute &/or feed frame) using PAT sensor. The proposed control strategy then uses this inlet concentration to determine a signal for the diversion strategy that can accurately be used to reject tablets that are out of tolerance limits at the outlet of the tablet press. Two strategies, i.e. â€˜fixed window based strategyâ€™ and RTD based strategy have been developed, compared and evaluated. In fixed window approach, the tablet diversion is facilitated through knowledge of time delays from the point of detection to the point of the affect (tablet press outlet gate) in the system. The sensor that detects the concentration is connected to a comparator block which decides if the said concentration is within the specifications. If it is not within specification, the experimentally derived time delay is applied and post this the diversion begins. The diversion stops when a concentration within spec is detected and the another time delay is applied. In RTD based approach, the RTD is used to predict the outlet concentration from the inlet concentration. The predicted signal is then used to initiate the diversion. The first approach is simpler to implement but may lead to lower production efficiency. The second approach is based on more advanced technique and will ensure more efficiency but is relatively complex to implement. In second phase of this work, the proposed strategies have been implemented into our pilot-plant facility for experimental demonstration.

The objective of this presentation is two-fold; first to highlight the Insilco studies for the development and evaluation of strategies for drug concertation assurance and diversion of non-confirming tablets and then demonstrate the performance of best strategies into our continuous pharmaceutical manufacturing pilot-plant facility

Reference

1. Singh, R. (2018). System engineering for a novel continuous pharmaceutical manufacturing process. Pharma. Issue 30. https://www.pharmafocusasia.com/manufacturing/system-engineering-pharmaceutical-manufacturing-process.
2. Barros, F. N., Bhaskar, A., Singh, R. (2017). A validated model for design and evaluation of control architectures for continuous tablet compaction process. Processes Journal, 5(4), 76. doi:10.3390/pr5040076

529911 Developing a System Based Model for Continuous Direct Compression â€“ Going Beyond Tablet Assay

Abstract:

The emergence of system based modelling for pharmaceutical manufacturing processes is providing an opportunity to holistically model the influence material and process parameters on intermediate and end product quality attributes. These models can be used to increase process understanding, identify critical process parameters, and support risk assessment and control strategy development. With an increase in the adoption of continuous processing, system models can also be used to understand system dynamics and serve as a basis for developing an integrated on-line control scheme.

For continuous direct compression (CDC) processes, where constituent formulation components are continuous fed, blended and compressed, system modelling efforts have focused on characterising the residence time distribution in separate unit operations and using this to model the influence of systematic changes and disturbances on intermediate blend uniformity and tablet assay. This type of model is very useful for simulating a relatively stable process and for developing and tuning a control scheme. In this paper we explore the development of additional models to complement this approach, extending prediction to additional quality attributes, such as tablet tensile strength. This type of system model can then be used holistically for process design and to explore sensitivity to material and process changes.

529932 *Twin-Screw Continuous Granulations: Technological and PAT Developments*

Abstract:

INTRODUCTION

Continuous pharmaceutical manufacturing, both at upstream (active pharmaceutical ingredient, API production) and downstream (formulation technology) processing, is rapidly gaining interest as a promising way to improve efficiency, reduce operating costs and ensure easier scale-up and shorter time-to-market. Twin-screw processes such as continuous granulations are especially in the focus given that the majority of the product manufacturing lines contains a granulation step. Nevertheless, conversion from batch to continuous manufacturing entails the necessity of continuous process monitoring and an integrated quality control strategy. Real-time quality assurance of continuously manufactured products must be developed in order to be able to implement continuous manufacturing processes, the future of pharmaceutical manufacturing.

In this work, continuous twin-screw wet and melt granulations are discussed. In case of wet granulation, the goal was to develop a proper granulation process during which a drug with low dose can be homogenized. Furthermore, different kinds of PAT tools have been created to support this process: NIR and Raman spectroscopies were applied to monitor the active ingredient content, while dynamic image analysis was employed to in-line monitor and control the particle size of the obtained granules. Continuous melt granulation is also a promising downstream process. In this case, our aim was two-fold: to apply twin-screw melt granulation to obtain a powder readily tabletable and for the flowability enhancement of amorphous solid dispersion containing powders. Dynamic image analysis could be used in this process as well for the in-line particle size determination.

Materials AND Methods

Carvedilol (CAR) was used as model API for the granulation experiments. As excipients, lactose (Granulac® 230, Meggle Pharma, Wasserburg, Germany) and potato starch (Roquette Pharma, Lestrem, France) was applied, while the granulating agent was polyvinylpyrrolidone K30 (Kollidon® 30, BASF, Ludwigshafen, Germany) in 96% ethanol (Merck Ltd., Budapest, Hungary). For melt granulation experiments, crospovidone (Kollidon® CL, BASF) and polyethylene glycol with different moleculare weights (2000, 6000, and 20000, Sigma Aldrich, Budapest, Hungary) were used.

For the discussed processes, a TS16 extruder equipped with a special granulation die was applied (ProCepT, Zelzate, Belgium). Conveying and kneading zones are varied in this machine. The length-to-diameter ratio is 25.

A Kaiser RamanRxn2® Hybrid *in situ* analyzer (Kaiser Optical Systems, Ann Arbor, USA) coupled with PhAT (Pharmaceutical Area Testing) probe was utilized in reflection mode to acquire the spectra of the homogenized powder and tablets. The focal length was 25 cm while spot size was 6 mm. A diode laser source of wavelength of 785 nm and 400 mW was applied. Spectra were collected in the range of 200-1890 cm^{-1} with a resolution of 4 cm^{-1}.

NIR spectra have been collected by a Bruker Optics MPAâ„¢ FT-NIR (Bruker OPTIK GmbH, Ettlingen, Germany) spectrophotometer. The instrument was run in reflection with the aid of a REFLECTOFLEX SS316L probe. The spectral range was between 4000 and 12500 cm^{-1}. The resolution was 32 cm-1 while the spectrophotometer was averaging 32 scans. The reflected beams arrived to a Te-InGaAs detector.

A process camera (Spark SP-5000M-GE2, JAI A/S, Tokyo, Japan) was applied for the in-line measurements. It was attached to the twin-screw equipment with a custom-made stainless steel mount along with a polished steel chute for the particles. The inclination of the metal chute allows the granules to flow freely, allowing random orientation of the particles to be captured.

RESULTS AND DISCUSSION

Technological development of continuous wet granulation

The starting materials possess weak flowability. Based on our results, granules with excellent flow properties could be obtained on the twin-screw granulator even with low, 3% of PVPK30. Tablets prepared from the granules had good hardness and disintegration time and appropriate friability. The model API (CAR) was pumped to the machine with the PVPK30 in ethanolic solution through a liquid addition port. Two different kinds of pumps were applied: a peristaltic pump and an infusion pump. Due to the pulsation of the peristaltic pump, CAR was not perfectly homogenized in the granules and in tablets based on our HPLC measurements (relative standard deviation was ~8%). This problem could be mitigated by the application of an infusion pump lowering the RSD of API content to ~2% (on a longer term, an HPLC pump could be used in a continuous manner).

PAT developments for monitoring the API content during continuous wet granulation

During this research, the challenge was that the model API was used in an extremely low dose (<0.05%), therefore spectroscopic determination of its content is impossible with limits of detection of todayâ€™s spectrophotometers. However, CAR was added to the system along with PVP which has a much higher concentration in the product (3-6%). Our idea was to measure the content of PVP and validate with HPLC measurements that the content of CAR can be estimated based on that. Both NIR and Raman spectroscopies were employed. In this case, the latter was found to be more precise. Naturally, a feedback control strategy could be developed on a long term based on these spectroscopic measurements.

Particle size control based on image analysis

A process camera was used for the in-line determination of particle size in this study. Firstly, the ability of it to determine the particle size was certified by comparing it to a microscope with certain particles. A software was developed in-house for the image analysis. With this system, the particle size enhancement with increasing liquid-to-solid ration could be monitored. With a P controller, it was possible to modify the liquid addition rate after disturbances in order to return to the set point of particle size. In the future, an even more advanced PI controller is planned to be developed.

Continuous melt granulation

Twin-screw melt granulation is a very promising process since it alloys effective bulk density and flowability enhancement (like wet granulation) with being solvent-free providing the possibility to omit drying (like roller compaction). Therefore, melt granulation fits perfectly into continuous manufacturing lines. Our goal was to prepare a directly compressible powder by melt granulation. This was successfully carried out since tablets could be compressed without the addition of any further excipient, the frictional forces did not increase (PEG acted as a lubricant). The process was monitored by thermal and process camera to observe the temperature and the particle size enhancement. This can be a base for controlling the process.

Melt granulation seems very suitable for the downstream processing of amorphous solid dispersions (ASDs). Spray drying and electrospinning as ASD manufacturing technologies usually result in a powder with low bulk density and weak flowability. Melt granulation can be the perfect way to enhance these properties via a solvent-free process. The temperature is the most critical process parameter as it must be elevated above the melting point of the granulating polymer but under the glass transition temperature of the ASD. According to our results, the selection of the granulating agent is of great importance. The amorphous Soluplus® gave better results in terms of dissolution of the subsequently obtained tablets.

CONCLUSION

Twin-screw granulations are certainly very promising techniques. Development of such processes can open up new ways for downstream processing. The monitoring and controlling of these processes via in-line techniques are of great importance since these are needed to ensure appropriate quality of the obtained product. The presented work showed promising results about monitoring and controlling critical quality attributes such as content uniformity and particle size and certainly can be of industrial interest.

536960 The Continuous Manufacturing of Pharmaceutical Capsules

Abstract:

New capabilities are being developed for continuous manufacturing systems as the use of this paradigm continues to become more widespread in the pharmaceutical industry. Continuous manufacturing allows for a high level of process control and knowledge. This enables process to be operated at a high level of quality while reducing cost. While most research and investment in the continuous manufacturing of oral solid dose (OSD) products has focused on the production of tablets, many products on the market are manufactured as capsules. In fact, most OSD products use encapsulation at some point during their development. Therefore, the development of a continuous operation with the capability to produce tablets at production scales (~200,000 capsules per hour) would allow the benefits of continuous manufacturing to be utilized by a wide range of products. This talk will describe the challenges and benefits inherent in the development and integrated of a full scale direct blend encapsulation process.

526001 Design and Implementation of a Control Strategy Using Risk-Based Probabilistic Design Principles: A Case Study in the Manufacturing an API with a Critical Impurity

Abstract:

In the quality by design (QbD) paradigm proposed by the FDA, the concept of risk analysis and risk management is a fundamental part of the development of the design space and control strategy for critical quality attributes (CQAs) in the manufacture of active pharmaceutical ingredients (APIâ€™s). However, there is not a standardized or well-adopted method for the quantification of risk in the development of processes for the manufacture of API. Herein, we present a case study highlighting the use of probabilistic (Bayesian) modeling techniques as (1) a tool to quantify the relative risk between different processing options, and (2), as a method for developing the control strategy for a specific CQA. We describe the process for manufacture of an API molecule that contains one critical impurity that is controlled as a CQA; the impurity is formed in the final chemical step and is primarily purged in a complex crystallization involving the use of a rotor-stator wet mill. In this case study, using probabilistic models to quantify variability, estimates of expected failure rates for different scenarios were tested, enabling the evaluation of process parameters and different options for control strategies. Lab-scale and pilot plant scale experiments were conducted to generate, test, and validate a series of models, which were used to develop a risk-based probabilistic design space and a control strategy for the CQA. Ultimately, we present the use of probabilistic modeling as a risk-mitigation and optimization tool for the engineer in the pharmaceutical industry.

532026 Proteomics Guided Process Engineering for Anti-HER2 Antibody Production

Abstract:

The high-quality monoclonal antibody (mAb) for cancer treatment is a fast growing market segment of biopharmaceutical industry. Chinese hamster ovary (CHO) cells have been widely used as host cells for mAb production. This study applied a comprehensive proteomics to develop an in-depth understanding of intracellular metabolism regulation and heterologous gene expression, which could benefit the production process optimization. The effect of four basal media and three cell culture operation modes on intracellular protein expression was evaluated in anti-HER2 mAb producing CHO DG44 cell line. A high-titer and high-quality mAb was produced from a fed-batch cell culture process in 2-L bioreactor with titer of >2.5 g/L. Moreover, the possible key host cell regulators of transcription, translation and post translational modification that correlated with important process parameters were identified. Several strategies of rational process engineering were also proposed to guide future cell culture process optimization.

520922 Formation Mechanism and Phase Transformation Behaviors of Polymorphs of Esomeprazole Sodium

Abstract:

Esomeprazole sodium is one kind of proton pump inhibitors. In literature, several polymorphic forms of esomeprazole sodium have been reported. However, no detailed information about the stability, formation mechanism and transformation behaviors of esomeprazole sodium has been reported. In this work, Powder X-ray diffraction (PXRD), thermogravimetric analysis (TGA), differential scanning calorimetry (DSC), Raman spectroscopy, and infrared spectroscopy (IR) were used to identify and characterize different polymorphs of esomeprazole sodium. Single crystal structure of form P was determined. The formation mechanism of form P was investigated and discussed from the molecular level. Presently, form J of esomeprazole sodium is the main commercial product in the market. However, it was found that form J can transform into form P and form L through solvent-mediated transformation and suspension transformation, respectively. Furthermore, selected area electron diffraction (SAED) was used to analysis the structure of different polymorphs of esomeprazole sodium.

525435 Process Modeling of a Continuous Drug Substance Manufacturing Process

Abstract:

Continuous manufacturing has the potential to increase the efficiency, flexibility, agility, and robustness of pharmaceutical processes delivering benefits to both patients and industry. With respect to drug substance manufacturing, there are examples of continuous unit operations such flow synthesis steps for API manufacturing. The current trend is towards increased process integration as opposed to isolated continuous unit operations in an overall batch process and more advanced control strategy approaches including PAT, diversion of non-conforming products etc.

The unit operation connectivity (steady state mode) and effect of disturbances (process dynamic) on process robustness and challenges of process control to produce â€œin-rangeâ€ quality product could be simulated by process modeling. The simulations will be used for case studies and sensitivity analysis to define the sensitivity of the process to particular process parameters, and to investigate the effects of Critical Process Parameters (CPPs) on Critical Quality Attributes (CQAs).

Two steady-state and dynamic process simulation models were developed for an integrated drug substance continuous manufacturing process. The manufacturing process risk analyses were performed and effect of various CPPs on the process robustness and final CQAs were evaluated. With the dynamic model sensitivity analysis and disturbance effects were studied as part of the system dynamic investigation. The system dynamic case studies demonstrate the response of the system to a various disturbances of CPPs (temporal and spatial distributions of disturbances). As a result of this simulation development, process control and manufacturing stability risk analyses were performed and response of the process to different disturbances evaluated under different control strategies.

526637 Green Process Development and Characterization of an Alkylation Reaction to Improve API Impurity Control

Abstract:

Anacetrapib is a CETP inhibitor which was developed to prevent cardiovascular disease, via control of cholesterol levels, by Merck. Extensive efforts have been made to develop an alkylation reaction as the last synthesis step for this drug substance to deliver high conversion, robust impurity control, as well as green chemistry elements such as low Process Mass Intensity (PMI) and reduced waste. Compared to the previously developed NaHMDS-based chemistry, the use of catalytic tetrabutylammonium iodide (TBAI) and potassium carbonate, as a base, not only minimize the epimerization of the final API into the Trans impurity, but also eliminates two process unit operations that were in the previous version of this step. Thorough process characterization was executed with the end-game chemistry to better understand the alkylation reaction, extraction and crystallization procedures for good product quality control. The combination of Design of Experiments (DOE) with statistical analysis and one factor at a time (OFAT) type experiments surrounding multiple process performance-impacting parameters further confirmed the reaction pathway and mechanism. It was found that an undesired hydrolysis reaction and an epimerization of final API were competing against the target pathway. The by-product formation was largely impacted by reactant stoichiometry and TBAI charges, solvent volume and reaction temperature. During the extraction sequence using selected extractants, the hydrophobic nature of API resulted in the formation of a 3rd solution phase (an oily phase), which was found to have no impact on the downstream crystallization process. The risked posed by the oil phase formation in the crystallization process was well controlled by slowing down the batch transfer rate and the final cooldown rate. The optimized reaction, extraction and crystallization process which were performed within the proposed Proven Acceptable Ranges (PAR) for the regulatory filing, showed high conversion, robust impurity control, and improved operability and productivity of the entire process step, as well as decreasing the environmental impact of the step by decreasing the PMI and waste volumes.

529180 Leveraging PAT for Efficient Process Development Workflows

Abstract:

Leveraging PAT for Efficient Process Development Workflows

James C. Marek*, Eric G. Moschetta, Travis B. Dunn,

AbbVie, Inc. Process R&D, 1401 Sheridan Road North Chicago, IL 60064

*speaker: james.marek@abbvie.com

Timelines to develop process understanding and controls for pharmaceutical processes are universally short, necessitating efficient workflows to design and execute data-rich experiments leading to sound conclusions. Leveraging on-line Process Analytical Technology (PAT) is an integral component of efficient workflows. We present a case study that illustrates the power of *in situ* ATR/FTIR to elucidate fundamental kinetics for a complex reaction system, focusing in this talk on the development of the approach. Measuring the relevant chemical species concentrations was essential to modeling the kinetics, and on-line measurements with rapid sampling frequencies collected many data points over the course of the reaction.

The PAT approach was taken after careful consideration of the potential benefits and risks compared to the HPLC method intended for the manufacturing process. Simple calibration models constructed using FTIR measurements of solutions with a single reaction component yielded accuracy sufficient for the purposes of the study. The models were further used to derive species concentrations from FTIR profiles of the reaction run in batch mode. The concentration profiles were readily exported to standard software for kinetic modeling. Because the reaction was relatively fast, and the HPLC method involved a lengthy derivatization step, PAT via on-line ATR/FTIR was clearly the better approach for this investigation. ATR/FTIR was further used to monitor the reaction chemistry in the pilot plant process using the same calibration models developed for the lab study, with results demonstrating excellent scalability of the reaction.

This abstract presentation was sponsored by AbbVie. AbbVie contributed to the design, study conduct, and financial support for the research. AbbVie participated in the interpretation of data, writing, reviewing, and approving the publication. All authors are AbbVie employees.

529073 On-Line FTIR for Accurate Fundamental Kinetic Analysis, Real-Time Process Monitoring, and Process Controls Justification in Pharmaceutical Manufacturing

Abstract:

On-Line FTIR for Accurate Fundamental Kinetic Analysis, Real-Time Process Monitoring, and Process Controls Justification in Pharmaceutical Manufacturing

Eric G. Moschetta,* James C. Marek, and Travis B. Dunn

AbbVie, Inc. Process R&D, 1401 Sheridan Road North Chicago, IL 60064

*corresponding author: eric.moschetta@abbvie.com

Fundamental kinetic analysis of complex reactions is critical to process understanding and development in the pharmaceutical industry. With aggressive timelines for active projects, it is imperative to design and execute data-rich experiments to glean as much kinetic information in as few experiments as possible. However, existing analytical techniques may not be fully developed for complete kinetic analysis, nor may they be amenable to collecting many data points in a single experiment. For example, reaction samples that require labor-intensive workup or processing to analyze quantitatively do not lend themselves well to data-rich experimentation because of the significant burden on the development scientist and the available instrumentation. An alternative approach is to implement on-line process analytical technology (PAT) tools for monitoring the concentrations of the relevant chemical species participating in the reactions. On-line measurements are amenable to collecting many data points over the course of an experiment because the instruments often have rapid sampling frequencies. PAT tools that rely on spectroscopy offer the additional benefit of structural characterization and can distinguish between different chemical species. Careful identification and assignment of characteristic peaks and bands from the on-line signals can provide time-dependent concentration data simultaneously for multiple chemical species. These data can be used not only to build a robust chemometric kinetic model for the process, but for simple identification of logical process endpoints or even to monitor critical quality attributes (CQAs) in real time. As such, implementing on-line PAT tools to improve process understanding at a fundamental level and streamline the workflow in the laboratory is critical to developing robust manufacturing processes in the pharmaceutical industry.

We present a business case in which we implemented an on-line FTIR probe for *in situ* operation in data-rich kinetic experiments. Our business objective was to justify the process controls for the rate of addition of a reagent into a chemical reactor such that the starting material in the reactor was always consumed to the justified level. We hypothesized that a kinetic model describing the decomposition of the starting material would provide excellent justification for the process. However, the kinetic analysis was complicated by two factors: (1) the reagent being dosed to the reactor could further react with the desired product to form an impurity and (2) the current HPLC method required a time- and labor-intensive derivatization process, limiting the amount of data points that could be collected and analyzed in any given experiment. All of our relevant kinetic species contained different functional groups bound to a carbonyl group, so we pursued on-line FTIR as a possible PAT tool to acquire data. We will show how the key experimental observations were more easily elucidated by the on-line IR than the off-line HPLC method and how we used these observations to design and execute the minimum number of data-rich experiments to build a robust kinetic model of the process. Specifically, we will explain how we were able to observe a subtle reaction effect using the on-line FTIR that proved our working hypothesis of the reaction mechanism was wrong. Finally, we show how our chemometric model can be used to monitor pilot plant scale processes in real time based on the fundamental principles developed in the laboratory. This level of understanding allows for predictive power in scalable processes and provides excellent details for process control justification.

The design, study conduct, and financial support for this research was provided by AbbVie. AbbVie participated in the interpretation of data, writing, reviewing, and approving the publication. All authors are AbbVie employees.

530753 Integration of Orthogonal PAT Tools to Facilitate API Process Development

Abstract:

Integration of Orthogonal PAT Tools to Facilitate API Process Development
Jeffrey Nye, Dimitri Skliar, Gregory Beutner

The Reaction Science and Engineering group at BMS is a multidisciplinary team of chemists and chemical engineers that utilizes first principles-based understanding of reactions and engineering fundamentals to optimize chemical processes, generate knowledge and enable robust manufacturing strategies for active pharmaceutical ingredients (API). A focus area for the group is in supporting data-rich experimentation, wherein during experimental execution, data streams from many disparate sources are captured, aggregated and contextualized in real time. By expanding the scope of easily-accessible data streams, lab scientists are able to maximize the knowledge generated from individual experiments, achieving the goals of increased understanding with shortened development timelines.

Here we present a series of case studies wherein process development is enabled by the incorporation of two or more orthogonal process analytical technologies (PAT). These real-time tools provided critical insights into complex chemical systems enabling knowledge generation on lab scale, and process control on large scale. In one case study we describe the simultaneous use of FTIR, NIR and hydrogen uptake to map the kinetics and understand diastereoselectivity of a reductive amination using anhydrous ammonia. In the second case study we couple Raman spectroscopy and real-time microscopy with off-line techniques to elucidate the chemical behavior of inorganic carbonate bases in a metathesis reaction involving a poorly-soluble intermediate with complex form behavior. The data-rich platforms provided by PAT in both case studies allowed for the rapid understanding and decision-making that are critical to enable a dynamic and complex API portfolio.

531180 Online LC in Support of Data Rich Experimentation

Abstract:

Online Liquid Chromatography analysis is becoming increasingly important for studying the kinetic profiles of chemical reactions, from which valuable mechanistic information can be obtained. Historically, online sample collection was performed *via* either flow of the reaction medium through a sample loop, followed by injection, or by metered withdrawal of a defined volume flowing via a sampling tube or capillary. Both of these approaches work well for homogeneous solutions, but are not well suited to the investigation of heterogeneous reactions where solids are present or where active precipitation or crystallization is taking place. The recently introduced Mettler Toledo EasySampler 1210 addresses this limitation, allowing for reliable sampling of heterogeneous reactions using a mechanically actuated, 20 μL fixed-pocket based sampling mechanism. However, this tool is limited to the collection of a carousel of 12 samples, limiting the utility for detailed kinetic profiling with frequent, and extending sampling needs, especially when there is poor sample stability of the diluted solution. In this work, we demonstrate expansion of the EasySampler's application range to include direct injection into an LC stack showcasing the EasySampler as a robust sampling solution enabling high data density experimentation.

534400 Data Reconciliation in the Quality-By-Design (QbD) Implementation of Pharmaceutical Continuous Tablet Manufacturing

Abstract:

Pharmaceutical continuous manufacturing is now entering the arena of big data with more and more critical quality attributes (CQA) and critical process parameters (CPP) readily available in real time using process analytical technology (PAT). Although the management and integration of process operational and product quality data have been receiving extensive interests, [1] the transition from data-rich to information and knowledge-rich operation has not received much attention thus far. [2] Under the conventional implementation of multivariate statistical process control (MSPC), the nominal variation in the *process systems* (equipment\operation\material) and *sensor measurements are monitored by tracking a small number of principal components and* testing them against pre-specified statistical control limits, e.g., Hotelingâ€™s T^2 and square prediction error (SPE). However, while simplifying monitoring, this approach does not directly use the valuable information on the true product quality variations, which are more directly-related to in-process and final product quality attributes. [3] Furthermore, there are some critical material attribute (CMA) or CQA variables that cannot be measured in line and thus cannot be captured by MSPC methods. Finally, an adequate database of nominal operations for MSPC development is not always available, especially not during the early-stages of product/process development.

Data reconciliation (DR) is an important process systems engineering tool for accommodating process and measurement variations and estimating the true state of the manufacturing system. Based on process mathematical models, DR seeks to minimize the errors between model predicted system states and the state information provided by sensor measurements that inherently have different levels of accuracy and precision. Effectively by reconciling the model predictions of system states with the nominal sensor measurements, the differences in sensor measurement errors and offsets are taken into account. This is in consistent with the US FDA Quality by Design (QbD) guidelines, which highlights the importance of process understanding in process development and product quality control. The role of DR is especially important given the significant differences which normally occur between CQA and CPP variable variations. For example, most CQA variables are now indirectly measured using in situ spectroscopy probes, e.g., active pharmaceutical ingredient (API) composition measured by Near Infrared (NIR) probe. Such measurements are often subject to variations/drifts due to fouling, inherent errors in chemometric model predictions, material property changes (particle size, bulk density, etc.), environmental factor changes (humidity, temperature, etc.), and so on. On the other hand, the CPP variables are commonly and directly measured using reliable mechanical or electrical sensors, e.g., the compression force in a tablet press, which tend to have smaller measurement errors. It is worth mentioning, that the robust design of traditional manufacturing equipment (tablet press, roller compactor, etc.) which provided minimum variation in CPPs, has allowed pharmaceutical manufacturing to continue to operate in batch mode while handling CQA variations via end of line statistical quality control (SQC) methods. The challenge in continuous operation is to effectively integrate the noisier and possibly biased CQA measurements into the process control system so as to effectively supervise the control of CPPâ€™s while minimizing the need for end-on-line SQC. However, this integration does impose additional dynamics on the process, potentially amplifying variations in CPPs and thus in product quality attributes.

In this study, a data reconciliation framework is proposed for implementation in pharmaceutical continuous manufacturing of oral solid dosage. The special features of using the Welsch robust estimator for the errors between model prediction and sensor measurement, [4] and the updating of model parameters to address the model-plant or model-material mismatches are highlighted. The Welsch robust estimator is shown to be capable of rejecting the measurement gross errors without affecting the reconciled model predictions and the robust operation of process control systems. The role of the estimator is especially important when the sensors were undergoing maintenance or replacement due to failures. The real-time model parameter updating is useful when there was a change in raw material properties or process operations. Moreover, the resulting changes in model parameters can be further used to diagnose the system variations in process operation and material properties, as well as to quantify their effects on product quality attributes. The proposed data reconciliation framework is demonstrated on a continuous tablet manufacturing process via direct compaction, which uses a Natoli BLP-16 tablet press. The Kawakita model which captures the relationship between main compression force (CPP) and tablet weight (CQA) is incorporated in the DR framework. [5] The Kawakita model parameter are automatically updated to accommodate the changes in material compactibility, e.g., bulk density, particle size, etc. The reconciled main compression force and tablet weight values are used in a hierarchical three-level control design to maintain consistent tablet weight production. [6] A tablet weight measurement based on a load cell is used to provide real time but higher variability measurement. The reconciled model predictions of tablet weight are periodically verified and corrected using at-line Sotax Auto Test 4 tablet weight measurements.

The performance of the data reconciliation framework using our developed control system design for continuous tablet manufacturing is quite promising in offering a robust tablet weight control.

References

1. Markl D, Wahl PR, Menezes JC, et al. Supervisory control system for monitoring a pharmaceutical holt melt extrusion process. *AAPS PharmSciTech*. 2013;14(3):1034-1044.

2. Ierapetritou M, Muzzio F, Reklaitis G. Perspectives on the continuous manufacturing of powder-based pharmaceutical processes. *AIChE Journal*. 2016;62(6):1846-1862.

3. Moreno M, Liu J, Ganesh S, et al. Steady-state data reconciliation of a direct compression tableting line. Paper presented at: AIChE Annual Meeting, 2017; Minneapolis, US.

4. Liu J, Su Q, Moreno M, Laird C, Nagy Z, Reklaitis G. Robust state estimation of feeding-blending systems in continuous pharmaceutical manufacturing. *Chemical Engineering Research and Design*. 2018:Accepted.

5. Su Q, Bommireddy Y, Gonzalez M, Reklaitis GV, Nagy ZK. Variation and risk analysis in tablet press control for continuous manufacturing of solid dosage via direct compaction. Paper presented at: The 13th International Symposium on Process Systems Engineering PSE 2018, 2018; San Diego, CA.

6. Su Q, Moreno M, Giridhar A, Reklaitis GV, Nagy ZK. A systematic framework for process control design and risk analysis in continuous pharmaceutical solid-dosage manufacturing. *Journal of Pharmaceutical Innovation*. 2017;12:327-346.

536638 Near Infrared Hyperspectral Imaging to Monitor Drug Content and Coating Thickness Uniformity of Oral Disintegrating Films during a Continuous Manufacturing Process

Abstract:

Continuous manufacturing of oral disintegrating films (ODF) involves several unit operations such as mixing of raw materials such as active pharmaceutical ingredient (API), film forming polymer, plasticizer and other excipients to form a uniform mixture/slurry, coating the mixture followed by drying, die cutting to desired shape and size and packaging. Coat thickness and drug content uniformity are two critical in-process variables which directly impact the critical quality attribute (CQA) of ODF such as drug assay. Non-uniformity in the coating thickness or drug content can lead to deviation from target drug load, which in turn may affect the in vivo drug delivery rate. Hyperspectral Imaging using near infrared spectroscopy (NIRS) is a potential technology that integrates spectroscopy and conventional imaging to obtain both spectral and spatial information of a material while it is in motion. Objective is to develop and validate the use of Near Infra-red (NIR) hyperspectral imaging for its ability to monitor drug content and coat thickness uniformity of ODF. Hyperspectral Imaging system equipped with a full range (1000-2500 nm) short wavelength infrared (SWIR) camera was employed for this study. ODF samples of varying coating thickness and drug content were prepared and divided into training and test sets to develop and validate the chemometric model for coat thickness and drug content. The results and findings of this study will be presented.

527116 Data-Rich Reaction Profiling of Copper-Catalyzed Methoxylation of an Aryl Halide

Abstract:

In the development and scale-up of viable synthetic pathways for active pharmaceutical ingredients (APIâ€™s), it is advantageous to understand a reactionâ€™s sensitivities to various process conditions. Reaction profiling is a useful way to characterize these sensitivities but can be very time, labor, and resource intensive. The use of small volume parallel reactors with automated sampling strategies enables data rich experimentation in a resource-sparing manner. Inclusion of in-situ analytical techniques, such as FTIR, provides further insight to the reaction dynamics. This data can provide mechanistic insight to the reaction and can be used to develop a robust kinetic model of the reaction.

We employ this approach in a case study of the copper-catalyzed methoxylation of an aryl halide. The resulting profiles were used to construct a kinetic model in agreement with a proposed reaction mechanism. Laboratory profiling results were also used as a comparison to those obtained in the pilot plant during scale-up.

530489 Parameter Estimation of Reaction Kinetics from Spectroscopic Data - Developments and Applications

Abstract:

The development of drug manufacturing processes involves dealing with spectroscopic data. When dealing with spectroscopic data, the identification of parameters and variances still remains a challenging task. In many cases kinetic parameter identification from spectroscopic data has to be performed without knowing the absorbing species in advance, such that they have to be estimated as well. However, kinetic parameter estimation is important in order to lower production costs, i.e. save measurements and equipment. Furthermore, scaling up from laboratory to industrial level relies on accurate kinetic parameter values.

That is why, we take a closer look at the development of optimization-based procedures in order to estimate the variances of the noise in the system variables and spectral measurements. Then, with the estimated variances we determine the concentration profiles and kinetic parameters simultaneously using adequate strategies. The work is based on the approach proposed by Chen et al. (2016) using maximum likelihood principles for simultaneous estimation of reaction kinetics and curve resolution from process and spectral data. For this a new software environment was developed which is continuously enhanced. This environment is based on Pyomo with different discretization options for the processes described by ordinary differential/differential algebraic equations, such as collocation using Pyomo.DAE and an efficient implementation of variable-order and variable-coefficient BDF methods, implemented in IDA by sundials. The further investigations regarding this software environment, the identification of absorbing species and challenges arising from pharmaceutical processes are presented within this talk. The outcomes are illustrated by several case studies of pharmaceutical processes. The nonlinear programming (NLP) problems, formulated for these applications, are solved using IPOPT and using sIPOPT for the determination of confidence regions for the kinetic parameter estimates.

We gratefully acknowledge Pfizer Inc.â€™s funding.

References

W. Chen, L. T. Biegler, and S. G. MuÃ±oz. An approach for simultaneous estimation of reaction kinetics and curve resolution from process and spectral data. *Journal of Chemometrics*, 30:506â€“522, 2016. doi:10.1002/cem.2808. URL http://dx.doi.org/10.1002/cem.2808.

529885 An Application of Portable Raman Spectroscopy for Rapid and Intact Polymorphism Quantitative Analysis of Multi-Component Pharmaceutical Tablets

Abstract:

An Application of Portable Raman spectroscopy for Rapid and Intact Polymorphism Quantitative Analysis of Multi-component Pharmaceutical Tablets

Yufeng Quan [1, 2, ¡ì], Zhixuan Huang [3, ¡ì], Lina Jia [1, 2], Yun Cao [1, 2],Junbo Gong [1, 2, *], Da Chen [3, *]

[1] National Engineering Research Center of Industry Crystallization Technology, Tianjin, China;

[2] Collaborative Innovation Center of Chemical Science and Engineering (Tianjin), Tianjin, China;

[3] State Key Laboratory of Precision Measurement Technology and Instruments, Tianjin

University, Tianjin, China

¡ì These authors contributed equally to this work and should be considered as co-first authors.
*junbo_gong@tju.edu.cn

*dachen@tju.edu.cn

Keywords: portable Raman spectroscopy, polymorphism quantitative analysis, pharmaceutical tablets

Polymorphism phenomenon is common and important among pharmaceutical products. On the one hand, changes from one polymorph to another may have a big effect on the bioavailability of the final products. One the other hand, a particular form of an active pharmaceutical ingredient (API) represents an opportunity to prolong the patent protection period which will bring a certain economic benefit. In this study, a detailed application of portable Raman spectroscopy (a weight of only 1.5kg), equipped with a micro sample platform, was performed on Vortioxetine hydrobromide tablets in the presence of five different excipients for qualitative and quantitative analysis. API is usually under the interference of the auxiliary materials and the substance so that a chemo metrics algorithm was applied to extract the API¡¯s Raman signal from tablets. The algorithm adopted a prior knowledge-oriented feature extraction, which gets the main pure excipient spectral information through the removal of fluorescence interference. Then, the spectrum fitting method was used to analyze the information of the interference components, such as the auxiliary materials, from the complex spectrum of the tablets, and the intensity information of the API in the tablets was extracted by the differential spectrum method. The results (shown in Fig.1 and Fig.2) demonstrates that this method is more effective than the traditional fitting algorithm to overcome the interference of different contents of the auxiliary material to the spectrum. Furthermore, combined with a movable platform, the whole surface of each tablet can be detected to reduce uncertainty.

Fig.1 Raman spectroscopy of ¦Á form, ¦Â form and excipients tablets

Fig.2 Raman spectroscopy of polymorph mixtures (bottom line) and reconstructed spectrum

536158 Liposome Processing Using a Turbulent Jet in Co-Flow: Experiments and Multi-Scale Models

Abstract:

PURPOSE

Continuous processing administers considerable advantages over batch manufacturing to improve scalability, automated control and incorporation of process analytical technology. Liposomes are complex dosage forms with multiple FDA approved drug products presently available. Conventional methods to form liposomes do not provide for precise control of the particle size and particle size distribution both of which can directly impact encapsulation efficiency, cellular uptake and biodistribution. The continuous manufacturing platform provides precise control over liposome formation including monitoring of particle size and size distribution. The present study focuses on coaxial turbulent jet flow design to control liposome properties and quantify them as a function of the flow conditions (velocity, temperature, Reynoldâ€™s number). Computational Fluid Dynamics (CFD) is further performed to verify the dynamics of the environment at site of liposome formation. Molecular Dynamics (MD) simulation are being further performed to analyze the lipid-lipid interactions and lipid-salt interactions at the liposomal formation site.

METHODS

Liposomes were formed using an ethanol injection method with a custom-built continuous processing system that is controlled using a program developed using National Instruments LabVIEW. Briefly, the continuous processing system consists of multiple tanks that contain lipid dissolved in ethanol and another set of tanks for the aqueous components (*e.g.* DI water or buffer salts). At-line particle size measurements were performed with a Malvern Zetasizer Nano using a flow-through cell. Particle size measurements were measured off-line and compared with the at-line measurements. Fluid flow profiles (i.e. Flow Velocity ratio (FVR) vs Reynolds number (Re) were related to the liposomal size distribution (*i.e.* Z-Average particle size and PDI). CFD geometry similar to the continuous liposome manufacturing unit was developed using ANSYS 19.0. Species Transport Equations were implemented to predict the fluidic microenvironment at the site of liposome formation and was validated with experimental data by image analysis and fluid flow dimensions (FVR and Re) of the co-axial jet formed from ethanol water mixture.

RESULTS

The continuous processing system at UConn can form monodispersed liposomes. Process parameters can be controlled to regulate the velocity, density, and viscosity at the co-axial jet. The liposome mean particle size can also be controlled and has dependence on the Reynolds number of the mixed fluid. The CFD studies have successfully be used to predict the experimental flow properties observed at the co-axial jet. Based on the FVR and Re, the model provides insight on the alteration of the micro-environment at the liposome formation site. In addition, MD modeling can be used to understand the differences in liposome formation based on factors such as type of lipid and aqueous phase additives.

CONCLUSION

The liposome size and PDI can be systematically controlled using a co-axial turbulent jet in co-flow. The non-linear increment of the ethanol-water mixture viscosity at the site of liposome formation (predicted from the CFD studies), plays a significant role to alter the fluid properties and can be directly related to the liposome dimensions and distribution. MD simulations are being performed to predict the mechanism of liposome formation due to variability of fluid flow properties and physical properties of the formulation.

ACKNOWLEDGEMENTS

This work was supported by the FDA (1U01FD005773-01).

DISCLAIMER

This article reflects the views of the authors and should not be construed to represent FDAâ€™s views or policies.

524387 Optimal Design of Separation Systems for Continuous Manufacturing of Nevirapine, an Active Pharmaceutical Ingredient (API) for H.I.V. Treatment

Abstract:

Affordability and accessibility of essential medicines remains a pressing issue for the treatment of diseases prevalent in developing countries. The treatment of HIV continues to be one of the most prominent global health challenges, with 36.7 million people living with HIV by the end of 2016, and approximately 70% of those being in low- and middle-income countries [1]. The development of efficient, cost-effective manufacturing routes towards APIs for HIV treatment is paramount to ensure global, affordable access to such medicines [2].

Continuous pharmaceutical manufacturing (CPM) has emerged as a new production paradigm for its promise of enhanced efficiency and greater economic viability over currently implemented batch protocols [3]. The utility of CPM platforms for the development of active pharmaceutical ingredients (APIs) for HIV treatment [4] has been demonstrated in the literature. While experimental demonstration of feasible API continuous flow synthetic routes is the foundation of any CPM campaign [5], the design of essential continuous separation processes for integration into upstream CPM plants for HIV API production is essential to realise the benefits of end-to-end continuous manufacturing [6].

Nevirapine is a widely-prescribed API for HIV-1 treatment, whose continuous flow synthesis was recently demonstrated [7]. Process intensification of the synthetic route compared to commercial reaction pathways allowed enhanced yields, improved material efficiency (reduced process mass intensity, PMI) and reduced number of unit operations. Systematic comparison of continuous separation processes is essential to further aid the development of leaner manufacturing routes towards this societally-important API. Process modelling and optimisation methodologies can be used to establish cost-optimal design configurations [8].

This paper presents a combined experimental and modelling study for the upstream continuous flow synthesis and separation of nevirapine. Reaction kinetic parameter estimation from API continuous flow synthesis experiments is performed, followed by continuous separation process model validation and subsequent nonlinear optimisation of the integrated upstream CPM plant. Optimal process configurations for CPM of nevirapine are presented with a critical analysis of experimental and modelling methodologies implemented. This work illustrates potential advantages attainable via end-to-end CPM and the importance and utility for modelling and optimisation studies at the early stages of design.

REFERENCES

[1] HIV.gov, "People living with HIV on antiretroviral therapy," 2017. [Online]. Available: https://www.hiv.gov/hiv-basics/overview/data-and-trends/global-statistics. [Accessed: 08-Mar-2018].

[2] J. M. Fortunak, R. O. M. A. de Souza, A. A. Kulkarni, C. L. King, T. Ellison, and L. S. M. Miranda, "Active

pharmaceutical ingredients for antiretroviral treatment in low- and middle-income countries: a survey.," *Antivir. Ther.* **19**(S3): 15–29 (2014).

[3] S. D. Schaber, D. I. Gerogiorgis, R. Ramachandran, J. M. B. Evans, P. I. Barton, and B. L. Trout, "Economic Analysis of Integrated Continuous and Batch Pharmaceutical Manufacturing: A Case Study," *Ind. Eng. Chem. Res.* **50**(17): 10083–10092 (2011).

[4] R. O. M. A. de Souza and P. Watts, "Flow processing as a tool for API production in developing economies," *J. Flow Chem.* **7**(3–4): 146–150 (2017).

[5] M. B. Plutschack, B. Pieber, K. Gilmore, and P. H. Seeberger, "The Hitchhiker's Guide to Flow Chemistry," *Chemical Reviews* **117**(18): 11796–11893 (2017).

[6] I. R. Baxendale, R. D. Braatz, B. K. Hodnett, K. F. Jensen, M. D. Johnson, P. Sharratt, J.-P. Sherlock, and A. J. Florence, "Achieving continuous manufacturing: technologies and approaches for synthesis, workup, and isolation of drug substance," *J. Pharm. Sci.* **104**(3): 781–791 (2015).

[7] J. Verghese, C. J. Kong, D. Rivalti, E. C. Yu, R. Krack, J. Alcázar, J. B. Manley, D. T. McQuade, S. Ahmad, K. Belecki, and B. F. Gupton, "Increasing global access to the high-volume HIV drug nevirapine through process intensification," *Green Chem.* **19**(13): 2986–2991 (2017).

[8] H. G. Jolliffe, S. Diab, and D. I. Gerogiorgis, "Nonlinear optimisation via explicit NRTL model solubility prediction for antisolvent mixture selection in artemisinin crystallisation," *Org. Process Res. Dev.* **22**(1): 40–53 (2018).

537278 Utilizing Mechanistic Modelling to Assist in the Process Development of Pharmaceutical Drug Substance Processes

Abstract:

The development of a pharmaceutical API/drug product requires a large number of experiments, the design and interpretation of which is often based on statistical modelling tools. This approach achieves the definition of parameter ranges that delivers products of high quality in a robust manner. However, the drawback to this approach is that the data only covers the processing conditions investigated and cannot be extrapolated outside of these conditions. Also, this approach does not guarantee a full understanding of all the phenomena involved in a given process and hence, it may yield a viable but not an optimised processes or sequence of operations. For the above reason, mechanistic modelling approaches would be the preferred choice, although sometimes not the most practical, yielding a higher degree of understanding of the chemical and physical phenomena occurring in a given pharmaceutical process. This approach can be applied to the development of both single unit and multi-stage operations, for both batch and continuous processes. Coupling models across multiple unit operations would lead to the optimisation not only of a single operation but the entire process, thus leading to reduced costs and maximising product yields.

In this context, the work presented models batch unit operations, including reactions, separations, and links them together following a common pharmaceutical process structure. In this case, gPROMS FormulatedProducts software was applied to meet this mechanistic modelling need. This tool includes a wide array of mechanistic models for drug substance applications. The kinetic parameters for the reaction mechanisms considered were validated using a small number of data rich experiments, allowing for reaction model and mechanism discrimination to achieve the most descriptive mechanistic model for the batch reaction step. In the future, the validated process models will be utilized to describe the integrated drug substance processing steps. This end-to-end mechanistic modelling approach will help to both optimize a full pharmaceutical manufacturing process and determine its overall robustness against disturbances.

520278 System-Model Development for Continuous Drug Substance Manufacturing Process

Abstract:

Continuous manufacture (CM) of small molecule drug substance offers several advantages over traditional batch processes, especially when the chemistry is very complex and/or process requires fine control of process conditions. The advantages of CM have been highlighted elsewhere [1, 2], and as demonstrated and noted by these authors, precise understanding of system dynamics via system/process modeling and/or residence time distributions is crucial in CM.

In this presentation, GlaxoSmithKlineâ€™s CM process and system model development approach will be discussed for an asset which utilizes three continuous drug substance unit operations. The focus will be on biphasic reaction and continuous evaporator models. System-level understanding via global sensitivity analysis and modelâ€™s usage in various stages of process development will be discussed.

References

[1] Cole et al., 2017, Science, 356(6343), 1144-1150.

[2] Lee et al., 2015, J. Pharm. Innov., 10, 191-199.

521891 An Industrial Application of Parameter Estimation of Biocatalytic Transaminase Reactions

Abstract:

2018 American Institute of Chemical Engineers (AIChE) Annual Meeting

An industrial application of parameter estimation of biocatalytic transaminase reactions

Javier Magano, David Damon, John Wong, Steven Guinness, Ke Wang, Jason Mustakis, **Lu Han**

Abstract

In the pharmaceutical industry, there is growing interest to use enzymes to perform highly selective chemical reactions. Through advancements in protein engineering, the enzyme properties can be tuned and optimized, to enhance their selectivity and activity for desired reactions. Biocatalysis presents a valuable tool that complements existing chemical synthesis approaches. The kinetic models that describe the enzyme-catalyzed reactions consist of several parameters, which are often correlated. Typically, many experiments are required in order to effectively estimate these kinetic parameters [1-3]. Utilizing automation for these types of experiments at small-scale has the potential to save resources as well as provide comprehensive data for modeling.

In this work, we demonstrate an application of a high-throughput reaction screening platform at Pfizer to estimate the kinetics of a biocatalytic transaminase reaction. The transaminase experiments were performed at a resource-sparing scale under well-controlled conditions. The generated data was fitted to a kinetic model, and coupled to a reactor model to simulate the effects of different reaction and process conditions. The process model was applied to experimental data generated from lab and plant scale. The results from the current study reveal that a simple kinetic model can be developed from a small-scale screening platform and leveraged for predicting data at the larger scale.

References

[1] R. H. Ringborg and J. M. Woodley. *React. Chem. Eng.*, 2016, 1, 10-22.

[2] N. Al-Haque, P.A. Santacoloma, W. Neto, P. Tufvesson, G. Rafiqul, J.M. Woodley. *Biotechnol. Prog.*, 2012, 28(5), 1186-1196.

[3] B. Chen, E. Hibbert, P. Dalby, J.M. Woodley. *AIChE J.*, 2008, 53(8), 2155-2163.

527937 Investigating the Impact of Water on the Energetics and Kinetics of a Reductive Amination Reaction – a Computational and Experimental Approach

Abstract:

Reductive amination reactions constitute up to 14% of the total reactions used for drug synthesis in pharmaceutical industry [1]; in particular, the reductive amination of carbonyl compounds is a highly useful, robust, and broadly employed transformation for the synthesis of further functionalised amines and has become an important work tool of pharmaceutical industry. [2] A key emerging consideration within pharmaceutical research is the use of better candidate solvents in these reactions as replacements for those that are less sustainable, pose adverse health and safety issues or demonstrate a detrimental environment impact. Therefore, studying the detailed mechanism, energetics and kinetics of imine formation and sequential reductive amination reactions is central to improving our understanding of the critical role that these reactions hold in pharmaceutical processes. The mechanism of such reactions follows a two-step approach in which an imine is formed via a carbinolamine intermediate (step 1) and the subsequent elimination of a water molecule (step 2). Previous computational studies have suggested the catalytic effect of water molecules on reaction energetics and kinetics. [3-5] In this study, we investigate the impact of water molecules on the energetics and kinetics of the reaction between benzaldehyde and 4-trifluoromethyl aniline, using quantum-mechanical and molecular-mechanical methods. Our efforts are focused on studying the microsolvation process in the gas and liquid phase, incorporating a varied number of water molecules around the reagents of the reaction. We demonstrate that addition of water molecules monotonically reduces the energy barrier of both reaction steps and accelerates the reaction kinetics, suggesting solvent effect importance on the energy and kinetic properties of the selected reaction. We additionally perform different water-content kinetic experiments, through ^{19}F-NMR and ^{1}H-NMR spectroscopy, which strongly support our computational findings regarding the trend in barrier heights and reaction kinetics.

References

1. Carey J. et al., ***Biomol. Chem.*** 2006, 4, 2337.
2. McGonagle F. et al., ***Green Chem.*** 2013, 15, 1159.
3. Williams, I. H. ***Am. Chem. Soc.*** 1987, 109, 6299.
4. Hall and Smith, ***Phys. Chem. A*** 1998, 102, 4930
5. Louie M. K. et al., ***Phys. Chem. A*** 2016, 120, 1358

533712 Importance of the Rheological Characterization on the Spray Performance of Pharmaceutical Formulations

Abstract:

Importance of the rheological characterization on the spray performance of pharmaceutical formulations

T. Porfirio[1,2*], R.C. Silva[1], J.Vicente[1], V.Semião[2]

[1] Hovione Farmaciência SA, Estrada do Paço do Lumiar, 1649-038 Lisbon, Portugal; *tporfirio@hovione.com

[2]LAETA, IDMEC, Mechanical Engineering Department, Instituto Superior Técnico, Universidade de Lisboa, Av. Rovisco Pais 1, 1049-001 Lisbon, Portugal

Spray drying is one of the most used particle engineering technologies in the pharmaceutical industry with several enabling applications in modern drug delivery. The theoretical description of the phenomenon involved in this technology typically considers two key events – atomization and droplet dying / particle formation. Droplet size, during the atomization, is the key parameter governing the control of particle size and bulk density of the spray dried material. The droplet formation process is strongly influenced by the atomizer design and operating conditions as well as the feedstock properties viz. viscosity, surface tension and density [1].

The determination of feed properties is crucial for an accurate estimation of the droplet size. However, in the case of non-Newtonian fluids – which are typically the case of the pharmaceutical formulations used in spray drying – the properties may vary during atomization due to the high shear at the nozzle level. Therefore, the rheological behavior of the fluid plays an important role especially in swirl pressure nozzle in which the shear rate is high by default. Besides that, an accurate estimation of the shear rate depending on the nozzle design and operating conditions is paramount to establish a suitable droplet size correlation.

The purpose of the present work is to examine the fundamental principles of the liquid break-up occurring during the atomization process to establish the relationship between the rheology of the most common organic solutions used in the spray drying of pharmaceutical formulations and the spray performance and particle size of the spray dried product. The rheological curves are established using a rotational rheometer with a cone-plate spindles to achieve shear rate around 10^5 s^{-1}. Suitable models are applied to the rheological curves depending on the fluid behavior and formulation variables i.e. solvent(s) and solid concentration.

In order to increase upon the existing knowledge in shear rate estimation for swirl pressure nozzles, a Computational Fluids Dynamics (CFD) model is benchmarked in this work and compared with literature data. A velocity and wall shear stress profile is drawn using such methodology (Figure 1), in tandem with the experimental feed rheological curves of Newtonian and non-Newtonian fluids, which in turn is used to establish an understanding and correlation between the shear rate and the nozzle design / operating conditions.

The method of the present work can be incorporated in atomization models to increase the accuracy of spray drying scale-up methods.

Figure 1 – Velocity profile inside a pressure nozzle

[1] A.H. LEFEBVRE, X.F. WANG, Mean drop sizes from pressure-swirl nozzles, J. Propuls. Power. 3 (1987) 11–18. doi:10.2514/3.22946.

534570 Scaling up and Optimization of Process Parameters for a Spray Drying Plant Using Mechanistic Modelling

Abstract:

Scaling up and optimization of process parameters for a spray drying plant using mechanistic modelling

Sridevi Challa, Mohammed Yakoob Sardar, Veera Pratap Reddy Kasina, RaviChandra Palaparthi, Srividya Ramakrishnan, Ravikumar Gorle

Generic pharmaceutical industry has to operate in tight timelines of product development and scale up from lab to plant where time to market makes or breaks the viability of the business case. Getting it right the first time demands detailed product & process understanding and implementing appropriate mechanistic modelling and Quality-by-design (QbD) approaches in the product life cycle, to minimize downstream risks especially with the complexity of the formulations increasing exponentially in the current decade for the generic industry. This work describes an example case of such efforts on a model system comprising of excipients/ API involving the scale up of spray drying operation from lab to plant scale.

Typical requirements in spray drying operations involve: identifying the process conditions at one scale where one can minimize the right residual solvent content and maximize the yield of the spray dried powder; translating the process conditions as the product goes from lab to plant scale spray dryers. A combination of experimental studies that characterize the relevant material and lab/ plant equipment properties, and mechanistic models that link these properties, and the process conditions to the final product quality (figure 1) is expected to facilitate the scale-up. The broader spray dryer configuration (closed loop/open loop, Figure 2) determines the type of process conditions to be accounted for (i.e., those of the atomizer, the actual spray drying section, cyclone separator, condenser and utilities).

Figure 1: Interactions of different process and product parameters impacting spray dryer product quality

Figure 2: Spray dryer configuration (a) Closed loop; (b) Open loop

Ref: Modelling of a spray drying process, Poul Bach, Martin Norby, Mark Pinto, Sean Bermingham, 2011, AICHE Annual Meeting ,Minneapolis, MN

The first part of this work shows how detailed experimentation can help quantify the relevant material properties and the spray dryer equipment characteristics. The second part focuses on leveraging a combination of process flow sheeting type of a model, along with detailed computational fluid dynamic simulations of the spray dryer chamber for understanding the interactions of the various inputs on the product quality (figure 3). The CFD simulations help in addressing challenges due to changes in geometry going from lab to plant scale spray dryers. The last part shows how this understanding helped in scaling up from 1 l/hr to 50 l/hr evaporating capacity spray dryer.

Figure 3 Coupling framework of CFD-Process flow sheeting model

522099 Industrial Application of Mechanistic Model for Fluid-Bed Granulation for Technology Transfer and Design Space Exploration

Abstract:

Fluid bed granulation is a frequently-used formulation step whose benefits typically include improved flowability and compressibility of powders. The process consists of two main stages: a spraying phase, during which fluidized particles are coated with an aqueous binder and agglomerate, and a drying phase used to evaporate away the water, resulting in a dry powder.

While the exact process outcome depends on a large variety of factors, making this process difficult to model in its entirety, it is found that the moisture-level profile experienced by the powder during the granulation is a useful and scale-independent measure of the process performance within a sizeable region of the parameter space [1]. Hence, here, a mechanistic model for this moisture-level profile is presented.

The tool, a parameterized heat- and mass-balance with a customized graphical user interface, has been used extensively during a technology transfer of multiple products from an old to a new site in Shaanxi province, China. Specifically, the model was used to plan pilot-scale experiments at different conditions, used to estimate model parameters and to study the effects of different parameters on outputs such as tablet hardness and dissolution. Subsequently, this information was used together with the model to design a set of target conditions and scale-up the process to full-scale.

In this work, we give an overview of the results obtained at the different scales, together with the challenges faced during development and implementation. Furthermore, we provide insight into how such a model can be leveraged in order to assess the effects of input fluctuations and disturbances to derive a more meaningful set of acceptable parameter ranges, founded on detailed process understanding.

[1] Lyngberg, O., et al. (2016). Applications of Modeling in Oral Solid Dosage Form Development and Manufacturing. Process Simulation and Data Modeling in Solid Oral Drug Development and Manufacture. M. G. Ierapetritou and R. Ramachandran. New York, NY, Springer New York: 1-42.

529890 Practical Application of a Mechanistic Model for Twin Screw Granulation for Pharmaceutical Process Development

Abstract:

The scientific understanding of twin screw wet granulation processes has advanced in recent years, allowing for the development of mechanistic models that quantify the effects of design decisions on granule attributes. Beyond capturing the scientific understanding of these processes, mechanistic models can be used to solve practical problems in pharmaceutical process development. However, the workflows and best practices for model validation are not always clear, and the modelâ€™s capabilities for answering targeted, meaningful questions are not well tested.

In twin screw granulation process development, finding the optimal design and operational space for a new formulation can consume significant time and material due to the highly configurable screw layout in addition to the main process parameters. In this work, a case study will be presented that utilizes a mechanistic model of a twin screw granulation process to improve on the current, experimentally driven process development workflow. Model validation against experimental data will be presented, and data requirements will be discussed. Further, the ability of the model to reduce the experimental burden during process development will be explored. Finally, the incorporation of downstream processing models, such as dryers, mills, and tablet presses, will be investigated.

534558 IbranceÂ® Capsule Commercial Process: Designed for Robustness

Abstract:

The commercial manufacturing process of Ibrance® (palbociclib) capsules was redesigned to ensure the quality attributes met or exceeded the Quality Target Product Profile (QTPP). The optimized process was established based on parameter ranges from process understanding achieved through multivariate and univariate analysis of small scale development trials, and computational modeling tools. The registration stability and clinical supply manufacturing campaigns, were utilized to broaden the process parameter ranges, establish process understanding, and ultimately define the proven acceptable ranges for the commercial manufacturing process.

During the registration stability and clinical supply manufacturing campaigns, an edge of failure was identified for the quality attribute of stratified capsule dissolution. The impact on dissolution was only observed at the end of the encapsulation run. It was determined that the edge of failure was an encapsulation processing tail-end effect which was associated with the original design/operation of the encapsulator hopper and powder feed system. Three engineering design modifications were identified and implemented to remediate the issue, and subsequently tested in a number of additional clinical and commercial scale capsule batches. Optimization of the encapsulator hopper and feed system defined the equipment design modifications that ensure consistent product performance throughout the entire batch.

531957 Application of Integrated Modeling Approach for Quality-By-Design (QbD) Process Development of Lyophilization

Abstract:

Lyophilization has been one of the most important unit operations in drug-product processes involving biologics as it affords low temperature drying to produce stable product that is amenable to storage and transportation. In order to achieve successful scale-up and technology transfer of a lyophilization process from laboratory to manufacturing, it is imperative to understand the scale-up issues/challenges, specifically the equipment performance (capabilities and limitations) and the differences in manufacturing environment. These need proactive considerations in the process design during the development stages. Computational fluid dynamics (CFD) provides inexpensive option to assess process equipment performance across scales.

In this work, CFD modeling was employed to study laboratory and manufacturing scale lyophilizers to understand the equipment performance differences arising from design differences and their impact on the process performance. CFD models were utilized to study and predict chamber pressure and temperature as well as flow patterns and velocities. Coupling CFD with the one-dimensional heat and mass balance was used to predict product temperature for a given formulation (Tgâ€™/collapse temperature) for both laboratory scale as well as manufacturing scale lyophilizers. A safe operational design space was constructed for a lyophilizer and product combination and utilized for quality-by-design (QbD) of lyophilization cycle scale-up and tech-transfer. We present here the CFD study on four lyophilizers at different scales coupled with two different formulations. The model predictions were verified using appropriate experiments. The design space provides flexibility to change/operate without revalidation as well as explanations for parameters excursions during operations.

Kushal Sinha, Ehab Moussa, Deliang Zhou, Madeleine Witting, Nandkishor Nere, Sherwin Shang, Mario Hirth, Martin Bultmann, Nupur Dutta, Ted Tharp, Feroz Jameel and Martin Gastens are employees of AbbVie. The design, study conduct, and financial support for this research was provided by AbbVie. AbbVie participated in the interpretation of data, review, and approval of the publication.

521464 Drug Product Process Modeling

Abstract:

During the past decade, computational process models have become more common place in applying to pharmaceutical development of drug products. This new state of the art has been forged predominantly by the chemical and mechanical engineering community, members of PD2M in AIChE. In this presentation, recent examples of modeling techniques being applied to drug product processing operations are discussed as support to the design, development, and scale-up for solid oral drug products. These process modeling techniques are discussed and exemplified with case studies ranging from powder flow and segregation, encapsulation, and film coating mapped to process parameter predictions. The major processing to produce capsules/tablets include i) powder processing; ii) dry and wet granulation; iii) encapsulation or tablet compression; and iv) film coating. Specific consideration has been given to transfer and scale-up issues along with general process design related challenges to pharmaceutical process R&D.

530526 Greenness By Design for Pharmaceutical Synthetic Processes

Abstract:

Designing efficient and green approaches to complex molecules is a challenge faced by any organization seeking to deliver modern pharmaceutical compounds to patients in a prompt manner. To measure process greenness, Process Mass Intensity (PMI) is a basic metric of the environmental efficiency for a synthetic step or sequence of steps. The PMI for synthetic sequence generally decreases as the process becomes optimized and environmentally friendly, however in selecting the synthetic route to a drug substance, the environmental impact of the final optimized process is uncertain. Therefore, choosing between different routes based on environmental impact becomes challenging and can be difficult for project teams to justify.

In this presentation we incorporate a data science approach to classify historical processes according to the sequence of chemical transformations to support a predictive analytics framework capable of quantifying the probable efficiency of a proposed synthesis. This method leverages real-world data to predict PMI green chemistry scores for proposed synthesis options, acting as a decision tool during the route selection process, predicting greenness metrics for any proposed, potential, or any unoptimized synthetic route, and enabling the direct comparison of the greenness score of an optimized synthesis to all comparable chemistry. Monte Carlo simulation provides teams with a realistic expectation of typical and best-case process performance, and a calculator interface facilitates prediction and comparison of PMI metrics. We envision that this rational approach for route selection will deliver significant impact to the chemistry community enabling greener decisions to be made during the route selection and process development.

531094 Data Rich Experimentation Methods Towards Immobilized Biocatalysis for Drug Substance Manufacture

Abstract:

Enzymes are increasingly used as highly specific and selective catalysts for small molecule API production. In many biocatalytic processes it is desirable to immobilize the enzyme on or within a solid support. Immobilization can improve enzyme stability, activity and catalyst loading and enables catalysis in environments which normally denature enzymes (e.g. extreme pH, high temperatures, and significant organic solvent). From a manufacturing perspective, immobilization greatly simplifies product isolation, reduces aqueous waste streams, and enables reactions to be telescoped into continuous flow processes.

Despite widespread use in chemical production, immobilized processes are generally developed by manually examining how the activity of a given enzyme varies when immobilized onto a small subset of resins. Here we discuss development of robust immobilized biocatalytic process at Merck. We focus on the development of flow-based reactors for characterizing enzyme immobilization on different resin classes and for monitoring biocatalytic processes in flow. We further discuss different biophysical methods and screening techniques which we are implementing to enable high throughput process development and improve our fundamental understanding of enzyme-resin interactions with an ultimate goal of developing robust, commercially viable manufacturing processes.

531894 Enzyme Deactivation Probed By Non-Isothermal Continuous Activity Assay

Abstract:

Enzymes have narrow temperature operating windows and complex deactivation mechanisms compared to chemical catalysts. Knowledge of the mechanism of deactivation is important for designing continuous processes incorporating biocatalysts. Improved enzyme performance can be achieved without protein engineering if the process is designed with the enzymeâ€™s deactivation mechanism and kinetics in mind.

A novel approach for determination of a kinetic model of enzyme deactivation is presented incorporating time and temperature effects into a single, continuous assay. Unique temperature profiles enhance model descrimination with fewer experiments and man-hours compared to linear temperature scans typical of techniques such as differential scanning calorimetry. Using the model enzyme penicillin G acylase (PGA), three models are examined in detail, however, the technique can be applied to any deterministic model of enzyme deactivation. Using the Akaike information criterion, the Lumry-Eyring mechanism was found to best capture PGA deactivation behavior and the corresponding kinetic parameters are presented for the first time. Simulated experiments are also used to better inform real experiments and add more confidence to the model discrimination. The results were consistent with conventional, but tedious, isothermal batch experiments. [1]

The development of this assay provided insight into continuous enzymatic catalysis. PGA demonstrated tolerance to small perturbations in temperature, which enables the operation of unique reactor designs. One design selected for further study is a reactive crystallizer [2] with a fines dissolution loop. A fines dissolution loop is a bypass at elevated temperature that destroys small crystals, increasing the average crystal size for easier downstream handling of solid products. The process tolerance to perturbations in temperature also depends heavily on enzyme stability. Knowledge of how an enzyme will react to fluctuations in temperature is necessary for design tolerances.

References:

[1] McDonald, M. A., et al., 2018. Kinetic Model Discrimination of Penicillin G Acylase Thermal Deactivation by Non-Isothermal Continuous Activity Assay. *Chem. Eng. Sci.*, IN REVISION

[2] McDonald, M. A., et al., Enzymatic reactive crystallization for improving ampicillin synthesis *Chem. Eng. Sci.*, 2017. 165, 81-88

537609 Removing Endotoxins from *E. coli*-Based Cell-Free Systems: Towards Enabling on-Demand Distributed Production of Therapeutics

Abstract:

Very recently researchers demonstrated the ability to lyophilized the reagents for *E. coli*-based cell-free protein synthesis. The resulting long-term stability of these reagents at room temperature has greatly simplified the use of this system for the on-demand on-site production of protein therapeutics. Indeed, this technology has been combined with DNA synthesis technology such that small robotic systems have been developed where an amino acid sequence is digitally inputted and a product protein is produced in a matter of hours using *E. coli*-based cell-free protein synthesis. This rapid turnaround time would be especially important for personalized cancer therapeutic and pandemic threat response applications. The low cost, robust, and rapid nature of *E. coli*-based cell-free systems makes it one of the most promising systems for on-demand therapeutic production. However, the presence of endotoxins at the extremely high concentrations of millions of endotoxin units per milliliter is a challenge for on-site distributed production. Here we report the utilization of liquid-liquid phase separation, affinity chromatography, and genetically engineered *E. coli* to remove endotoxins from cell-free reagents prior to protein production to reduce the risk of endotoxin contamination. This technology would eliminate additional endotoxin purification steps following protein production as well as additional quality control steps which would be otherwise necessary for on-demand on-site protein therapeutic production applications.

532727 Mechanistic Modeling and Parameter-Adaptive Nonlinear Model Predictive Control of a Microbioreactor

Abstract:

Microbioreactors are a promising innovative technology to accelerate biologic drug development. In aerobic cellular respiration, one potential limit to the productivity of such systems is the transport of oxygen from an external gas to the most oxygen-deficient cells, whereas another limit is the potential for excessive spatially localized dissolved oxygen which can result in cellular damage. In this work, a mechanistic model is constructed for the spatiotemporal transport of oxygen through a gas-permeable membrane to the cells within a microbioreactor. An analytical solution to the partial differential equations for oxygen transport is derived using the Finite Fourier Transform method. A parameter-adaptive extended Kalman filter is shown to produce highly accurate estimates of the oxygen uptake rate of the cells, with some fluctuation in estimates of the specific cell growth rate and the specific oxygen uptake rate. The estimates are fed to a model predictive control formulation that improves the spatial control of dissolved oxygen during cell growth by more than 30% compared to a traditional controller. The rest of this extended abstract describes the opportunities and challenges of microbioreactors, and the objectives addressed in this work, while also providing a review of the most closely related literature.

Conventional bench-scale stirred-tank bioreactors are well established for bioprocess development, but are labor-intensive and expensive, especially when operated in parallel for the optimization of media and the determination of optimal operating protocols during startup and transition to perfusion mode. Microbioreactor systems with embedded sensors for control and automation have been proposed as a more efficient alternative [1,2]. The primary fast dynamics that need to be controlled in a bioreactor in general and microbioreactor in particular are associated with the dissolved oxygen (aka DO) concentration. DO concentration either too high or too low causes cell damage or death [3].

The conventional method of bubble sparging is not feasible for aeration at the small volumes in microbioreactors, where bubbles would cause clogging and flow disruptions. An alternative aeration method that is feasible at small volumes is to select one or more walls of the microbioreactor as a gas-permeable membrane for oxygen to diffuse through from an oxygen source [4,5,6,7]. This surface aeration provides oxygen to the cells but creates an oxygen concentration gradient in the microbioreactor, rather than having perfect mixing. Due to physical limitations with implementing the sensors, the DO concentration is only measured at the bioreactor wall opposite of the gas-permeable membrane. One of the objectives of our work is to derive and validate a distributed parameter model for DO concentration in the microbioreactor.

In industrial practice, the pH, temperature, and DO concentration within bioreactors are typically controlled using Proportional-Integral-Derivative (PID) controllers. Many academic studies have been published on the open- and closed-loop control of bioreactors, which include gain-scheduled, nonlinear-inverse-based, singular, adaptive, and model predictive control systems [8,9]. Nonlinear model predictive control (NMPC) is of interest due to its ability to explicitly address dynamic nonlinearities and constraints. Among NMPC schemes, one of the earlier proposals for bioreactor control employed a nonlinear autoregressive with exogenous input model [10]. Later, an adaptive NMPC strategy was evaluated in simulations for the maximization of productivity of a continuous fermenter [11]. Numerous later studies considered NMPC with state estimation. For example, one study investigated the potential of using NMPC with an unscented Kalman filter to control starvation-induced cell death in Chinese hamster ovary (CHO) cells in the bioreactor [12]. Another study considered NMPC with an extended Kalman filter (EKF) for control of nitrogen and oxygen concentrations for a biological nitrogen-removal process, to make the effluent organic concentrations below regulatory limits [13]. Very recently, an NMPC implementation that includes dynamic flux balance models was investigated for fed-batch fermentation [14]. Also, NMPC has been experimentally implemented in recent years, including a study that demonstrated increased biomass and lipid productivity in a microalgal photobioreactor system [15]. Another objective of our work is to propose and evaluate an NMPC algorithm for the control of DO concentration throughout the cell-containing spatial domain within the microbioreactor. The algorithm employs parameter-adaptive EKF algorithm for simultaneous parameter and state estimation. The spatial variation of the DO within the microbioreactor results in different considerations than reported in the literature for stirred-tank bioreactors.

References

[1] J. W. Kim and Y. H. Lee, "Development of microfermenter chip," Journal of the Korean Physical Society, vol. 33, pp. S462-S466, 1998.

[2] Y. Kostov, P. Harms, L. Randers-Eichhorn, and G. Rao, "Low-cost microbioreactor for high-throughput bioprocessing," Biotechnology and Bioengineering, vol. 72, no. 3, pp. 346-352, 2001.

[3] A. Baez and J. Shiloach, "Effect of elevated oxygen concentration on bacteria, yeasts, and cells propagated for production of biological compounds," Microbial Cell Factories, vol. 13, no. 1, p. 181, 2014.

[4] A. Zanzotto, N. Szita, P. Boccazzi, P. Lessard, A. J. Sinskey, and K. F. Jensen, "Membrane-aerated microbioreactor for high-throughput bioprocessing," Biotechnology and Bioengineering, vol. 87, no. 2, pp. 243-254, 2004.

[5] N. Szita, P. Boccazzi, Z. Zhang, P. Boyle, A. J. Sinskey, and K. F. Jensen, "Development of a multiplexed microbioreactor system for high-throughput bioprocessing," Lab on a Chip, vol. 5, no. 8, pp. 819-826, 2005.

[6] Z. Zhang, P. Boccazzi, H.-G. Choi, G. Perozziello, A. J. Sinskey, and K. F. Jensen, "Microchemostat-microbial continuous culture in a polymer-based, instrumented micro-bioreactor," Lab on a Chip, vol. 6, no. 7, pp. 906-913, 2006.

[7] H. L. Lee, P. Boccazzi, R. J. Ram, and A. J. Sinskey, "Microbioreactor arrays with integrated mixers and uid injectors for high-throughput experimentation with pH and dissolved oxygen control," Lab on a Chip, vol. 6, no. 9, pp. 1229-1235, 2006.

[8] J. R. Banga, E. Balsa-Canto, C. G. Moles, and A. A. Alonso, "Dynamic optimization of bioprocesses: Efficient and robust numerical strategies," Journal of Biotechnology, vol. 117, no. 4, pp. 407-419, 2005.

[9] K. Y. Rani and V. R. Rao, "Control of fermenters - A review," Bioprocess Engineering, vol. 21, no. 1, pp. 77-88, 1999.

[10] T. Hong, J. Zhang, A. Morris, E. Martin, and M. Karim, "Neural based predictive control of a multivariable microalgae fermentation," in Proceedings of the IEEE International Conference on Systems, Man, and Cybernetics, vol. 1, pp. 345-350, 1996.

[11] P. Saha, S. Patwardhan, and V. Ramachandra Rao, "Maximizing productivity of a continuous fermenter using nonlinear adaptive optimizing control," Bioprocess and Biosystems Engineering, vol. 20, no. 1, pp. 15-21, 1999.

[12] L. Simon and M. N. Karim, "Control of starvation-induced apoptosis in chinese hamster ovary cell cultures," Biotechnology and Bioengineering, vol. 78, no. 6, pp. 645-657, 2002.

[13] H. El Bahja, P. Vega, O. Bakka, and F. Mesquine, "Non-linear GPC of a nutrient removal biological plant," in Proceedings of the IEEE Conference on Emerging Technologies & Factory Automation, pp. 1-7, 2009.

[14] L. Chang, X. Liu, and M. A. Henson, "Nonlinear model predictive control of fed-batch fermentations using dynamic ux balance models," Journal of Process Control, vol. 42, pp. 137-149, 2016.

[15] S. J. Yoo, D. H. Jeong, J. H. Kim, and J. M. Lee, "Optimization of microalgal photobioreactor system using model predictive control with experimental validation," Bioprocess and Biosystems Engineering, vol. 39, no. 8, pp. 1235-1246, 2016.

527880 Automated *in silico* Crystallization Process Design Using Solubility Models; Web Applications for Visualization and an Overview of a Solvent Selection Workflow

Abstract:

The accuracy of solubility models for realistic active pharmaceutical ingredients (API) has improved to the point that qualitative and quantitative predictions in aqueous and organic solvents are now becoming routine. In crystallization process design, such solubility predictions can be leveraged as a first step for selection of realistic solvent systems or isolation method (e.g. thermal, antisolvent or reactive crystallization) before going into the laboratory. In this contribution, a workflow is presented to show how such predictions are leveraged to aid in crystallization process design at Lilly. Additionally, custom Python and R code is described that takes in the modeled solubility data and simulates ~10000 possible crystallization processes for consideration as a starting point for process design. A user interface built upon web applications (Bokeh and Rshiny) for the visualization and process selection is also described. Application examples including R-UNIFAC from dynochem and Cosmotherm solubility models will be discussed.

521858 Computational Fluid Dynamics Boosted Stochastic Modelling for Integrated Quantitative Understanding of API Crystalline Product Manufacturing Process

Abstract:

A large number of APIs are marketed in dried crystalline form and have to comply with tight specifications on crystal morphology and crystal size distribution. Manufacture of these APIs involves several processing steps which include reaction, crystallization (anti-solvent and/or cooling), filtration and drying. Each operation significantly modifies crystal morphology. Considerable efforts are needed towards meeting the market and regulatory specifications on a commercial scale which is often a resource intensive task. However, resources expended, especially time and cost, can be greatly reduced by employing simulation tools which help in the tuning of production parameters to meet the final specification in an optimized manner.

Simple mathematical models provide information on gross behavior (e.g. rates of precipitation) but may not be effective towards generating more detailed and granular information, for example, distribution of morphology or particle size. The availability of inexpensive computing power opens up the opportunities to efficiently execute stochastic models. Such models, using â€œkinetic Monte-Carloâ€ methods, were used to simulate diffusive processes such as growth of crystal defects, dissolution of photolithography polymers, etc. A further advantage entailed by this approach lies in identifying a full crystal size distribution using a minimal set of parameters describing nucleation, growth, combination and break up.

Our approach used data from various sources: crystal size distributions from dynamic light scattering analyzer; phase diagram from gravimetric experiments; nucleation and growth kinetics from particle track technologies; and crystal breakup from shear contours of crystallizer using computational fluid dynamics. Similar models were deployed to understand crystal breakup and/or lump formation in filtration, drying and milling equipment.

This approach demonstrates an *end-to-end* quantitative model for API processes using quality-by-design (QbD) framework allowing engineers and scientists to develop robust API commercial manufacturing processes.

537627 Engineering the Surfaces of Fluorescently-Labeled Polymeric Nanoparticles for Drug Delivery

Abstract:

The use of nanoparticles in drug delivery can address many of the difficulties faced by traditional methods such as systemic toxicity, transport limitations, and drug solubility. By loading drugs into biocompatible polymeric nanoparticles, drugs that were previously unusable due to their solubility or toxicity can now be used. Fluorescently labeling the nanoparticles with a small molecule like 6,13-Bis(triisopropylsilylethynyl)pentacene (TIPS pentacene), the particles can be tracked for biodistribution studies in biological systems. TIPS pentacene is a fluorophore that has been primarily used in semiconductors, LEDs and solar cells.[1, 2] However, its high quantum yield and photostability make it a promising candidate as a fluorescent imaging agent for biomedical applications.

Using the Flash Nanoprecipitation process[3], different loadings of this fluorophore and a filler homopolymer of poly(D,L-lactic acid) (PDLLA) have been incorporated into amphiphilic diblock copolymer nanoparticles of poly(ethylene oxide)-b-poly(D,L-lactic acid) (PEO-PDLLA). We investigated the effects of varying the TIPS pentacene loading as well as the homopolymer to diblock copolymer weight:weight ratio on the characteristics of the resulting nanoparticles such as size, zeta potential, and fluorescent intensity.

The PEO-PDLLA nanoparticles made with varying amounts of fluorophore and homopolymer resulted in particles with sizes ranging from 60-200 nm. The zeta potentials were typically ~-15 to -20 mV. Previous work done by Pansare et al. analyzed a synthesized pentacene derivative Et-TP5 showing promising optical properties.[4] Recent work from our group showed TIPS pentacene loaded nanoparticles readily taken up by macrophages and the TIPS pentacene visible through imaging cytometry even at weight percent loadings in particles as low as 1 wt%.[5] The presentation will delve into the nanoparticle fabrication and engineering of these particles, looking at the fluorescence and quenching properties of the loaded TIPS pentacene. Initial studies showed that even at TIPS pentacene loadings of 1 wt%, self-quenching due to Forster Resonance Energy Transfer (FRET) may be present. Through similar analysis to that of Pansare et al., the work presented shows that the TIPS pentacene may be a comparable fluorophore appropriate for these applications but with the advantage that it is commercially available.

This research also tests the hypothesis that the surface chemistry of the particles can easily be altered by the addition of acid end capped PDLLA homopolymer to increase carboxylate groups on the surface while also changing the size properties. The carboxylate groups provide a useful functional group for coupling streptavidin to the surface. Streptavidin is a glycoprotein from the bacteria *Streptomyces* avidinii and with biotin, form on of the strongest non-covalent interactions with a dissociation constant of K_D â‰ˆ 10^{-15} M. By coating the nanoparticles with streptavidin, it provides a facile method for functionalizing the surface with biotinylated ligands for targeted drug delivery.

Flash nanoprecipitation shows promise in its scalability and efficacy in producing well-defined nanoparticles and this talk explores a potential design and processing framework of engineering polymeric nanoparticles appropriate for biological studies in a variety of different applications within the drug delivery field.

References

1. Anthony, J. E., Silylethyne-Substituted Pentacenes. *Material Matters: Organic and Molecular Electronics* **2009,** *4* (3), 58-60.
2. Platt, A. D.; Day, J.; Subramanian, S.; Anthony, J. E.; Ostroverkhova, O., Optical, Fluorescent, and (Photo)conductive Properties of High-Performance Functionalized Pentacene and Anthradithiophene Derivatives. *Journal of Physical Chemistry C* **2009,** *113* (31), 14006-14014.
3. Gindy, M. E.; Panagiotopoulos, A. Z.; Prud'homme, R. K., Composite block copolymer stabilized nanoparticles: Simultaneous encapsulation of organic actives and inorganic nanostructures. *Langmuir* **2008,** *24* (1), 83-90.
4. Pansare, V. J.; Bruzek, M. J.; Adamson, D. H.; Anthony, J.; Prud'homme, R. K., Composite Fluorescent Nanoparticles for Biomedical Imaging. *Mol. Imaging. Biol.* **2014,** *16* (2), 180-188.
5. McDaniel, D. K.; Jo, A.; Ringel-Scaia, V. M.; Coutermarsh-Ott, S.; Rothschild, D. E.; Powell, M. D.; Zhang, R.; Long, T. E.; Oestreich, K. J.; Riffle, J. S.; Davis, R. M.; Allen, I. C., TIPS pentacene loaded PEO-PDLLA core-shell nanoparticles have similar cellular uptake dynamics in M1 and M2 macrophages and in corresponding in vivo microenvironments. *Nanomedicine: Nanotechnology, Biology and Medicine* **2017,** *13* (3), 1255-1266.

537333 Optimizing the Surface Property-Activity Relationship of Nanoscale Hydrogel Drug Delivery Systems

Abstract:

Ovarian cancer is the deadliest of all gynecological cancers and the fifth leading cause of death due to cancer in women, resulting in more than 15,500 deaths each year [1]. Screening for ovarian cancer in the general population is not recommended by any medical group as many other benign conditions also exhibit expression of the same biomarkers, including endometriosis-induced irritation, pregnancy, and liver disease, leading to a significant rate of false diagnosis [2]. As a result, the development of improved therapies remains a key challenge. Despite broadly increased research into the development of new anti-cancer therapies, there has been no change in the mortality rate for ovarian cancer in the last 30 years [3].

Current treatment plans typically include chemotherapy with radiation or surgical removal of the tumors. However, traditional chemotherapy does not distinguish between healthy and cancerous cells and typically leads to significant side effects. The use of nanoparticles as delivery vehicles has shown promise for enhancing the systemic delivery of chemotherapeutic agents and solving some of the limitations that conventional administration presents [4]. There are three major requirements for nanoparticles to deliver their payload to the tumor. They need to: (i) be stable in the circulation without releasing drug prematurely, (ii) evade opsonization and accumulate in the tumor efficiently, and (iii) intelligently release drug inside the tumor cells. This means the nanogel must have sufficient time in blood circulation to reach intended sites of the body and accumulate in and around the tumors via leaky vasculature. Ultimately, localizing and controlling drug release at the disease site would offer advantages over the current chemotherapeutic regimens because it limits the toxicity to healthy tissues and decreases the side effects for the patient.

In this work, we have developed long-circulating, biocompatible nanoscale hydrogels (nanogels) that respond to the acidic intracellular environment for controlled delivery of chemotherapeutics. The nanogels are comprised of: (i) hydrophilic, cationic 2-(diethylamino)ethyl-methacrylate, (ii) tetraethylene-glycol-dimethacrylate crosslinker, and (iii) surface-grafted poly(ethylene-glycol). First, we show that the nanogel molecular architecture can be tailored to carry a variety of cargos with varying physicochemical properties, and the release kinetics can be tuned to a desired stimuli. The physical properties of the resulting nanogels were compared using dynamic light scattering, zeta potential, titration, pyrene fluorescence, and red blood cell hemolysis to elicit the influence of polymer composition on swelling ratio, surface charge, pKa, hydrophile-hydrophobe phase transition, and erythrocyte membrane disruption capability. The delivery capacity was analyzed using carboplatin and paclitaxel.

Next, we optimized the nanogel surface properties to avoid the body's natural clearance mechanisms and increase circulation time, while still maintaining the necessary characteristics to promote cellular uptake. Stealth coatings shield the particle surface charge and thereby reduce opsonization by blood proteins and uptake by macrophages of the mononuclear phagocyte system. Yet, after localizing at the tumor site, the coating may hinder drug release and target cell interaction and can therefore be an obstacle in the realization of the therapeutic response. To this extent, we developed a novel stimuli-responsive coating on the particle surface that unmasks a ligand that binds to cell surface receptors overexpressed in ovarian cancer cells. Using hyaluronic acid, PEG, and dextran, a library of nanoparticles with varying degrees of surface density was synthesized, and the impact to particle physicochemical properties, protein binding, drug release, and binding affinity was investigated. The influence of the surface properties to cellular cytocompatibility, uptake, degree of nanogel-cell specificity, and therapeutic efficacy was evaluated with OVCAR-3 cells as a function of time and concentration.

We demonstrated that both the surface-grafted and hydrophobic monomer composition can be varied to enable precise control over the nanogel surface and core characteristics to enable effective, controlled delivery. Varying both chain length and steric bulk allowed for precise control over the thermodynamic response (swelling ratio), dynamic behavior (pKa, membrane disruption), and drug-polymer interaction (delivery capacity). Ultimately, we demonstrated that the tunability of our multicomponent nanogel can be exploited to enable long-circulation and effective transport of the drugs to the tumor site.

References:

[1] Ovarian Cancer National Alliance. wwwovariancancerorg. 2015.

[2] Nolen BM, Lokshin AE. Biomarker testing for ovarian cancer: clinical utility of multiplex assays. Molecular diagnosis & therapy. 2013;17(3):139-46. Epub 2013/04/05. doi: 10.1007/s40291-013-0027-6. PubMed PMID: 23552992; PMCID: PMC3670781.

[3] National Cancer Institute. SEER Cancer Statistics Factsheets: Ovary Cancer. http://seercancergov/statfacts/html/ovaryhtml. 2015.

[4] Kirtane AR, Kalscheuer SM, Panyam J. Exploiting nanotechnology to overcome tumor drug resistance: Challenges and opportunities. Advanced drug delivery reviews. 2013;65(13-14):1731-47. Epub 2013/09/17. doi: 10.1016/j.addr.2013.09.001. PubMed PMID: 24036273; PMCID: PMC3849460.

534117 Paramagnetic Cations-Loaded Polydopamine Nanoparticles Cytotoxicity

Abstract:

Melanins are natural occurring biopolymers with different functions in the organisms in which they are present. Among these functions, skin UV protection is perhaps the most recognized. However, they participate in many other biological processes, as the regulation of the cellular concentration of free metallic ions, being able of sequestering them or acting as an ion reservoir [1,2].

Taking into account this fact, polydopamine, a synthetic melanin analogous, has acquired an important role in material science [3]. Since its ability to chelate paramagnetic cations is known, it has been considered to device MRI contrast agents or more advanced theranostic systems. In this line, several studies have demonstrated polydopamine biocompatibility during the first 24 hours of treatment, but its long-time cytotoxicity has not been studied yet [4].

Thus, in the present work, the interaction of polydopamine particles with different cations (Fe^{2+}, Fe^{3+}, Ca^{2+}, Na^{+}, Zn^{2+} and Cu^{2+}) has been carried out, analyzing their metal loading capacity and affinity. A tentative assignation for such metal binding sites has been performed by IR spectroscopy and, finally, polydopamine cytotoxicity has been studied in a cancer (BT474) and a healthy cell lines (NIHT3T3). As result, it has been found that this melanin analogous is toxic for both cell lines, possibly due to a ROS production process. So, in this manner, there would be a concern in the use of these metal-enriched polydopamine particles in some biomedical applications, but they could represent a potential antineoplastic agent [5].

References

[1] Hong L, Liu Y, Simon JD. Binding of metal ions to melanin and their effects on the aerobic reactivity. Photochemistry and Photobiology. 2004;80:477-81.

[2] Hong L, Simon JD. Current understanding on the binding sites, capacity, affinity and biological significance of metals in melanin. The Journal of Physical Chemistry B. 2007;111:7938-47.

[3] Lee H, Dellatore SM, Miller WM, Messersmith PB. Mussel-inspired surface chemistry for multifunctional coatings. Science. 2007;318:426-30.

[4] Cho S, Park W, Kim D-H. Silica-coated metal chelating-melanin nanoparticles as a dual-modal contrast enhancement imaging and therapeutic agent. ACS Applied Materials & Interfaces. 2017;9:101-11.

[5] Vega MA, Nieto C, Marcelo G, Martín del Valle EM. Cytotoxicity of paramagnetic cations-loaded polydopamine nanoparticles. Colloids and Surfaces B: Biointerfaces. 2018. doi: 10.1016/j.colsurfb.2018.04.027

536485 Two-Photon Microscopy for Deep-Tissue Imaging of Dopaminergic Neuromodulation in the Brain

Abstract:

Non-invasively imaging the brain neuronal structure and neuro-chemical activity stands to inform our understanding of psychiatric and neurological disorders and their pharmacological therapies. However, the high optical density of brain tissue makes imaging disease-relevant regions at its center a challenge using conventional fluorophores and microscopy techniques, particularly for intact brains in living animal models. We present a microscopic approach that could enable whole-brain, through-skull, and continuous imaging of neuromodulation. To this end, we have combined two imaging modalities that have emerged in parallel with opposing imaging strengths and weaknesses: infrared microscopy, and two-photon microscopy.[1] Recently, functionalized single walled carbon nanotubes (SWNTs) have emerged as promising near-infrared (NIR) fluorescent nanosensors for the sensitive and rapid detection of dopamine, a neuromodulatory neurotransmitter of clinical importance to brain function and disease.[2] SWNT nanosensors fluorescence in the NIR-II window (1000-1700 nm), where absorption and scattering of light is minimal, making them optimally suited for non-invasive imaging in brain tissue. However, to overcome the challenge of absorption and scattering of excitation photons in deep-tissue imaging, we show that intrinsically-near-infrared fluorescent SWNT nanosensors can be efficiently imaged with two-photon excitation (in lieu of one-photon excitation) through photoexcitation by a NIR-II femtosecond pulsed erbium fiber laser. We measure an effective, average quantum yield of 0.2% and a two-photon cross section of 216,000 GM. Furthermore, we demonstrate that two-photon photoexcitation improves the imaging quality of nanosensors in brain-tissue phantoms at depths approaching 2 mm and confirm that non-linear excitation preserves the molecular recognition of dopamine. Finally, we show that this imaging paradigm can be expanded to a laser-scanning system paired with a NIR photomultiplier tube, enabling layer-by-layer 3D imaging, motivating future efforts to achieve real-time imaging of dopamine neurotransmission in awake and behaving animals.

1 J. T. Del Bonis-Oâ€™Donnell, R. H. Page, A. G. Beyene, E. G. Tindall, I. R. McFarlane and M. P. Landry, *Adv. Funct. Mater.*, 2017, **27**, 1702112.

2 A. G. Beyene, I. R. McFarlane, R. L. Pinals and M. P. Landry, *ACS Chem. Neurosci.*, 2017, **8**, 2275â€"2289.

538019 Ultrafast Post-Formulation Core Radiolabeling of Biodegradable Nanoparticles for PET Contrast Agents

Abstract:

Among almost all cancers cases, diagnostic accuracy and speed remains one of the major factors for patient prognosis. Positron emission tomography (PET) is among the most sensitive of imaging modalities but requires accumulation of radionuclide at the targeted site. ^{18}F-FDG (Fludeoxyglucose) is the current gold standard for PET diagnostics but still has challenges with accessibility, sensitivity and specificity. The short half-life of ^{18}F limits PET usage to only institutions with easy access to a cyclotron. Furthermore, since resolution is dependent on concentration of positron emitters at the imaging site, smaller and low metabolism cancers might not be detectable using FDG. Lastly, non-tumor areas with high metabolism, such as brain tissue or inflammation sites, produce high background signals that limit the specificity of FDG-PET. Nanoparticle-based PET modality has the potential to overcome all these limitations. Targeted particles can both localize PET signal and drastically improve sensitivity by clustering the PET imaging agents in one place. Additionally, nanoparticles have more flexibility in labeling with longer half-life PET isotopes such as Cu-64 (12.7 hrs) and Zr-89 (3.3 days), suitable for long-time imaging and drug biodistribution studies. However, the lack of a clinically relevant formulation and radiolabeling method is the current bottleneck for nanoparticle contrast agent development. Traditional surface labeling methods are not industrially scalable and can lead to off target effects *in vivo*. Alternatively, radiolabeling the nanoparticle core requires careful selection of biocompatible chelators that can rapidly bind to radionuclides to form stable complexes. To this end, we here present the use of Flash Nanoprecipitation (FNP), a continuous and scalable self-assembly process, to form porphyrin encapsulated polymeric nanoparticles (NPs) at various sizes ranging from 50 to 200 nm. The core component, hematoporphyrin derivative, is a biocompatible small molecule that has been FDA approved for photodynamic therapy (Photofrin). Its iron(III) complex precursor is Heme, a major component of hemoglobin. When incubated in a solution of [^{64}Cu]CuCl2, a long half-life positron emitter, the porphyrin core rapidly chelated copper in a stable and irreversible manner, generating PET active nanoparticles. Under very mild temperature and pH conditions (25C and pH 5.5), the porphyrin nanoparticles' chelation to ^{64}Cu was complete in less than 15 minutes with essentially all the initial ^{64}Cu activity (1.1 MBq) localized within the nanoparticle core. This core radiolabeling method is comparable or even faster than currently used nanoparticle surface labeling methods. Additionally, the porphyrin:^{64}Cu complex in the nanoparticle core was found to be stable for at least 24 hours and resistant to leaching from external chelators such as ethylenediaminetetraacetic acid (EDTA). Chelation rate modeling was done based on absorbance changes of the porphyrin nanoparticles when incubated with non-radioactive copper and was found to be consistent with experimental results. Drugs could also be encapsulated into the core of these nanoparticles to facilitate a theranostic approach to cancer diagnostic and treatment. This development of a formulation and rapid radiolabeling method to generate stable and biocompatible PET nanoparticles facilitates more accessible, sensitive and specific cancer diagnostics.

A close up of text on a white background Description generated with high confidence

539532 Voltage-Sensitive Ultrasound Enhancing Agent: *In Vitro* and *In Vivo* analysis

Abstract:

We have created a voltage-sensitive ultrasound enhancing agent comprised of perfluorocarbon nano-droplets coated in a monolayer of polymeric surfactant that is then nested inside a negatively charged phospholipid bilayer. The particle, termed Electrast, is responsive to an applied electric field allowing for its internal components to become acoustically activated at ultrasound intensities that are inadequate to cause activation by ultrasound alone. Electrast is novel because activation depends on the presence of an electric field rather than the intensity of ultrasound. This talk will discuss key results from the bench top and the clinic, while commenting on the role of water transport in nesting structures, lipid relocation and reorientation, and lateral phase separation. Calcein and cobalt chloride leakage studies were performed to quantify the amount of water transported across the phospholipid bilayer when exposed to high frequency ultrasound, an applied electric field or a combination of the two. A custom-built, tissue-mimicking gel phantom coupled with a function generator was used to determine the contrast-to-tissue ratio of two different nesting architectures at varying electric field strengths and imaging parameters. Electrast has shown to be selectively activated in coronary circulation in two different animal models (rat and swine) showing an increase in brightness (dB) in the myocardium while leaving the ventricles largely unchanged. This talk will show that Electrast has the potential to be used clinically for myocardial perfusion imaging, assessing heart failure, and evaluating left ventricle wall motion.

537768 Polymorphic Selection of Biominerals By Anionic Polyelectrolytes

Abstract:

In most biomineralization processes many soluble biopolymers exist in minute quantities; these control the nucleation, growth, orientation, and polymorph selection, leading to the formation of minerals with diverse biological functions and distinct properties but also to pathological mineralization, e.g., bone disease, dental calculus, and renal stone formation. In this work, we study the effects of supersaturation, temperature, and additives on the nucleation and growth of calcium oxalate (the major constituent of kidney stones) in aqueous solutions. We have demonstrated that certain anionic polyelectrolytes can be very effective toward kidney stone inhibition by reducing the growth rate, altering the crystal morphology, and favoring the formation of the metastable dihydrate (COD) over the thermodynamically stable monohydrate (COM) which is mainly responsible for kidney, gall, and bladder stones. We compare our results to citrate, a naturally occurring inhibitor in the urinary system and currently the standard pharmacological treatment for stones. Our findings offer a better understanding of the calcium oxalate crystallization kinetics and provide information on the viability of synthetic polymers as potential calcium oxalate stone therapies.

538910 Coupling Experimental Results with Molecular Dynamics Simulations to Describe Polymorphism Obtained Using Solution Shearing

Abstract:

During pharmaceutical drug development, it is important to carry out an extensive search to identify all drug forms. Stabilizing metastable polymorphs is necessary to ensure that the correct drug form is selected for development to balance form stability and drug bioavailability. However, searching for rare forms is a nontrivial process. Here, we show that a combination of fluid dynamics and surface chemistry can be used to overcome some of the challenges associated with searching this extensive landscape of polymorphism. In this study, multiple forms of glycine and acetaminophen were isolated, including metastable forms that are difficult to obtain using conventional methods. The mechanisms through which multiple polymorphs are obtained are described herein. Molecular dynamics simulations and in-situ grazing incidence wide angle X-ray diffraction lend insight to the phase transitions that occur during the crystallization process for fundamental understanding of solution shearing.

533104 Investigation of the Polymorphy Phenomenon of the Carotene Lutein

Abstract:

Carotene is a special group of hydrocarbon pigments characterized with π-π conjugated double bonds in the molecular backbone. This group of compounds is supposed to have great potential in pharmaceutical, healthcare and optical materials because of its excellent properties in antioxidative, light-harvesting, photo-protective, and conductive. Carotene is unstable, due to the conjugated double bonds intend to abstract electron by photo-oxidation. Several methods such as encapsulation have been employed to enhance the stability of carotene. However, seldom studies concerned crystal engineering of carotene to improve its stability has been reported. Nonetheless, H and J aggregation phenomenon of carotene moleculars have been investigated with specific solvents. Recently, β-carotene were reported to have the ability to bind decanuclear homo-and heterometal chain through μ_{10}-briding π-coordination. These findings suggested that the π-π conjugated bond in carotene molecular might have strong plasticity in molecular self-assembling process. As a member of the carotene family, lutein also posseses two hydroxy group (which is easy to occurs polymorph due to the H-bonds between molecular chains) in the molecular structure besides the conjugated double bonds.Thus, cooling crystallization, diffusion zone process and vapor diffusion crystallization methods with selected solvents were tested for lutein, and crystal powders of lutein with needle like, platelet and pyramids morphylogy were obtained. By XRD, TG and DSC analyzing, three crystalline polymorphic forms of lutein were identified. The three crystalline forms of lutein exhibited distinct characters in morphylogy and stability, and the correlation between the stability and the crystal form will be investigated further.

A Kinetic Study of Crystallization Process of Imatinib Mesylate with Polymorphic Transformation Phenomenon

Mengxing Lin, Yuanyi Wu, Sohrab Rohani *
The University of Western Ontario, Department of Chemical and Biochemical Engineering, London, Ontario N6A 5B9, Canada
* Corresponding author: srohani@uwo.ca

Abstract

For pharmaceutical crystallization design and control, the kinetics of nucleation and growth of crystals are significant parameters, especially for the system exhibiting polymorphic transformation. In this study, imatinib mesylate, whose solubility, nucleation and growth kinetics are lacking in the literature, was thoroughly explored in the aspects of characterization, solubility and polymorphic transformation, and evaluation of nucleation and growth rate. Two forms of imatinib mesylate, α and β, were characterized by X-ray powder diffraction, scanning electron microscopy and differential scanning calorimetry. The solubility measurement of the two forms was performed with ultraviolet–visible spectroscopy in three solvents from 278.15 to 333.15 K at atmospheric pressure. The results indicated the β-form is more stable in methanol,1-propanol, and 2-propanol under the experimental condition. The solubility order is methanol >1-propanol >2-propanol. The solvent-mediated polymorphic transformation (SMPT) of imatinib mesylate from α-form to β-form was studied with the in-situ Raman spectroscopy and conductivity meter. It is found that the SMPT process of imatinib mesylate from α-form to β-form in methanol is controlled by nucleation and growth of β-form. Finally, the nucleation and grow rate of β-form of imatinib mesylate were estimated by minimizing the difference between the calculated and experimental solution concentration with MATLAB optimization function.

Keywords: A1. Crystal morphology, A1. Solubility, A2. Growth from solutions, B2. Organic compounds

1. Introduction

Depending on the arrangement of molecules, a solid can be a crystal, a quasicrystal or an amorphous. A phenomenon called polymorphism, exists widely in the crystalline solids, especially in organic compounds[1]. Even though polymorphs have the same molecular formula, they are different in molecular conformations and/or packing arrangements in the solid state, resulting in the distinct physical and chemical properties, such as morphology, melting point, flow properties, mechanical properties, solubility, and Gibbs free energy[2]. In terms of active pharmaceutical ingredients (API), polymorphism would affect the processability, bioavailability and stability of the API[3]. Therefore, the identification and control of polymorphs of a given API are essential for the pharmaceutical industry.

Comparing with the stable polymorph, the metastable polymorph has higher solubility and Gibbs free energy. The divergence in Gibbs energy drives polymorphism to transit from the metastable form to the stable state. Generally, there are two ways to realize the transition: solid-state polymorphic transformation (SST) and solvent-mediated polymorphic transformation (SMPT)[4]. The SMPT process consists of three steps: dissolution of the metastable form, nucleation of stable form, and growth of stable form. The slowest step of the SMPT process limits the whole transition rate. Based on the relative rate of dissolution, nucleation, and growth, O'Mahony et al. rationalized the SMPT process into four possible scenarios: dissolution-controlled, nucleation-dissolution-controlled, growth-controlled, nucleation-growth-controlled [5].

In order to model and control a crystallization process, the kinetic parameters, namely nucleation rate, growth rate, dissolution rate, agglomeration, and breakage have to be determined first through experiments. For the nucleation and growth rate, there are three experimental methods commonly used to determine the nucleation and growth rate. The first one relies on the crystal size distribution of the final product obtained from a Mixed-Suspension, Mixed-Product Removal (MSMPR) crystallizer. The slope and intercept of the "particle number density vs. crystal size" curve provide the information necessary to extract the nucleation and growth rates [2]. The second method which is somewhat approximate, is based on the count and chord length distribution measured by the Focused Beam Reflectance Measurement (FBRM) device [6]. The change in the count of fine crystals presents the nucleation rate, while the change in average chord length refers to the growth rate. The last method uses a rigorous parameter estimation algorithm and works by finding a set of proper constants in the kinetic equations to minimize the difference between the experimental data (measured solute concentration and/or crystal size distribution) and modeling results [7].

The compound studied in this work is imatinib mesylate. Imatinib mesylate (Figure 1), chemically 4-[(4-methylpiperazin-1-yl) methyl]-N-[4-methyl-3-[(4-pyridin-3-ylpyrimidin-2-yl) amino]-phenyl] benzamide methanesulfonate, was developed by Novartis under the trade name "Gleevec/Glivec®" in the late1990s. It was designed as a protein tyrosine kinase inhibitor and shows a remarkable therapeutic effect in Philadelphia chromosome positive chronic myelogenous leukemia (CML) and metastatic malignant gastro intestinal stromal tumors (GISTs) [8]. Many polymorphs of crystalline imatinib mesylate have been observed by precipitating from a solution that consists of an organic solvent, imatinib base and methanesulfonic acid. The different solvents lead to different polymorphs of the final product[9]–[16].

Among the dozens of polymorphs of imatinib mesylate, the α-form and the β-form proposed in WO99/03854 [9] are the most frequent and commercialized. The cubic-shaped β-form has the lower hygroscopic ability, higher flow properties, and higher stability at temperatures below 140° C than the needle-shaped α-form. Therefore, the β-form shows better performance than α-form in terms of processability, manufacturing, pharmaceutical preparation, and storage [17].

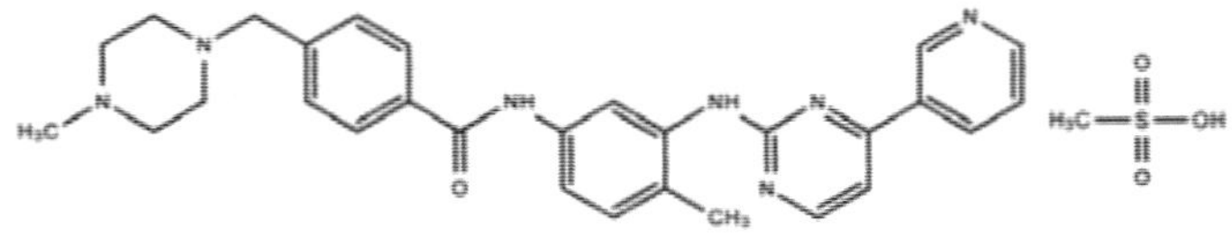

Figure 1. Schematic of imatinib mesylate

Grillo et al. revealed the crystal structure of α-form and β-form imatinib mesylate and found these two polymorphs have significantly different molecular conformations. By grinding them, the amorphous imatinib mesylate was produced that converted to a crystalline phase by either thermal treatment or aging [18]. Veverka et al. modified the morphology of α-form imatinib mesylate with various organic solvent mixtures by precipitation [19]. Srivastava et al. presented a comprehensive study of the vibrational spectroscopies of β-form with the FT-Raman and FT-IR spectra [20]. The polymorphic composition of two polymorphs mixture was determined quantitatively with PXRD, ATR-FTIR, and DSC in the literature [21]. The results show that PXRD technique is more accurate than ATR-FTIR and DSC.

Despite the mentioned studies, the investigation of the imatinib mesylate properties, particularly those related to the crystallization process, was still lacking. The solubility data has not been published up to date. In this study, we measured the solubility of two forms of imatinib mesylate in three solvents (methanol, 1-propanol, and 2-propanol) from 278.15 to 333.15 K at atmospheric pressure to study their relative stability. Subsequently, the solvent-mediated polymorphic transformation (SMPT) from metastable α-form to stable β-form was studied in-situ with Raman spectroscopy and conductivity meter. Finally, the nucleation and grow rate of β-form imatinib mesylate were estimated by minimizing the difference between the calculated and experimental solute concentration with MATLAB optimization function.

2. Material and Experiments

2.1 Material

The α-form imatinib mesylate was supplied by Apotex Pharmachem Inc. (Brantford, ON, Canada) and used without any further processing. The β-form was obtained by the solvent-mediated polymorph transformation from α-form in methanol. The solvents (methanol, ethanol, isopropanol, n-propanol, 1-butanol, acetone, acetonitrile, and tetrahydrofuran) with HPLC grade were purchased from Sigma-Aldrich (Milwaukee, WI,) and no further purification was performed.

2.2 Characterization

2.2.1 Powder X-Ray Diffraction (PXRD)

The polymorphs of solid-state α-form and β-form imatinib mesylate were identified with powder X-ray diffraction (Rigaku, Miniflex, Tokyo, Japan) with the Cu-Kα source ((λ for Kα = 1.54059 Å) at room temperature. The conditions of voltage (30kV), current (15mA), scan angle (2° to 50°) and scan speed (2°/min) were kept the same for all samples.

2.2.2 Differential Scanning Calorimetry (DSC)

Thermal properties of crystals, such as melting point and enthalpy, were measured by differential scanning colorimeter ((DSC, Mettler Toledo, Chicago, United States). The samples (5-10 mg) were placed in a 40-μl aluminum crucible with a pierced lid for gas escaping and then heated from 25°C to 250°C with 10°C/min heating rate under nitrogen purge at a flow rate of 40 ml/min.

2.2.3 Optical microscopy and Scanning Electron Microscopy (SEM)

The optical microscopy (Axioskop 40, Carl Zeiss Light Microscopy, Germany) with a magnification of 500 times was used as an off-line method to give a rough judgement of the polymorphic composition of solid-state imatinib mesylate in suspension. Scanning electron microscope (LEO 1450XB , Zeiss, Oberkochen, Germany) was employed for detailed morphologic information. The samples have been coated with coated with 5 nm of osmium metal in Osmium Plasma Coater (OPC80T, Filgen Inc. Japan) before SEM testing.

2.2.4 Raman spectroscopy

A Raman RXN Analyzer spectroscopy (Kaiser Optical Systems, Inc., Ann Arbor, MI, USA) equipped with a diode laser (784.8-nm) and a fiber optic probe was applied to monitor the polymorphic composition of solid-phase during the SMPT process from α-form to β-form. The characteristic peaks of α-form and β-form of imatinib mesylate are at 1664 cm^{-1} and 1656 cm^{-1}, and the single point baseline is at 1640 cm^{-1}.

2.3 Solubility measurement

To study the stability of two forms, we used Ultraviolet-visible spectroscopy (UV-Vis, Cary Bio 100 spectrophotometer, Varian, Mississauga, ON) to measure the solubility of imatinib mesylate in methanol, 1-propanol, and 2-propanol. The calibration curve of "Absorbance-Concentration" should be determined with a series of known concentration solutions at room temperature at first. The characteristic UV-Vis absorption band at 271-nm and 275-nm was applied for quantification in methanol and propanol, respectively. Then, an excessive sample was added to the selected solvent in a 5-ml glass vial. The vials were immersed in a constant temperature bath with an accuracy of 0.1 °C and agitated for several hours at a specific temperature to reach equilibrium. After 30 mins standing, the supernatant was taken by a syringe and filtered with a 0.45-μm membrane syringe filter (VWR, Mississauga, ON), followed by proper dilution and concentration measurement with UV-Vis. The excess solids in the vials were filtered, dried, and checked with PXRD.

As the α-form imatinib mesylate transformed to β-form in methanol during the solubility measurement, the solubility of α-form imatinib mesylate was determined during the transition process (see section 2.4 for more details).

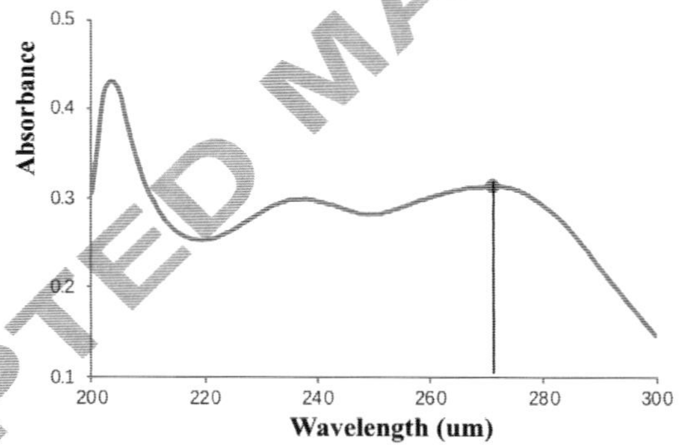

Figure 2. UV-Vis absorption spectra of imatinib mesylate in methanol with a characteristic peak at 271-nm

2.4 SMPT experiments

Eight solvents (methanol, ethanol, isopropanol, n-propanol, 1-butanol, acetone, acetonitrile, and tetrahydrofuran) were screened for mediating the polymorphism transformation. An excessive amount of α-form imatinib mesylate was added to the solvents and agitated in a thermostatic shaker for 2 weeks. The start time of transition was determined by the morphology of the crystals remaining in the solution with the microscope. After filtration and drying, the crystal polymorph was determined by PXRD.

Subsequently, the SMPT process from α-form to β-form imatinib mesylate was investigated in methanol at different temperatures. A carefully excess weighed amount of α-form was added to 35-ml methanol in a 50-mL double-jacketed crystallizer. The crystallizer temperature was kept constant by a water bath circulator (FP-50, Julabo LABORTECHNIK GMBH, Germany). The solute concentration and solid-phase polymorphic composition were monitored by a conductivity meter (Pinnacle Series: M541P) and Raman spectroscopy. The start and end time of the transition process was determined through the morphology of the undissolved crystals with the optical microscope. The solubility of α-form was detected during the SMPT process when the transition did not happen.

3. Theory

3.1 Solubility models

In this work, the solubility of two forms of imatinib mesylate in the selected solvents was fitted with the modified Apelblat equation, as shown in Eq. (1)

$$lnx = A + \frac{B}{T} + ClnT \quad (1)$$

where x is the molar fraction of solute in the solution, T is the absolute temperature in Kelvin, and A, B, and C are dimensionless parameters. The value of three parameters were estimated with MATLAB curve fitting toolbox.

The solubility, in molar fraction, can be calculated using Eq. (2):

$$x = \frac{m/M}{m/M + m_s/M_S} \quad (2)$$

where m and m_s are the mass of solute and solvent, M and M_S are the molecular mass of solute and solvent, respectively.

Solubility reflects the Gibbs energy and stability. The more stable a polymorph is, the more stable a polymorph is, the lower the solubility it is at a given temperature. The difference of Gibbs free energy between two polymorphs relates to the solubility by Eq. (3)

$$\Delta G = RTln\left(\frac{a_1}{a_2}\right) \approx RTln\left(\frac{x_1}{x_2}\right) \quad (3)$$

where ΔG is the difference of Gibbs free energy (J/mol), and a is the solute activity in the solution[22].

3.2 Transition process

The SMPT process consists of three steps: the dissolution of metastable form, nucleation of stable form and growth of stable form. The empirical equations for secondary nucleation and growth rate are expressed as Eq.(4)-(5) [23]:

$$B_{sn} = \frac{dn_0}{dt} = k_{sn}{M_T}^j S^b \quad (4)$$

$$G = \frac{dL}{dt} = k_g S^g \quad (5)$$

where B_{sn} is secondary nucleation rate [#/(m^3· s)], G is growth rate (m/s), S is relative supersaturation, M_T is suspension density (kg solute /kg solvent). k_{sn}, k_g, j, b and g are parameters, representing the nucleation rate coefficient, growth rate coefficient, nucleation order on suspension density, nucleation order on supersaturation, and growth order on supersaturation.

$$S = \frac{C - C^*}{C^*} \quad (6)$$

where C is the solution concentration, while the $C^*$is the equilibrium concentration, namely solubility. In this work, their units are kg solute /kg solvent.

3.3 Parameter estimation

The population balance equation (PBE) coupled with solute mass balance provides a mathematical framework of the crystallization processes. The PBE with the Lagrangian approach[23] describes the change in crystal population density with respect to time and space, as shown in Eq. (7).

$$\frac{\partial n}{\partial t} + \nabla.(v_e n) + \nabla.(v_i n) = B - D \tag{7}$$

where n is population density, v_e is external velocity vector, v_i is internal velocity vector, B is crystal birth rate, and D is crystal death rate. Assuming that a crystallizer has a uniform distribution of suspended solids, a size-independent growth rate, and negligible breakage and agglomeration, Eq. (7) can be simplified to a 1-D PBE as following when $L \neq 0$:

$$\frac{\partial n(t,L)}{\partial t} + G(t)\frac{\partial n(t,L)}{\partial L} = 0 \tag{8}$$

The boundary condition at $L = 0$ is

$$n(t,0) = \frac{B(t)}{G(t)}|_{L=0} \tag{9}$$

Hu et al.[24] proposed a methodology by converting Eq. (7) from a partial differential equation (PDE).to a set of ordinary differential equations (ODE). The method was applied in this study to solve Eq. (7) in MATLAB software. The single objective to minimize the sum of squares of the difference between the calculated and experimental solute concentration of all independent experiments is defined:

$$min\ \ F = \sum_{k=1}^{k}\sum_{i=1}^{N}\left(C_{i,k}^{exp} - C_{i,k}^{cal}\right)^2 \tag{10}$$

where N is the sample number and k is the experiment number. $C_{i,k}^{exp}$ and $C_{i,k}^{cal}$ experimental and calculated solution concentration, respectively. In this work, the last step of SMPT process at different temperature was used to estimate the parameters of nucleation and growth rate of β-form imatinib mesylate. The single objective function was solved with the MATLAB optimization function, $fmincon$, based on the sequential quadratic programming (SQP) method.

The solute mass balance is

$$\frac{dC(t)}{dt} = -3k_v\rho_c \int_0^\infty G(t,L)\,n(t,L)L^2 dL \tag{11}$$

where k_v is the crystal volume shape factor (k_v=1 for cube), and ρ_c is crystal density.

The average relative deviation (ARD) was introduced to evaluate the accuracy of estimation,

$$ARD = \frac{1}{N}\sum_{i=1}^{N}\left|\frac{C_{i,k}^{cal} - C_{i,k}^{exp}}{C_{i,k}^{exp}}\right| \qquad (12)$$

4. Results and discussion

4.1 SEM, PXRD and DSC

α-form imatinib mesylate supplied by Apotex Pharmachem Inc, and β-form transformed from α-form in methanol at 293K were used for SEM, powder XRD, and DSC test.

Figure 3 shows the SEM images of the α-form and β-form imatinib mesylate. Obviously, the α-form is needle-shaped while β-form is cubic-shaped. According to the morphology, we can distinguish the crystal polymorph easily.

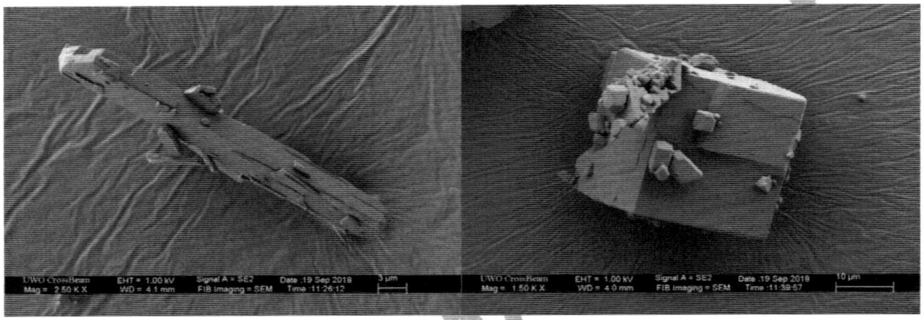

Figure 3. SEM images of imatinib mesylate. Left: α-form (3-μm), Right: β-form (10-μm)

Figure 4 displays the powder XRD patterns of α-form and β-form imatinib mesylate. The results are consistent with the α-form and β-form single-crystal XRD patterns, which were collected at 293K by Grillo et al.[18] and deposited in the Cambridge Crystallographic Data Center with publication numbers CCDC 821869 and 821871. α-form shows strong peaks at scattering angle (2θ) of 4.8°, 10.4°, 14.8°, 19.0°, 24.8° and 28.5°, whereas β-form has characteristic peaks at 9.7°, 13.9°, 18.2°, 20.0° 21.1° and 30.8°. The distinct differences on the XRPD spectrum can be used for polymorph identification of imatinib mesylate.

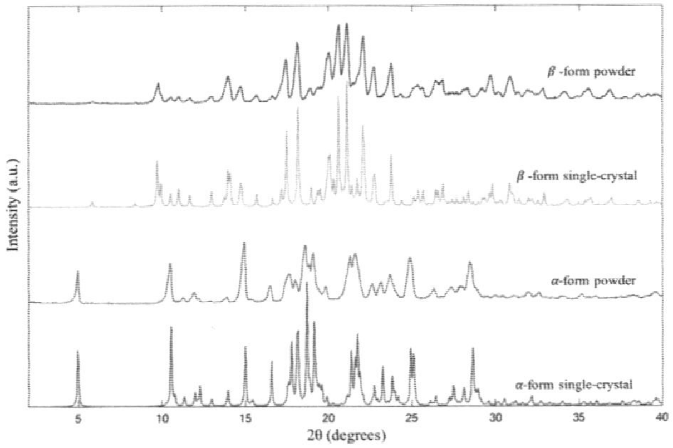

Figure 4. The XRD patterns of α-form and β-form imatinib mesylate

For both of α-form and β-form imatinib mesylate, there is only one endothermic peak observed in the DSC curves (Figure 5). The peaks are related to the melting event, which was confirmed with hot-stage microscopy and thermogravimetric analysis. The onset (start point) of the peak corresponds to the melting point, and the heat of fusion is equal to the integral of heat flow over time divided by sample mass. According to the heat of fusion rule [25], α-form and β-form imatinib mesylate are enantiotropic, as the α-form imatinib mesylate has higher melting point and lower heat of fusion than β-form.

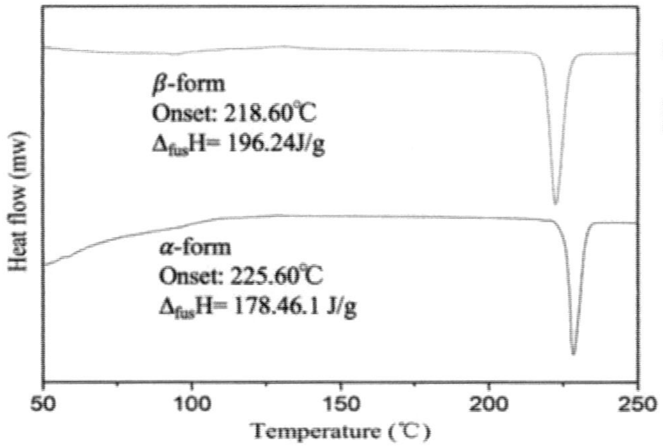

Figure 5. The DSC curves at a heating rate of 10 K/min of α-form and β-form imatinib mesylate

4.2 Solubility

The PXRD results of the residual solids demonstrated that no transformation or conversion occurred during the solubility experiments, except for the α-form in methanol. Figure 6 shows the solubility of α-form and β-form imatinib mesylate in methanol, 1-proponal and 2-proponal. As shown in Figure 6, the α-form has higher solubility than β-form in methanol,1-proponal and 2-proponal. The solubility order for both forms is methanol>1-propanol >2-propanol. The fitting parameters are listed in Table 1 and the results indicate that the modified Apelblat equation matches the solubility well with all R^2 greater than 0.99.

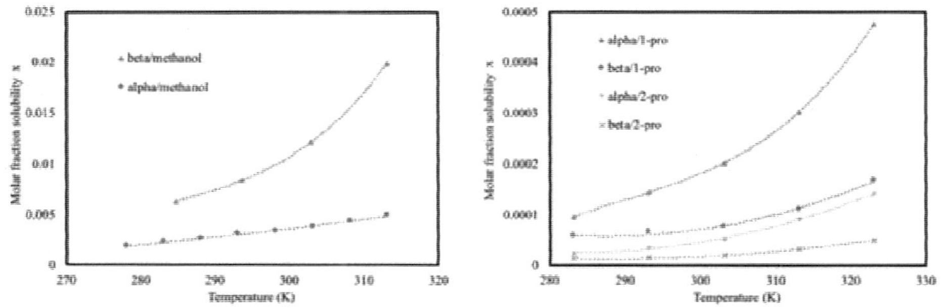

Figure 6. The molar fraction solubility of α-form and β-form imatinib mesylate in methanol, 1-proponal and 2-proponal. Dashed line: calculated curve with modified Apelblat equation

Table 1. Parameters of the modified Apelblat equation for imatinib mesylate in three solvents

		A	B	C	R^2
Methanol	α-form	-685.8	2.73E+04	103.5	1
	β-form	144.2	-8.61E+03	-21.24	0.9938
1-proponal	α-form	-279.2	9116	42.11	0.9983
	beta	-762.2	3.18E+04	113.4	0.9972
2-proponal	α-form	-455.2	1.66E+04	68.38	0.9982
	β-form	-933.4	3.89E+04	139	0.9949

As concluded above, α-form and β-form imatinib mesylate are enantiotropic. Figure 7 presents that the difference of Gibbs free energy between α-form and β-form imatinib mesylate increases with temperature. So, the transition temperature should be lower than 283.15K, which means that the β-form is always more stable than α-form above 283.15K.

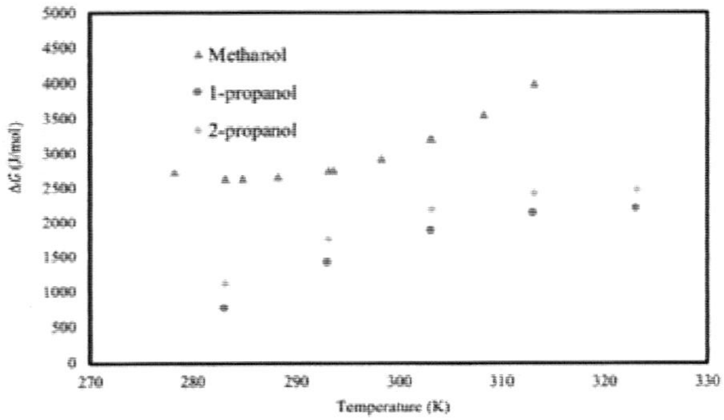

Figure 7. The difference of Gibbs free energy between α-form and β-form imatinib mesylate in methanol, 1-proponal and 2-proponal

4.3 Solvent-mediated polymorphic transformation (SMPT)

Through the solvent screening, we found that the SMPT from α-form to β-form imatinib mesylate could occur in methanol, ethanol, isopropanol, n-propanol, and 1-butanol. The transition time increased with the solvent in the order of methanol, ethanol, 1-propanol, 2-propanol and 1-butanol. Thus, methanol was chosen as the solvent to investigate the SMPT process at different temperatures by in-situ Raman spectroscopy and conductivity meter.

Due to the apparent fluctuations in the baseline, the relative heights of the Raman spectrum were used for quantitative analysis to reduce the effect produced by disturbance variables [26]. Here, the peaks at 1664 cm^{-1} and 1656 cm^{-1} were selected as the characteristic peaks of α-form and β-form imatinib mesylate, and the single point baseline was at 1640 cm^{-1}. Generally, the conductivity of a binary conducting solution is a function of temperature and solute concentration. For an isothermal process, conductivity increase in direct proportion to concentration, so the conductivity meter can be used for real-time measurement of the concentration.

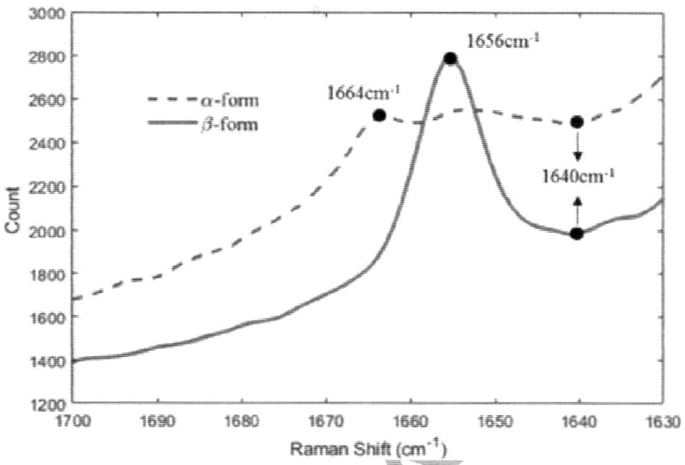

Figure 8. Raman spectra of α-form and β-form imatinib mesylate

Figure 9 displays how the concentration and solid composition change during the SMPT process in methanol at 293.15K. The process can be divided into four steps. At the beginning, the concentration increased rapidly as α-form dissolved, and the undissolved solid was pure α-form. The second step involves the nucleation of β-form. Meanwhile, the conductivity and the Raman intensity for both forms kept constant. The dissolution of α-form and growth of β-form happened during the third step, resulting in a decreased Raman intensity of α-form and an increased intensity of β-form. In the last step, the Raman intensity of α-form did not change, illustrating that all α-form crystals had been dissolved and the solid-phase were pure β-form. The nucleation and growth of β-form led to the drop of conductivity and a gradual increase of Raman intensity of β-form.

During the SMPT process, namely the third step, the concentration was close to the solubility of α-form, which indicated dissolution rate of α-form is much faster than growth rate of β-form. Therefore, it can be concluded that the SMPT from α-form to β-form imatinib mesylate is nucleation-growth controlled of β-form.

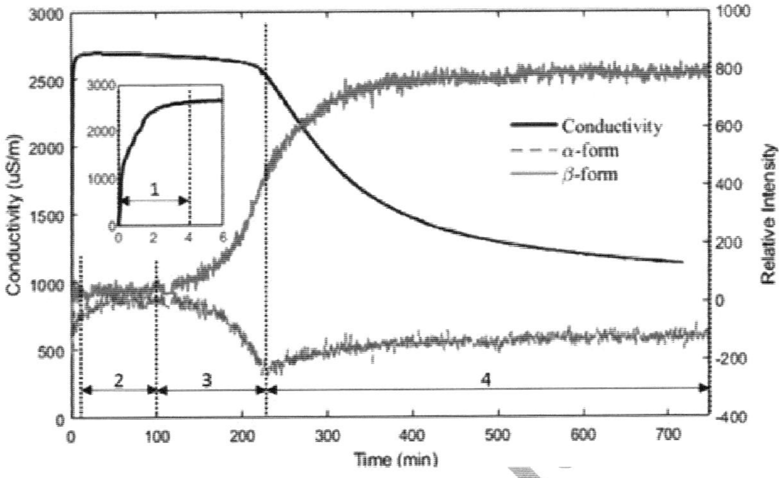

Figure 9. Solvent-mediated polymorphic transformation from α-form to β-form imatinib mesylate at 293k

4.4 Kinetic parameter estimation

As mentioned in section 4.3, all α-form crystals had dissolved and only β-form crystals existed in the solid-phase in the last step (step 4) of the solvent mediated polymorphic transformation. Therefore, the last step of the SMPT process can be regarded as a seeded isothermal batch crystallization. During the process, the solute concentration was measured by the conductivity meter and the suspension density was calculated by solute mass balance. Hence, the nucleation and growth can be estimated through minimizing the difference between the calculated and experimental solute concentration measured in all independent experiments. The estimated parameters are listed in Table 2.

The nucleation order with respect to supersaturation (b) is much higher than the nucleation order with respect to suspension density (j), indicating that supersaturation affects the secondary nucleation rate more significantly than slurry density does. $b > g$ shows that the nucleation rate has a stronger relationship with supersaturation than the growth rate, hence the mean crystal size will decrease with supersaturation. These estimated kinetic parameters would be very useful for optimizing and controlling the cooling crystallization process of β-form imatinib mesylate.

Table 2 Kinetic parameter estimated from optimization

Kinetic parameter	Estimated value	Units
k_b	9.22×10^{11}	$[(\#/m^3 \cdot s)/(kg/kgsolvent)^j]$
j	0.754	$[-]$
b	3.12	$[-]$
k_g	9.12×10^{-11}	$[m/s]$
g	1.23	$[-]$

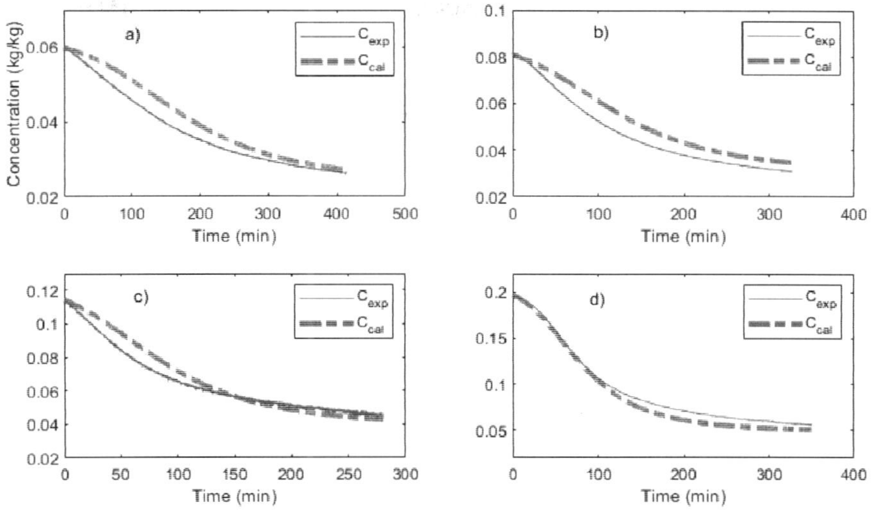

Figure 10. Calculated and experimental solution concentration
a)284.75 K, b)293.65 K, c)302.95 K and d) 313.15K

The calculated and experimental solute concentration at different temperatures are plotted in Figure 10. The ARD of the four experiments are 7.6%, 12.19%, 6.44 %, and 8.62%, respectively, which indicates the empirical equations (Eq. 4, 5, 7) used in this work are suitable for β-form imatinib mesylate. The PBE combined with agglomeration and breakage would give better estimation performance with considerable increase in computation time.

5. Conclusions

In this work, the α-form and β-form imatinib mesylate, were characterized by powder X-ray diffraction, scanning electron microscopy and differential scanning calorimetry firstly. Then, we measured their solubility in methanol,1-propanol, and 2-propanol with UV-Vis. The modified Apelblat equation was used to fit the solubility data. According to the DSC and solubility results, we found that α-form and β-form imatinib mesylate are enantiotropically related and the transition temperature is lower than 283.15K. Above 283.15K, α-form always has a higher solubility and lower stability than β-form in methanol, 1-proponal and 2-proponal. The solubility order in selected solvents is methanol>1-propanol>2-propanol. After solvent screening, the solvent-mediated polymorphic transformation (SMPT) from α-form to β-form imatinib mesylate was investigated in methanol. The solute concentration was monitored by a conductivity meter, while the solid phase was monitored by Raman spectroscopy. As the concentration was close to the solubility of metastable form during the SMPT process, it can be concluded that SMPT process of imatinib mesylate from α-form to β-form in methanol is controlled by nucleation and growth of β-form. Finally, the nucleation and grow rate of

β-form of imatinib mesylate were estimated by minimizing the difference between the calculated and experimental solution concentration with MATLAB optimization function.

6. Nomenclature

a	solute activity	-
b	nucleation order on supersaturation	-
B	crystal birth rate	$\#/m^3 \cdot s$
B_s	secondary nucleation rate	$\#/m^3 \cdot s$
C	solution concentration	kg solute /kg solvent
C^*	equilibrium concentration--solubility	kg solute /kg solvent
C^{cal}	calculated solution concentration	kg solute /kg solvent
C^{exp}	experimental solution concentration	kg solute /kg solvent
D	crystal death rate	$\#/m^3 \cdot s$
F	objective function	
g	growth order on supersaturation	-
G	crystal growth rate	m/s
ΔG	difference of Gibbs free energy	kJ/mol
j	nucleation order on suspension density	-
k	number of experiments included in the objective function F	-
k_g	growth rate coefficient	m/s
k_{sn}	secondary nucleation rate coefficient	$(\#/m^3 \cdot s)/(\text{kg/kg solvent})^j$
k_v	the crystal volume shape factor	
L	characteristic size of crystal	m
m	mass of solute	kg
m_s	mass of solvent	kg
M	molecular mass of solute	kg/mol
M_s	molecular mass of solvent	kg/mol
M_t	suspension density	kg solids /kg solvent
n	population density	$\#/m^3 \cdot m$
N	number of samples in each experiment included in the objective function F	-
R	universal gas constant	kJ/mol·K
S	relative supersaturation	-
t	time	s
x	molar fraction of solute	-
ρ_c	crystal density	kg/m^3

7. Reference

[1] G. P. Stahly, "Diversity in Single- and Multiple-Component Crystals. The Search for and Prevalence of Polymorphs and Cocrystals," *Cryst. Growth Des.*, vol. 7, no. 6, pp. 1007–1026, Jun. 2007.

[2] A. Myerson, *Handbook of Industrial Crystallization*. Butterworth-Heinemann, 2002.

[3] A. Y. Lee, D. Erdemir, and A. S. Myerson, "Crystal Polymorphism in Chemical Process Development," *Annu. Rev. Chem. Biomol. Eng.*, vol. 2, no. 1, pp. 259–280, 2011.

[4] J. W. Mullin, *Crystallization*. Elsevier, 2001.

[5] M. A. O'Mahony, A. Maher, D. M. Croker, Å. C. Rasmuson, and B. K. Hodnett, "Examining Solution and Solid State Composition for the Solution-Mediated Polymorphic Transformation of Carbamazepine and Piracetam," *Cryst. Growth Des.*, vol. 12, no. 4, pp. 1925–1932, Apr. 2012.

[6] M. Trifkovic, S. Rohani, and M. Sheikhzadeh, "Kinetics Estimation and Polymorphic Transformation Modeling of Buspirone Hydrochloride," *J. Cryst. Process Technol.*, vol. 02, no. 02, pp. 31–43, 2012.

[7] G. Morris, G. Power, S. Ferguson, M. Barrett, G. Hou, and B. Glennon, "Estimation of Nucleation and Growth Kinetics of Benzoic Acid by Population Balance Modeling of a Continuous Cooling Mixed Suspension, Mixed Product Removal Crystallizer," *Org. Process Res. Dev.*, vol. 19, no. 12, pp. 1891–1902, Dec. 2015.

[8] B. M. H. Al-Hadiya, A. H. H. Bakheit, and A. A. Abd-Elgalil, "Chapter Six - Imatinib Mesylate," in *Profiles of Drug Substances, Excipients and Related Methodology*, vol. 39, H. G. Brittain, Ed. Academic Press, 2014, pp. 265–297.

[9] J. Zimmermann, B. Sutter, and H. Buerger, "Crystal Modification of a N-Phenyl-2-Pyrimidineamine Derivative, Processes for Its Manufacture and Its Use," WO9903854 (A1), 28-Jan-1999.

[10] R. B. Parthasaradhi, R. K. Rathnakar, R. R. Raji, R. D. Muralidhara, C. R. K. Subash, and H. D. Limited, *Novel polymorphs of imatinib mesylate*. 2003.

[11] K. Amala, R. T. Srinivasa, K. S. B. R. Adibhatla, S. Rachakonda, C. N. Venkaiah, and K. Podili, "Novel polymorphic form of imatinib mesylate and a process for its preparation," WO2005077933A1, 25-Aug-2005.

[12] S. L. Pathi, R. Puppala, R. N. Kankan, D. R. Rao, C. Limited, and C. Wain, *Stable crystal form of imatinib mesylate and process for the preparation thereof*. 2005.

[13] H. Patel, R. Jani, and R. Thennati, "Imatinib Mesylate Crystal Form and Process for Preparation Thereof," WO/2006/048890, 12-May-2006.

[14] A. K. Kompella, K. S. B. R. Adibhatla, K. Podili, and C. N. Venkaiah, "Polymorphic forms of imatinib mesylate," WO2006054314A1, 26-May-2006.

[15] Mutz M., "Delta and Epsilon Crystal Forms of Imatinib Mesylate," WO/2007/023182, 02-Mar-2007.

[16] M. Mutz, "F,g,h,i and K Crystal Forms of Imatinib Mesylate," WO/2007/059963, 01-Jun-2007.

[17] J. Zimmermann, B. Sutter, and H. M. Bürger, "Crystal modification of a N-phenyl-2-pyrimidineamine derivative, processes for its manufacture and its use," US6894051B1, 17-May-2005.

[18] D. Grillo, G. Polla, and D. Vega, "Conformational Polymorphism on Imatinib Mesylate: Grinding Effects," *J. Pharm. Sci.*, vol. 101, no. 2, pp. 541–551, Feb. 2012.

[19] M. Veverka, P. Šimon, J. Lokaj, and E. Veverková, "Crystal habit modifications of imatinib mesylate under various precipitation conditions," *Monatshefte Für Chem. - Chem. Mon.*, vol. 143, no. 1, pp. 65–71, Jan. 2012.

[20] A. Srivastava, B. D. Joshi, P. Tandon, A. P. Ayala, A. K. Bansal, and D. Grillo, "Study of polymorphism in imatinib mesylate: A quantum chemical approach using electronic and vibrational spectra," *Spectrochim. Acta. A. Mol. Biomol. Spectrosc.*, vol. 103, pp. 325–332, Feb. 2013.

[21] E. Bellur Atici and B. Karlığa, "Quantitative determination of two polymorphic forms of imatinib mesylate in a drug substance and tablet formulation by X-ray powder diffraction, differential scanning calorimetry and attenuated total reflectance Fourier transform infrared spectroscopy," *J. Pharm. Biomed. Anal.*, vol. 114, pp. 330–340, Oct. 2015.

[22] A. Grunenberg, J.-O. Henck, and H. W. Siesler, "Theoretical derivation and practical application of energy/temperature diagrams as an instrument in preformulation studies of polymorphic drug substances," *Int. J. Pharm.*, vol. 129, no. 1, pp. 147–158, Mar. 1996.

[23] A. Ranodolph, *Theory of Particulate Processes: Analysis and Techniques of Continuous Crystallization*. Elsevier, 2012.

[24] Q. Hu, S. Rohani, and A. Jutan, "New numerical method for solving the dynamic population balance equations," *AIChE J.*, vol. 51, no. 11, pp. 3000–3006, Nov. 2005.

[25] A. Burger and R. Ramberger, "On the polymorphism of pharmaceuticals and other molecular crystals. I," *Microchim. Acta*, vol. 72, no. 3–4, pp. 259–271, May 1979.

[26] A. Caillet, F. Puel, and G. Fevotte, "In-line monitoring of partial and overall solid concentration during solvent-mediated phase transition using Raman spectroscopy," *Int. J. Pharm.*, vol. 307, no. 2, pp. 201–208, Jan. 2006.

8. Acknowledgment

Thanks to Apotex Pharmachem Inc. (Brantford, ON, Canada) for supplying the raw imatinib mesylate, Dr. Tim Goldhawk from Western University Nanfabrication Facility for help with SEM imaging, and Chenxi Li for her assistance in the solubility measurement. The authors acknowledge the financial support provided by the Natural Science and Engineering Research Council (NSERC) of Canada and the China Scholarship Council.

Highlights

- Two forms of imatinib mesylate, α and β, were thoroughly characterized.
- Solubility of the two forms was measured in methanol, 1-propanol, and 2-propanol.
- Developed kinetics of polymorphic transformation from α-form to β-form in methanol.
- The nucleation and grow rates of β-form of imatinib mesylate were estimated.

532399 Polymorphism of D-Mannitol: Selective Nucleation and Crystal Growth Mechanism

Abstract:

Polymorphism of D-mannitol: Selective Nucleation and Crystal Growth Mechanism

Weiyi Su, Chunli Li, Honghai Wang, Jing Fang

School of Chemical Engineering, Hebei University of Technology, Tianjin 300130, PR China

Polymorphism has gained more and more attention in the field of pharmaceuticals[1]. D-mannitol is a natural hexahydric alditol which has been widely used as a nephropathy treatment medicine and also an excipient in the formulation of various tablets and granulated powders[3]. It has been reported that there are three anhydrous polymorphs[4-6] and a hemihydrate[7] of mannitol. However the nomenclature of these polymorphs is often diverse in different literatures[8, 9] in despite of similar unit cell parameters. Therefore a review of mannitol polymorphism is initially presented along with the crystal structure parameters. Then the influence of supersaturation on the polymorphic nucleation of mannitol is investigated in aqueous solutions and ethanol-water mixtures. Moreover, a method reported by Kuldipkumar[10] was extended to determine the crystal growth of metastable ¦Ä mannitol, which was found to follow the 2D nucleation-mediated mechanism.

1. Polymorphs of D-mannitol

It has been confirmed that there are only three pure anhydrous polymorphs of mannitol even though numerous names have been used[6, 8]. Hereby, for clarity and to avoid any confusion, the unite cell parameters of mannitol polymorphs in various literatures are summarized in Table 1. It is clear in Table 1 that the first six ¦Â references have a similar structure, which is named ¦Â in our work. While focusing on the polymorphs from the 7th to 9th in Table 1, the parameters were nearly the same disregarding what they were called in the original literature. So this form is referred to as ¦Á mannitol. Finally, the last three items are significantly distinct from the others, and they are referred to as the ¦Ä form of mannitol in this paper.

Table 1. The review of unit cell parameters and nomenclature of D-mannitol anhydrous polymorphs in literatures

* volume per molecule, and others are volume per unit cell.

¡è the author also mentioned the ¦Â form they obtained was in agreement with Marwick¡¯s result.

2. Selective Nucleation of D-mannitol Polymorphs

The influence of the supersaturation on mannitol polymorphic nucleation is investigated in cooling crystallization, and the results are displayed in Figure 1 with a comparison to the solubility curves. As illustrated in Figure 1, it is quite interesting to find that the nucleation sequence is ¦Á, ¦Ä, ¦Â as the decrease of concentration while the solubility order is always ¦Ä > ¦Á > ¦Â along with temperature.

Figure 1. Polymorphic nucleation results of mannitol (open symbols, labeled as ¡°Nucleation¡±) at different initial concentrations and temperatures shown in aqueous solution with the solubility curves of the three polymorphs (solid symbols, labeled as ¡°Solubility¡±).

To clarify the nucleation zone for different polymorphs, the nucleation parameters (*DG**, *r**, and *n**) at various initial concentrations at 10 ¡ãC are obtained as shown in Table 2 based on the measured induction time and supersaturation data. It is apparent that when the initial concentration is higher than 0.0305, the critical excess free energy forming the ¦Ä form nuclei is lower than that of the ¦Â form. It indicates that the kinetic properties play a more important role than the thermodynamic properties. While the concentration is lower than 0.0246, the supersaturation is the dominant factor, that is why ¦Â mannitol is easier to nucleate in this region.

Table 2. Nucleation properties of two mannitol polymorphs at 10 ¡ãC, the subscripts ¦Ä and ¦Â refer to the polymorph ¦Ä and ¦Â forms respectively while *s* stands for the supersaturation

The values highlighted in italics in Table 2 show lower critical excess free energy values, smaller radius and molecular number of the critical nucleus for either the *¦Ä* or *¦Â* form at certain concentrations, leading to the selective nucleation of this polymorph at those conditions. In this way, controlling to get the desired polymorphs can be fulfilled.

A reverse anti-solvent crystallization of adding saturated mannitol solution into cold ethanol is also performed. The nucleation results at -10¡æ are shown in Figure 2. It is similar to Figure 1 that high concentration favors the metastable ¦Á form and the least stable ¦Ä mannitol can only be obtained at relatively low concentration.

Figure 2. Polymorphic nucleation results of mannitol (open symbols, labeled as ¡°Nucleation¡±) at different concentrations in ethanol-water mixture shown with the solubility curves of the three polymorphs (solid symbols, labeled as ¡°Solubility¡±).

3. Crystal growth

After nucleation, the nuclei should grow to be crystals by consuming supersaturation. But in bulk aqueous solution, the solvent-mediated polymorphic transformation is quite obvious for mannitol metastable forms, which makes it even more difficult to clarify the growth mechanism. In order to enlarge the acquaintance to the growth of ¦Ä mannitol, the induction time are used to provide the growth mechanism following a reported method by Kuldiplumar et al[10]. In this method, a function F(s) is defined for normal, spiral, and volume diffusion-controlled growth mechanisms:

(1)

Similarly for 2D nucleation-mediated growth, the function is defined as:

(2)

According to equation 1 and 2, the calculated F(s) against $1/(\ln^2 s)$ for the normal, spiral, and volume diffusion-controlled growth mechanism, as well as F(s) against 1/(lns) following the 2D nucleation-mediated growth mechanism are shown in Figure 3.

Figure 3. Plots of *F(s)* versus $1/ln^2 s$ for *¦Ä* form of mannitol under the normal growth mechanism (a), the spiral growth mechanism (b), the volume diffusion controlled growth mechanism(c), and *F(s)* versus *1/lns* under the 2D nucleation-mediated growth mechanism (d).

It is clear in Figure 3 that the largest R^2 of 0.9912 appears when the data are fitted to the 2D nucleation-mediated mechanism. Therefore it is reasonable to believe that the growth mechanism of the metastable *¦Ä* form of mannitol in aqueous solution should be 2D nucleation-mediated.

Reference

[1] J. Bernstein, Polymorphism in molecular crystals, Oxford University Press, USA, 2002.

[2] S. Datta, D.J.W. Grant, Nat. Rev. Drug Discov. 3 (2004) 42-57.

[3] B. O'Sullivan, The application of in situ analysis to crystallization process development, University College Dublin, Dublin, 2005, p. 177.

[4] L. Walter-Levy, C. R. Acad. Sc. Paris, Ser. C 267 (1968) 1779.

[5] F.R. Fronczek, et al., Sect. C: Cryst. Struct. Commun. 59 (2003) o567-o570.

[6] A. Burger, etal., J. Pharm. Sci. 89 (2000) 457-468.

[7] C. Nunes, et al., J. Pharm. Sci. 93 (2004) 2800-2809.

[8] J. Cornel, et al., Industrial and Engineering Chemistry Research 49 (2010) 5854-5862.

525689 Can the Solvation Manner Leads to a Nucleation Diversity for Polymorphic System? the Case of D-Mannitol

Abstract:

Can the solvation manner leads to a nucleation diversity for polymorphic system? The case of D-mannitol

There has been much interest in the role of solution chemistry plays in determining the polymorphic nucleation process, especially for the field requires a precise control and even design of the final crystalline product, such as pharmaceutical and fine chemical industries. In this study we focused on a quite challenging subject: D-mannitol (D-man). An interesting solvent-dependent nucleation behavior was found by us: in pure methanol the alpha form can be easily harvested regardless of the supersaturation and temperature, in water the alpha can also be sporadically observed but the solution mainly crystallized beta form, whereas for mixtures of water and methanol, interestingly, the pure delta form was exclusively yielded instead of either alpha or beta. We wondered the reason for this solvent dependent crystallization outcomes, especially the unexpected appearance of delta form, however, the nearly identical molecular conformations and extensive hydrogen bonds throughout the crystal structures making D-man a system difficult to resolve [1], as shown in Fig. 1.

Two points are mainly concerned in this study : 1. exploring is there any structural feature could differentiate three anhydrous polymorphs of D-man properly; 2. based on such difference, whether a ¡°link¡± could be detected between the nature of intermolecular interactions present in solution and the crystallization outcomes, especially the unexpected appearance of delta form.

Though a Hirshfield Surface analysis [2], we found two important characteristics can be used to differentiate three polymorphs, namely the dimerization effect and terminal packing intensity, as illustrated in Fig.2 and TABLE 1.

In light of above comparison, the nature of interactions in different D-man solutions was investigated in terms of the radial distribution function (RDF) calculation. The solvation effect on D-man¡¯s body parts (-O3H3) and terminal parts (-O1H1) were both investigated. The RDF results together with snapshots for the molecular distribution within different solutions are shown below:

From above results, no link was found between the solvation effect on D-man¡¯s body part and the dimerization effect in three forms, indicating that the dimerization strength cannot be a probe for the nucleation behavior explanation. While a good correlation was observed for terminal packing characteristics: in pure methanol and water, the solvation intensities at the end part are 3 and 2.7, respectively, while in the mix solvent, much higher solvating power on terminals was observed, with the terminal-water in 4 for g(r) and terminal-methanol in 3. A double-layer solvent shell with water molecules in the inside track while methanol in outsides was formed, considering the higher probability of methanol-water assemblies than the homologous solvent aggregates (as illustrated in Fig. 5(c)), we have reason to believe that those outside methanol may be bonded to the inside water shell to some degree. In the following process of solute molecule assembly, this distribution will serve as a shield for solute aggregation, create limited space for solute-solute interactions, thus facilitate a head-to-tail connection due to the relatively small volume of both ends, and meanwhile bring about a lower local packing compactness in final structure, consequently leads to the appearance of delta form. On the other hand, the light solvation degree in other two solutions can provide favorable condition for more solute molecules to occupy the area adjacent to terminals. As lots of solutes accumulate around the ends, one exposed hydroxyl group at the molecular terminal may be likely to bond with others in more orientations, contribute a lower selectivity for terminal groups, which could be supported by the separate peaks of C1 and C6 (two terminal carbon atoms) obtained from solid state NMR characterization, as shown in Fig. 6. The broader space in solution together with the optional bonding orientation finally stabilized an approximate herringbone linkage network and a dense packing motif exist in alpha and beta.

Thus we have found that the packing characteristics in the lattice of D-man should be a direct reflection of the solvation state of D-man in solution, the solvating power of different solvents on the both ends of D-man can provide distinct selectivity for terminal ¨COH groups to interact with each other. To verify the assumption, same experiments were carried out in the acetone-water system. The results showed that in water-acetone solvent, the delta form was not crystallized exclusively, instead a mixture of alpha and delta was usually yield. This phenomenon to a certain extent supports our conclusion, as acetone possesses lower HB donor and acceptor abilities than methanol, cannot construct a firm double-solvent-shell as observed in methanol-water, as a result a local environment being suitable for both alpha and delta was created.

This study, for the first time, investigated the solvent-dependent nucleation behavior of D-man which is one of the most important excipients in pharmaceutical and food industry [3, 4], and revealed the significant roles played by solvation manners in determining polymorphic outcomes of D-man. In this regard, the use of local packing compactness as a probe might constitute a potential way could be extended to other systems for polymorphic nucleation research.

Reference:

[1] Carespacheco, M. G., Vacamedina, G., Calvet, R., Espitalier, F., Letourneau, J. J., & Rouilly, International Journal of Pharmaceutics, 475 (2014) 69.

[2] M.A. Spackman, J.J. McKinnon, CrystEngComm. 4 (2002) 378.

[3] Wagner, C.M., Pein, M., Breitkreutz, Pow. Tech. 270 (2015) 470.

[4] Rowe, R.C., Sheskey, P.J.S., Owen, S.C., 2006. Handbook of pharmaceutical excipients, 449.

524390 Process Modelling, Simulation and Optimisation for Continuous Biopharmaceutical Manufacturing

Abstract:

Batch processing is the traditional mode of manufacturing for a wide range of products by the pharmaceutical industry. While batch methods have advantages such as flexible equipment usage and the ability for specific batch recall, issues with product quality variations and process reliability necessitate technological innovation to maintain profitability and sustainability. Continuous pharmaceutical manufacturing (CPM) is a promising alternative to the batch paradigm [1], offering remarkable operating and economic benefits [2], attracting strong interest from academia, industry and regulatory bodies [3]. Continuous biopharmaceutical manufacturing also shows promise for technoeconomic benefits [4]; the elucidation of end-to-end continuous biopharmaceutical manufacturing campaigns, including essential continuous purification and separation processes, is essential for successful implementation.

Process systems engineering (PSE) methodologies are valid alternatives to laborious experimental campaigns at the early stages of process development, allowing screening for promising candidate process configurations prior to further development and scale-up [5], as well as in retrofitting [6]. For sterile filling of biopharmaceuticals, a systematic decision-making procedure has been developed to support the selection of equipment technology, in order to choose between conventional multi-use or novel single-use options [7]. Process modelling and nonlinear optimisation studies have also been used to establish cost optimal operating configurations for the CPM of pharmaceutical [8] and biopharmaceutical products [9]. However, plantwide modelling, simulation and technoeconomic optimisation of a continuous biopharmaceutical manufacturing plant has yet to be implemented to establish such cost optimal operating configurations.

This paper presents the nonlinear optimisation for total cost minimisation of the upstream continuous manufacturing of a biopharmaceutical product. Kinetic parameter estimation for biopharmaceutical synthesis modelling has been performed on the basis of available literature data for a batch culture of hybridoma cell line [10] and for a continuous culture of budding yeasts [11]. Plantwide models for both upstream synthesis and downstream separation process models are then discussed, and the formulation of the objective function (plant cost) for total cost minimisation is presented, along with the illustration of minima for various process configurations, followed by a critical discussion of the implemented methodology. This work illustrates potential advantages attainable via end-to-end continuous biopharmaceutical manufacturing, and the importance of process modelling and optimisation studies at the early stages of design.

REFERENCES

[1] B. Gutmann, D. Cantillo, and C. O. Kappe, "Continuous-Flow Technology: A Tool for the Safe Manufacturing of Active Pharmaceutical Ingredients," *Angew. Chemie-International Ed.* **54**(23): 6688–6728 (2015).

[2] D. Dallinger and C. O. Kappe, "Why flow means green – Evaluating the merits of continuous processing in the context of sustainability," *Curr. Opin. Green Sustain. Chem.* **7**: 6–12 (2017).

[3] S. L. Lee, T. F. O'Connor, X. Yang, C. N. Cruz, S. Chatterjee, R. D. Madurawe, C. M. V. Moore, L. X. Yu, and J. Woodcock, "Modernizing Pharmaceutical Manufacturing: from Batch to Continuous Production," *J. Pharm. Innov.* **10**(1): 191–199 (2015).

[4] M. S. Croughan, K. B. Konstantinov, and C. Cooney, "The future of industrial bioprocessing: Batch or continuous?," *Biotechnol. Bioeng.* **112**(4): 648–651, (2015).

[5] S. K. Teoh, C. Rathi, and P. Sharratt, "Practical Assessment Methodology for Converting Fine Chemicals Processes from Batch to Continuous," *Org. Process Res. Dev.* **20**(2) 414–431 (2015).

[6] A. Segawa, S. Yoshikawa, T. Toyama, H. Nakanishi, E. Kikuchi-Uehara, M. Hirao, and H. Sugiyama, "Method for reducing environmental, health, and safety risks in active pharmaceutical ingredient manufacturing based on multiobjective evaluation," *Process Saf. Environ. Prot.* **104**: 304–313 (2016).

[7] H. Shirahata, M. Hirao, and H. Sugiyama, "Decision-Support Method for the Choice Between Single-Use and Multi-Use Technologies in Sterile Drug Product Manufacturing," *J. Pharm. Innov.* **12**(1): 1–13 (2017).

[8] H.G. Jolliffe, S. Diab, and D. I. Gerogiorgis, "Nonlinear optimisation via explicit NRTL model solubility prediction for antisolvent mixture selection in artemisinin crystallisation," *Org. Process Res. Dev.* **22**(1): 40–53 (2018).

[9] P. Kroll, A. Hofer, S. Ulonska, J. Kager, and C. Herwig, "Model-Based Methods in the Biopharmaceutical Process Lifecycle," *Pharm. Res.* **34**(12): 2596–2613 (2017).

[10] C. Kontoravdi, E. N. Pistikopoulos, and A. Mantalaris, "Systematic development of predictive mathematical models for animal cell cultures," *Comput. Chem. Eng.* **34**(8): 1192–1198 (2010).

[11] H. N. Chang, N. J. Kim, J. Kang, C. M. Jeong, J. D. R. Choi, Q. Fei, B. J. Kim, S. Kwon, S. Y. Lee, and J. Kim, "Multi-stage high cell continuous fermentation for high productivity and titer," *Bioprocess Biosyst. Eng.* **34**(4): 419–431 (2011).

523979 Process Control Strategies for Pharmaceutical Drug Product Continuous Manufacturing – Where Are We Now and Where Do We Go from Here?

Abstract:

Drug product continuous manufacturing is approved for multiple products. It is no longer new but it is early in the technology lifecycle and is rapidly evolving. The talk will be organized around the concept of process control as a triad of "measure, model, and control" to assure the highest product quality at the lowest cost. It will review the different approaches being taken to process control currently and discuss opportunities for future innovations in measurement, real-time modeling, and control.

532537 Data Accuracy for Smart Manufacturing in Continuous Pharmaceutical Systems

Abstract:

The principles of Quality by Design (QbD) and the innovative applications of Process Analytical Technology (PAT) tools have enabled the adoption of continuous processing paradigms to advance pharmaceutical manufacturing. As the community progresses towards strategies for real-time release testing and implementing advanced process control strategies, it is important to ensure accuracy in process data. Process and material quality measurements using PAT tools are highly valuable, however, present multiple challenges such as extensive calibration, data handling, sensor fouling, measurement bias, measurement drifts etc. The utility of such tools in real time hence can become limited unless properly managed. However, by leveraging the extensive real time generated by robust equipment sensors and field devices along with predictive capabilities offered by process models, the application of data reconciliation and gross error detection can enable enhanced real-time process data accuracy and validation [1]. Improved accuracy in raw measurements enhances plant-enterprise integration with increased reliability.

Data reconciliation (DR) and gross error detection (GED) are systems engineering tools for further improving the accuracy and consistency of process data through systematic data rectification and checking [2]. DR can be posed as a constrained optimization problem, which has at its objective providing the best estimate of process variables, which satisfy the process model, including material and energy balances. Under DR random errors and gross errors associated with measurements are rectified through a model-based approach providing the sensor network has the necessary redundancy. Reliable state estimates of process variables not only enhances process monitoring capabilities but also improves the performance of control systems.

DR is a powerful tool for continuous processes whose application in industries such as oil and gas is well known. It is a powerful tool for leveraging large datasets for improving monitoring consistency in the pharmaceutical industry. In this paper, we demonstrate improved accuracy in real-time monitoring of relevant process variables in a continuous dry granulation tableting line. The dry granulation process exhibits nonlinear response and fast dynamics and hence poses interesting challenges for monitoring and control. In this paper, an earlier simulation study highlighting the application of DR for a partial dry granulation line [3] is extended to investigate experimental performance using the entire tableting line. The DR is conducted using a steady-state model in order to achieve fast computational times. The DR framework is experimentally demonstrated using real time measurements of the relevant CQA's on the continuous tableting pilot plant at Purdue University.

References

[1] M. J. Bagajewicz, Smart Process Plants: Software and Hardware Solutions for Accurate Data and Profitable Operations, New York: McGraw-Hill, 2010

[2] S. Narasimhan and C. Jordache, â€œThe Importance of Data Reconciliation and Gross Error

Detection,â€ in Data Reconciliation and Gross Error Detection, Elsevier, 1999, pp. 1â€"31.

[3] S. Ganesh, M. Moreno, J. Liu, M. Gonzalez, Z. Nagy, G. Reklaitis. Sensor Network for Continuous Tablet Manufacturing, Proceedings of 13th International Symposium on Process Systems Engineering, 2018 (accepted)

531857 Impurity Control in the Continuous Reactive Crystallization of Beta-Lactam Antibiotics

Abstract:

Beta-lactam antibiotics such as cephalexin and ampicillin can be synthesized and crystallized simultaneously with the use of penicillin G acylase (PGA) as a catalyst. However, PGA also catalyzes the degradation of the antibiotic to form a slightly soluble by-productâ€"phenylglycine in the case of cephalexin or ampicillinâ€"that can contaminate the solid product [1]. It is important that by-product concentration remain below the solubility limit so that pure antibiotic can be filtered immediately without the need for recrystallization or other further purification. In a continuous process, online detection of phenylglycine crystals is necessary to ensure product quality. Focused beam reflectance measurement (FBRM) has been used to observe the nucleation of the byproduct in real time. However, it is desirable to take preemptive action to avoid nucleation of the by-product at all. The combination of several process analytical technologies such as ATR-FTIR and inline polarimetry enable the detection of phenylglycine before the solubility limit is reached. Purity can then be enforced by changing the crystallizer conditions to increase enzyme selectivity or decrease enzyme activity, both at a cost to productivity and conversion, but without the need to stop the continuous process due to solid phase impurity. A model of the reactive crystallization system is used to inform controller actions. Experiments conducted in a fed-batch crystallizer, where phenylglycine accumulation is easier to control, are used to evaluate phenylglycine crystal detection by FBRM as well as model accuracy. A mixed-suspension mixed-product removal (MSMPR) crystallizer is used to evaluate different control actions, such as decreasing pH (which simultaneously decreases antibiotic solubility and increases PGA selectivity), changing temperature (increasing temperature decreases solubility of phenylglycine but increases PGA activity while decreasing temperature has the opposite effect), and changing feed reactant concentrations (changing the reactant ratio affects both selectivity and activity).

References:

[1] McDonald, M.A., Bommarius, A.S., and Rousseau, R.W. (2017) *Chem. Eng. Sci. 165*, 81-88.

532353 Continuous Processing of Complex Dosage Forms: Liposomes and Polymeric Micelles

Abstract:

PURPOSE

Complex dosage forms such as liposomes and polymeric micelles have tremendous potential to revolutionize the pharmaceutical industry and provide significant health benefits. One challenge is to process large quantities of these nanoparticles while also maintaining physicochemical properties to ensure a high quality formulation. One approach to overcome such challenge was to turn to continuous processing. Continuous processing can achieve improved process control (*via* process analytical technology), while avoiding inherent issues associated with batch processing to ultimately improve product quality and reduce manufacturing cost. Accordingly, we developed a continuous process for liposomes and polymeric micelles. In this work, we outline important steps in the particle formation process, in-line/ at-line measurements, in-line concentrating and drug-loading strategies.

METHODS

Risk Assessment

Ishikawa diagrams were formed to outline process parameters and material attributes that had the likelihood to effect quality attributes. Some common critical quality attributes (CQAs) for nanoparticle formulations include particle size distribution (z-average particle size and polydispersity index), zeta-potential, nanoparticle concentration and encapsulation efficiency. A custom-built system was designed to incorporate process control strategies to ensure CQAs were within user specifications.

Nanoparticle Formation Process

The nanoparticles were formed using an ethanol injection method with a custom-built continuous processing system that is controlled using a program written using National Instruments LabVIEW. Briefly, the continuous processing system consists of multiple tanks that contain lipid dissolved in ethanol and another set of tanks for the aqueous components (*e.g.* DI water or buffer salts). The system further consists of an injection port such that a turbulent jet is formed under controlled flow conditions with downstream conditioning modules and concentrating modules.

In-line/ at-line measurements include: (1) flow rates; (2) temperatures; (3) particle sizing *via* dynamic light scattering; (4) particle concentration *via* an NIR turbidity meter; (5) pH and (6) conductivity.

Active-loading Doxorubicin-HCl into Liposomes

Hydrogenated soy phosphatidyl-choline (HSPC), cholesterol and DSPE-mPEG2000 were prepared at a molar ratio of 5.6:3.8:0.52 and were dissolved in ethanol. The lipid solution was subsequently injected into the aqueous phase (250 mM Ammonium Sulfate) *via* the co-flow injection port. Liposomes were then concentrated downstream and doxorubicin-HCl was injected into the pre-formed liposomes under controlled conditions. Offline UV/HPLC measurements were performed to confirm loading percentage.

RESULTS

HSPC:Chol:DSPE-mPEG2000 liposomes were successfully formed under different processing conditions. The z-average particle size could be controlled from approximately 40.0 to >100 d.nm with a PDI of less than 0.10 under most conditions. For the doxorubicin loading studies, high encapsulation efficiency was achieved using the liposomes prepared on our continuous processing system.

CONCLUSION

Continuous processing of liposomes that incorporates a turbulent jet in co-flow enables a fine-control of nanoparticles. On-line/ at-line measurements techniques are important to characterize and provide feedback during the continuous process and can easily be implemented into a process control strategy. Lastly, we have provided groundwork on continuous processing of liposomes that can be used to make doxorubicin-loaded liposomes with similar quality attributes of existing drug products.

ACKNOWLEDGEMENTS

This work was supported by the FDA (HHSF223201610121C and 1U01FD005773-01).

DISCLAIMER

This article reflects the views of the authors and should not be construed to represent FDAâ€™s views or policies.

534546 Active Process Control in the Quality-By-Design (QbD) Implementation of Pharmaceutical Continuous Tablet Manufacturing

Abstract:

Continuous manufacturing has been routinely implemented and systematically improved in the chemical, petroleum and refinery industries. In the last decade, the stringently regulated pharmaceutical industry has undergone a shift in paradigm from the conventional batch manufacturing to continuous manufacturing. [1] Under this trend, real-time product quality assurance in pharmaceutical continuous manufacturing (PCM) has been widely investigated under the Quality-by-Design (QbD) guidelines recommended by the US Food and Drug Administration. Active process control was highlighted recently [2] as a key component to enable QbD implementation. Specifically, it was proposed that the active control of critical quality attributes (CQA), such as drug loading in oral solid dosage, can address variations in critical material properties (CMA), such as bulk density of active pharmaceutical ingredients (APIs) by manipulating critical process parameters (CPP), such as tablet press dosing position. Despite the feasibility of implementing active process control, which has been demonstrated for a number of continuous manufacturing facilities, including continuous direct compaction, there remains hesitation in widespread adoption of active process control strategies due to existing investments in established and mature batch technologies. [3] In this work we investigate the development of an efficient active process control design for a rotary tablet press.

A rotary tablet press has a number of punch stations (e.g., 16) mounted on a round turret. Each punch station undergoes the following steps: die filling and metering, pre-compaction, main-compaction, tablet ejection, and tablet take-off from lower punch. Conventional batch manufacturing employs pre-batch testing of in-process CQAs, such as API composition and mixing uniformity of the blend, and post-batch testing of tablet CQAs, such as weight and tensile strength. This operating procedure minimizes the importance of active process control and thus requires CPPs (such as main compression force, dosing position, turret speed) to be kept within a predefined design space during the entire run. By contrast, continuous manufacturing relies on active process control, under QbD guidance, to ensure consistent quality production in real time and with less trial and error testing. Thus, active process control is a key component to enabling widespread adoption of pharmaceutical continuous manufacturing. However, an efficient active process control system design for rotary tablet press has not been fully investigated and its full benefits have not been elaborated. Specifically, the advantages of active process control algorithms and structures for rotary tablet press over established and mature batch manufacturing strategies in terms of product quality variances or risks merit investigation. It is worth noting that the robust design of traditional manufacturing equipment (such as a rotary tablet press, a roller compactor, etc.) has resulted in minimum variation of CPPs and CQAs during operation and thus allowed batch pharmaceutical manufacturing to assess quality using post-batch statistical quality control (SQC) methods. An active process control system, by contrast, can use possibly noisy and biased CQA measurements to effectively supervise the control of CPPs and thus minimize the need for batch-end SQC. An important aspect to investigate is the extent to which this integration imposes additional dynamics on the process, and how this could potentially amplify variations in CPPs and thus in CQAs.

In this study, a hierarchical three-level active process control, designed according to the ANSI/ISA 95 standard, was implemented on a Natoli BLP-16 rotary tablet press integrated within a continuous direct compaction pilot plant. [4] A Level 0 control consisted of the CPP controls built-into the programmable logic control (PLC) panel provided by the equipment vendor for CPPs (turret speed, dosing position, etc.). A Level 1 control was employed at the OPC server level using cascaded PID close-loop control of CQAs (drug loading, tablet weight, and tensile strength) by means of tablet weight and API composition measurements, obtained with load cell and NIR spectroscopy in-house designed sensors, respectively. A Level 2 controller was designed, using data reconciliation (DR) and model predictive control (MPC) techniques, with the objective of reconciling and supervising the control of CQAs and CPPs at the lower levels. In addition, an at-line Sotax AT4 tester was employed to collect a string of 16 tablets every five minutes with the purpose of determining the variance of tablet CQAs over time. The Level 2 DR implementation was found to be necessary to reconcile noisy sensor measurements of CQA used in the process control systems. [5, 6] Furthermore, it was also demonstrated that, under predetermined nominal material variances, the Level 0 control commonly used in batch manufacturing showed promising control performance due to the reliable mechanical design of the equipment. Under conservative tuning, both Level 1 and Level 2 active process control schemes achieved comparable promising performance in the presence of inherent system and sensor noise. However, when materials were subject to significant variations in, e.g., bulk density, the active process control with Level 1 and 2 exhibited better performance. In addition, the Level 2 MPC control showed better performance than the Level 1 PID loops in process automation roles. For example, Level 2 MPC can rapidly reach a CQA set point during process start-up.

The results of this work conclusively demonstrate that an active process control strategy, with process knowledge [7] and advanced model-based control techniques, is indispensable to the QbD implementation of pharmaceutical continuous manufacturing, and that it ensures more robustness and efficiency than the techniques used in conventional batch manufacturing, providing the key ingredient towards the new paradigm of Quality-by-Control (QbC). [8] **References**

1. Ierapetritou M, Muzzio F, Reklaitis G. Perspectives on the continuous manufacturing of powder-based pharmaceutical processes. *AIChE Journal*. 2016;62(6):1846-1862.

2. Yu LX, Amidon G, Khan MA, et al. Understanding pharmaceutical quality by design. *The AAPS Journal*. 2014;16(4):771-783.

3. Diab S, Gerogiorgis DI. Process modelling, simulation and technoeconomic evaluation of crystallization antisolvents for the continuous pharmaceutical manufacturing of rufinamide. *Computers and Chemical Engineering*. 2018;111:102-114.

4. Su Q, Moreno M, Giridhar A, Reklaitis GV, Nagy ZK. A systematic framework for process control design and risk analysis in continuous pharmaceutical solid-dosage manufacturing. *Journal of Pharmaceutical Innovation*. 2017;12:327-346.

5. Moreno M, Liu J, Ganesh S, et al. Steady-state data reconciliation of a direct compression tableting line. Paper presented at: AIChE Annual Meeting, 2017; Minneapolis, US.

6. Liu J, Su Q, Moreno M, Laird C, Nagy Z, Reklaitis G. Robust state estimation of feeding-blending systems in continuous pharmaceutical manufacturing. *Chemical Engineering Research and Design*. 2018:Accepted.

7. Su Q, Bommireddy Y, Gonzalez M, Reklaitis GV, Nagy ZK. Variation and risk analysis in tablet press control for continuous manufacturing of solid dosage via direct compaction. Paper presented at: The 13th International

Symposium on Process Systems Engineering PSE 2018, 2018; San Diego, CA.

8. Koswara A, Nagy ZK. On-Off feedback control of plug-flow crystallization: a case of Quality-by-Control in continuous manufacturing. *IEEE Life Sciences Letters*. 2017;3(1):1-4.

538504 Exploring the Kinetics of API Degradation Under Hot Melt Extrusion Conditions

Abstract:

An understanding of API degradation in the formulation of drug product is crucial to develop a robust formulation process. This becomes especially challenging for hot melt Hot Melt Extrusion (HME) processes. The complexity is due to axially varying temperature and pressure profiles. Moreover, the transition from solids blending (at the start of HME) to slurry generation to solution (at the end of HME) also needs to be understood to predict thermally induced impurity generation. In this presentation, we will use a combined experimental and modeling framework to predict such impurity generation under conditions relevant to HME. The multivariate interactions between impurity generation kinetics, underlying transport phenomena, and flow patterns in the HME will be elucidated and a consistent modeling framework to predict such impurity generation will be presented.

The work presented here was sponsored by AbbVie. AbbVie contributed to the design, research, and interpretation of data, writing, reviewing, and approving the presentation. All authors are employees of AbbVie.

533137 Towards Predicting the Quality of HME Products

Abstract:

Towards Predicting the Quality of HME Products

Josip Matić*, Carolina Alva*, Hannes Bauer*, Johannes Khinast*,**

*Research Centre Pharmaceutical Engineering GmbH, Inffeldgasse 13/III, 8010 Graz, Austria

**Institute for Process and Particle Engineering, Graz University of Technology, Inffeldgasse 13/III, 8010 Graz, Austria

Email for correspondence: khinast@tugraz.at

Introduction

Hot melt extrusion (HME) is a continuous manufacturing process primarily facilitated by co-rotating intermeshing twin-screw extruders (TSE). The process is mostly used in order to produce amorphous solid dispersions of poorly soluble active pharmaceutical ingredients (APIs), by dispersing them in polymer carriers. The process setup is extremely modular, allowing for a tailor made HME process for every formulation. On the other hand, the process setup flexibility leads to problems in achieving the desired product quality, i.e. the HME process setup is extremely challenging. This is the reason that HME might not even be considered as a possible manufacturing route. In addition, during HME, the product is exerted to high temperatures and significant shear input. The temperature and shear history (thermo-mechanical history) of the product determines the product quality after HME. Knowing the actual thermo-mechanical load exerted on the product is challenging, as accurate measurements are difficult to obtained and in some cases not even possible (melt temperature during the whole process).

Methodology

As a response to the challenges, simulations have been increasingly used as means for process understanding. In general, HME simulations have been used for understanding individual extruder screw elements, the HME process as a whole and to relate HME to other unit operations in a continuous manufacturing line. Most notable among them is the individual extruder screw investigation using the Lagrangian based Smoothed Particle Hydrodynamics (SPH) approach for understanding the melt flow and mixing in fully and/or partially filled extruder screw elements, for Newtonian and non-Newtonian fluids [1]–[5]. The knowledge gained is applied in a less complex and faster 1D HME code allowing for the accurate representation of the process as a whole[6]. All of these approaches have a focus on understanding the process, either by investigating the melt flow in detail and/or by investigating the influence of the process settings on the process state variables like melt temperature and RTD. Building on the knowledge gained from process simulations, experiments and product investigations, the developments of tools for predicting the product performance *in-silico* is the next challenge. Hence, relating the process state variables (melt temperature, SMEC, RTD...) with the product performance, is the next step in understanding the HME process.

Results

In accordance, the foundations are laid for *in-silico* product performance prediction of extruded products. Various experiments were carried out on a 12mm prototype TSE from Leistritz with the goal of testing the limits of the equipment and the formulation used. The resulting products obtained, with various process settings, were quantified in terms of their crystallinity, degradation, content uniformity and release profile, as measures of the product quality. The quality of the product was then overlaid with the HME state variables obtained via the 1D HME simulation code developed in-house, with the goal to find process descriptors for the product quality. The knowledge used will be applied in the HME setup and scale-up of various other formulations.

Literature

[1] J. J. Monaghan, "Smoothed particle hydrodynamics," *Reports Prog. Phys.*, vol. 68, no. 8, pp. 1703–1759, Aug. 2005.

[2] K. Kohlgrüber and W. Wiedmann, *Co-Rotating Twin-Screw Extruders*. Munich: Carl Hanser Verlag, 2008.

[3] A. Eitzlmayr and J. Khinast, "Co-rotating twin-screw extruders: Detailed analysis of conveying elements based on smoothed particle hydrodynamics. Part 1: Hydrodynamics," *Chem. Eng. Sci.*, vol. 134, pp. 880–886, 2015.

[4] A. Eitzlmayr, J. Matić, and J. Khinast, "Analysis of Flow and Mixing in Screw Elements of Corotating Twin-Screw Extruders via SPH," *AIChE J.*, 2017.

[5] R. Baumgartner, J. Matić, S. Schrank, S. Laske, J. G. Khinast, and E. Roblegg, "NANEX: Process Design and Optimization," *Internaitonal J. Pharm.*, vol. 506, pp. 35–45, 2016.

[6] A. Eitzlmayr *et al.*, "Mechanistic modeling of modular co-rotating twin-screw extruders," *Int. J. Pharm.*, vol. 474, no. 1–2, pp. 157–176, 2014.

529352 A Predictive Transport Model for Drying of Polymer Strip Films

Abstract:

Polymer strip films are flexible platforms for oral delivery of drugs. They have shown superior performance compared to traditional solid dosage forms [1]. These films are extensively used as a delivery form for water-soluble drugs but they can be promising candidates for the delivery of poorly water-soluble drugs, which attracted attention in recent years [1]. Drying is an important unit operation in the manufacturing process of polymer strip films. Drying process has significant effect on the morphology of the films and consequently on their quality and performance. Developing predictive models is crucial for optimal design and control of pharmaceutical processes. Fast drying and gentle process conditions are necessary to ensure quality films with the desired moisture content and thickness [2]. The drying model should capture external and intra-film transport processes as well as the shrinkage of the film, as water evaporates. A critical part in the drying model is the calculation of the diffusion coefficient in the polymer film. Different models have been proposed in the literature. A good review of the models has been published by Masaro and Zhu [3]. A significant contribution to modeling of the diffusion in polymer solutions was made by Vrentas and Duda [4,5]. They improved and extended the model based on free-volume theory originally developed by Fujita [6]. Although their model requires the estimation of several parameters, it is applicable to various polymer–solvent systems at different temperatures and polymer concentrations [3].

This work contributes to the development of a robust drying process for the production of polymer strip films containing poorly water-soluble drugs. A two-dimensional simulation and parametric study were done using Comsol Multiphysics and MATLAB for convective drying of the films considering the diffusivity model developed by Vrentas and Duda [4,5]. Polymer strip films containing hydroxypropyl methylcellulose (HPMC; Methocel E-15), glycerin, water and silica-coated drug particles (griseofulvin) were used in the drying experiments. The model results were validated and the required transport parameters were estimated based on the experimental data obtained from a batch drying set-up. Overall, this presentation will demonstrate how time-wise evolution of the moisture and film thickness, two critical quality attributes (CQAs) of strip films, can be predicted using a transport-based modeling framework along with model validation using a batch drying set-up.

1. Krull, S.M., Susarla, R., Afolabi, A., Li, M., Ying, Y., Iqbal, Z., Bilgili, E. and Davé, R.N., 2015. Polymer strip films as a robust, surfactant-free platform for delivery of BCS Class II drug nanoparticles. ***International journal of pharmaceutics***, ***489***(1-2), pp.45-57.
2. Susarla, R., Sievens-Figueroa, L., Bhakay, A., Shen, Y., Jerez-Rozo, J.I., Engen, W., Khusid, B., Bilgili, E., Romanach, R.J., Morris, K.R. and Michniak-Kohn, B., 2013. Fast drying of biocompatible polymer films loaded with poorly water-soluble drug nano-particles via low temperature forced convection. ***International journal of pharmaceutics***, ***455***(1-2), pp.93-103.
3. Masaro, L. and Zhu, X.X., 1999. Physical models of diffusion for polymer solutions, gels and solids. ***Progress in polymer science***, ***24***(5), pp.731-775.
4. Vrentas, J.S. and Duda, J.L., 1977. Diffusion in polymer—solvent systems. I. Reexamination of the free volume theory. ***Journal of Polymer Science Part B: Polymer Physics***, ***15***(3), pp.403-416.

5. Vrentas, J.S. and Duda, J.L., 1977. Diffusion in polymer–solvent systems. II. A predictive theory for the dependence of diffusion coefficients on temperature, concentration, and molecular weight. ***Journal of Polymer Science Part B: Polymer Physics***, ***15***(3), pp.417-439.

6. Fujita, H., 1961. Diffusion in polymer-diluent systems. ***Fortschritte Der Hochpolymeren-Forschung***, pp.1-47.

524082 High Shear Wet Granulation Scale-up Study Using Discrete Element Modeling

Abstract:

High shear wet granulation is a complex unit operation, since there are several rate processes occurring simultaneously that affect physical characteristics of the granulation (e.g., granule size distribution, granule density, wet mass consistency, etc.). A poorly understood system and inappropriate scale-up strategy can result in a poor process at the commercial scale.

This work illustrates the application of Discrete Element Model (DEM) to gain an understanding of the flow and force field distributions across the small scale, pilot scale and commercial scale granulators. It is well-known that granule density plays an important role in tablet tensile strength and drug product performance [1]. The granule density is highly influenced by the binder liquid quantity, and the magnitude of force inside the granulator relative to the dynamic yield strength of the wet granules. The total force field includes the force imparted by the impeller, force due to gravity, or weight of the granulation mass, and force due to other particle-particle and particle-wall (equipment boundaries) interactions. It was anticipated that at the commercial scale the effects of increased granulation mass would yield granules of reduced porosity, which are unsuitable for tableting, if the process design did not consider the total force field contributions across the different scales.

The DEM of the granulators of different scales was developed. A qualitative analysis of the variation of force field across the different scales was conducted. The process knowledge obtained from the DEM was used in directional adjustment of the operating conditions at the commercial scale that moved the process towards a lower risk operating space, which was confirmed through regime map and dimensionless spray flux analysis.

Reference

[1] van den Ban and Goodwin, 2017. Pharmaceutical Research, 34, 1002-1011

535426 Predicting Shear, Energy Dissipation, and Blending in Bioreactors for Mammalian Cells

Abstract:

Mammalian cells, as used in the in vitro cultivation biotechnological products, are very sensitive to shear forces arising from turbulence and fluid motion. This behavior is particularly apparent in agitated reactors, wherein cells are exposed to various hydrodynamic environments that may reduce cell growth, cause cell death, or introduce other adverse reactions. Competing with this need to minimize shear, however, is the requirement that reactors remain homogenized and well-stirred to maximize throughput and batch efficiency.

In this work, we discuss how large eddy simulation can be used to perform time-accurate, three-dimensional numerical experiments of agitated bioreactors. After this theoretical overview, we use this approach to predict the time-evolution of the shear and energy dissipation fields across the fluid domain. Next, we show how this time-accurate model can be used to perform real-time blending and recirculation experiments on modest computational resources. We conclude showing how cell trajectories can be integrated in tandem with the fluid motion to predict the precise hydrodynamic environments experienced by many cells as they move about a tank.

522151 Simulation of Industrial-Scale Aerated Bioreactors

Abstract:

The design and scale-up of industrial scale aerated stirred bioreactors is mainly based on empirical knowledge although they are used in many industrial sectors. Biological processes including the conversion of carbon substrates and dissolved oxygen to carbon dioxide can be simulated with knowledge about the mass transfer and transport processes inside the reactor. Computational fluid dynamics provides a basis to model the species transfer and transport and hence lay the foundation for new reactor designs.

Graphical processing units are employed to achieve short simulation times. To approximate the Navies Stokes equations the lattice Boltzmann method is used to take advantage of the processorâ€™s parallelism. The bubble movement is simulated by solving the Newtonâ€™s equation of motion. Stochastic models are used for coalescence and breakup.

To achieve the goal of simulating the biological activity in the reactor a species transport solver for e.g. the dissolved oxygen and carbon dioxide is essential. Current solutions which can be efficiently parallelized suffer from negative concentration values and artificial inertia. Our novel solver is based on the lattice Boltzmann method which results in fast and efficient execution on GPUs while avoiding the unphysical effects mentioned before. The speed and memory consumption of this new solver is evaluated by comparing it to a particle based solver.

All models were validated with literature data. The simulation program is now applied for comparing reactor designs and scale-up concepts.

525048 Spatial Confined Capillary Flow and Precisely Controlled Crystallization Via 3D Printed Platform: A Comprehensive View

Abstract:

Spatial Confined Capillary Flow and Precisely Controlled Crystallization via 3D Printed Platform: A Comprehensive View

Mingguang Han, Gaohong He, Xiaobin Jiang*

State Key Laboratory of Fine Chemicals, Research and Development Center of Membrane Science and Technology, School of Chemical Engineering, Dalian University of Technology, Dalian, P. R. China

*Corresponding author: Email: xbjiang@dlut.edu.cn

Abstract

Precision crystallization control is crucial problem for crystallization progress design and industrial crystallization optimization in crystallizer designing, which was widely applied in pharmacy. Pacing the development of high precision manufacture method, micro-level or even nano level microfluidic chip and stable microwell array chip was developed and applied in recent state of the art crystallization processing research[1-4]. This chips-based crystallizer provides manageable environment control and repeatable crystal formation or nucleation results rapidly and accurately.

Here, we developed material-free micro pillar array chip as crystallizer utilizing high precise 3D printer (Digital Light Process (DLP)) with systematic dripping module (shown in Fig. 1) and dripped crystal solution drops onto the chip with different drop volume and different position in mild evaporation surrounding condition. We also found diversified initial contact angles when dripping different volume of drops, which possess different liquid-air interface initially and influence capillary flow inside system impactful for crystal formation (shown in Fig. 2).

For sessile drop, to understand evaporation condition and crystal formed position, concentration gradient, evaporation time and distance between formation and pillar center was encountered as variables to measure concentration change, vapor diffusion flux and nucleation energy barrier, and indicated the relationship of crystal form and capillary micro flow (shown in Fig. 3); For cover drops, we raised inside capillary micro flow by coffee ring inspiration to explain coffee-ring disappearance and guide the stable diffusion crystal formation in concave structure with observation of crystal form (shown in Fig. 3); For sitting drops and missing drip, some irregular and unexpected crystal form appeared without explanation, which need to be avoided. This method was well popularized in commercial crystal compounds with facile management and modification for high throughout crystal production. Meanwhile, lower concentration or even, polymorph system was probably suitable in crystallization process.

Fig 1: (a) The closed-up review of tunnel concave structure and (b) lattice concave structure of designed crystallizer. (a-1) SEM image of the cross section of (a). (b-1) SEM image of lattice concave from (b) (scale of the bar, 500 ¦Ìm). (c) The process of dripping drop system.

Fig 2: (d-1): The hydrophilic property of commercial 3D material shown in 50¡ãintrinsic angle. (d-2)~(d-6): gradient contact angle results when dripping numerous volume liquid drops on concave structure

Fig 3: The relationship of crystal form and capillary micro flow in sessile drop condition and cover drop condition

Acknowledgment

We acknowledge financial contribution from National Natural Science Foundation of China (Grant No. 21527812, 21676043, U1663223, 21606035), Changjiang Scholars Program (T2012049), the Fundamental Research Funds for the Central Universities (DUT16TD19, DUT17ZD203) and Education Department of the Liaoning Province of China (No. LT2015007).

References

1. Duffy DC, C MJ, A SOJ. Rapid Prototyping of Microfluidic Systems in Poly(dimethylsiloxane). *Anal. Chem.* . 1998;70:4974-4984.

2. Leng J, Salmon JB. Microfluidic crystallization. *Lab Chip.* Jan 7 2009;9(1):24-34.

3. Sultana M, Jensen KF. Microfluidic Continuous Seeded Crystallization: Extraction of Growth Kinetics and Impact of Impurity on Morphology. *Crystal Growth & Design.* 2012;12(12):6260-6266.

4. B Z, S RL, F IR. Screening of protein crystallization conditions on a microfluidic chip using nanoliter-size droplets. *Journal of the American chemical society.* 2003;125(37):11170-11171.

5. Bourgks-Monnier C, M. E. R. S. Influence of Evaporation on Contact Angle. *Langmuir.* 1995;11:2820-2829.

6. Shanahan MER, Bourg C. Effects of evaporation on contact angles on polymer surfaces. *INT. J. ADHESION AND ADHESIVES.* 1994;14(3):201-205.

7. Deegan RD, Bakajin O, Dupont TF, Huber G, Nagel SR, Witten TA. Capillary flow as the cause of ring stains from dried liquid drops. *Nature.* 1997;389(6653): 827-829

8. Cira NJ, Benusiglio A, Prakash M. Vapour-mediated sensing and motility in two-component droplets. *Nature.* Mar 26 2015;519(7544):446-450.

9. Mersmann A, ed *Crystallization technology handbook*. New York and Basel: Marcel Dekker, Inc.; 2001.

527642 Polymorph Screening of L-Glutamic Acid By Anti-Solvent Crystallization in Easy-to-Use Microfluidic Device

Abstract:

Most pharmaceuticals have polymorphs. Since different polymorphs exhibit distinct physical properties, such as solubility, bioavailability, density, mechanical properties, color and morphology, which result in difficulties in the downstream processing, including separation, handing and so on. Hence, it is crucial to control the crystallization of targeted polymorph. As an amino acid commonly used in pharmaceuticals and food, L-glutamic acid has two polymorphs: prismatic or granular shaped metastable Î± form and four different shaped stable Î^{2} form, including need-, rod-, lozenge- and plate-shaped. In practical applications, Î± form has higher utilization efficiency. In previous work, to obtain a specific polymorph form of L-glutamic acid, the commonly adopted methods includes control of larger crystallization temperature difference, in quiescent and flow state through membrane crystallizers, adding amino acid additives, sonocrystallization and so on.

In this work, based on anti-solvent crystallization, solutions and antisolvent were loaded into separate syringes which were connected to channels. By varying the initial concentrations of solutions, temperature, and velocities of solution and antisolvent, the resulted crystals were flitered directly, and dried in an air oven. Then crystals were characterized by SEM and PXRD for morphology evaluation and form confirmation. The polymorph selectivity of L-glutamic acid was investigated. This method was simple in operation and low consumption was required. And the crystallization temperature was kept constant throughout the process. The results showed that this method was more flexible in the selectivity of L-glutamic acid polymorphs without other assistive technologies or additives which might introduce impurities.

533156 Thermodynamics of Co-Crystal Systems

Abstract:

Thermodynamics of Co-crystal Systems

Dipali. Ahuja,[1] and Åke. C. Rasmuson[1]

[1]*Synthesis and Solid State Pharmaceutical Centre, Bernal Institute, Department of Chemical Sciences, University of Limerick, Limerick, Ireland*

Introduction.

The pharmaceutical industry has a significant interest in co-crystals for their ability to alter the physical properties of the active pharmaceutical ingredient (API) without affecting its chemical structure *[1]*. Co-crystallization may enhance a wide range of pharmaceutical properties like hygroscopicity, thermal stability, solubility and bioavailability. Most research in this field has been focused towards discovering new co-crystals. Much less work has been done on studying the co-crystal thermodynamic properties and manufacturing techniques. Ternary phase diagrams are crucial for the design of the manufacturing process *[2]* as they highlight the stability regions for various solid crystalline phases. Factors like solvent and temperature significantly affect the shape of the phase diagram and the solubility differences between the co-crystal components can cause a system to be congruent or incongruent *[3]*. It is most likely that a large solubility difference causes a system to show incongruent dissolution. The main interest in the ternary phase diagram is the size of the region where co-crystal is the most stable solid phase.

Methods and Experiments.

For this work, thermodynamic studies for 1:1 sulfamethazine (SMT)-salicylic acid (SA) co-crystal system were carried out. SMT is a sulfonamide drug and is used an antibacterial and anti-infective agent; SA is used to relieve pain and fever and is used for anti-acne treatments. Ternary phase diagrams were constructed in three solvent systems (methanol, acetonitrile and DMSO-methanol) at three temperatures. The methodology behind the phase diagram construction involved determination of the pure component solubilities by gravimetric technique. The invariant points were determined by mixing the solid and solvent and allowing sufficient time for equilibration. Solids were then analysed by PXRD (Powder X-ray Diffraction) or DSC (Differential Scanning Calorimetry) and the solution composition was determined by HPLC (High Performance Liquid Chromatography).

Results and Discussion.

Solvent dependent congruent and incongruent dissolution was observed, highlighting how solvent choice can affect the phase diagram. The co-crystal system showed congruent dissolution in acetonitrile but showed an incongruent behavior in methanol and DMSO-methanol. The effect of temperature over the range studied was found to be weak. Based on the solubility data for the pure components and the co-crystal, the Gibb's free energy of co-crystal formation was determined, and ranged from -5.5 to -7.7 kJ/mol, changing with temperature. The negative values reveal the thermodynamic stability of the co-crystal and spontaneity of the co-crystal formation from SMT and SA. The enthalpy and entropy associated with co-crystallization were determined to be +23 kJ/mol and +0.1 kJ/mol, , respectively. From this data, co-crystal formation seems to be an entropy driven process. A relationship between the solubility ratio between API and coformer and the size of the region where the co-crystal is the most stable phase has been observed. In addition, it was found that the solubility ratio between the pure components does not allow for confident prediction of whether a system will show congruent or incongruent behaviour. The co-crystal yields and volumetric productivities in the three solvent systems were determined using the slurry crystallization technique.

References:

[1] J. Urbanus, C. P. M. Roelands, D. Verdoes, P.J. Jansens, J.H. ter Horst, Co-Crystallization as a Separation Technology: Controlling Product Concentrations by Co-crystals, Cryst. Growth Des., 10, 1171 (2010).

[2] D.M. Croker, M. E. Foreman, B. N. Hogan, N. M. Maguire, C.J. Elcoate, B. K. Hodnett, A.R. Maguire, Å. Rasmuson, Understanding the p-Toluenesulfonamide/Triphenylphosphine Oxide Crystal Chemistry: A New 1:1 Cocrystal and Ternary Phase Diagram, Cryst. Growth Des., 12, 869 (2012).

[3] S. Zhang, Å. Rasmuson, Cryst. Thermodynamics and Crystallization of the Theophylline–Glutaric Acid Cocrystal, Cryst. Growth Des., 13, 1153 (2013).

525830 Thermodynamic Properties of Paracetamol Impurities and Their Impact on the Crystallisation of Paracetamol from Solution

Abstract:

The impact of additives and impurities on active pharmaceutical ingredients is an essential yet poorly understood area in crystallization research. Paracetamol, which is one of the most widely used model systems in crystallization, is typically produced with trace amounts of unwanted impurities. However, it remains unclear how the main impurities 4-nitrophenol (NP) and 4â€™chloroacetanilide (CA) affect the solubility and polymorphism of paracetamol.

This work describes the solid- and solution characterisation of the main impurities of paracetamol. Temperature-dependent solid-liquid properties of 4-nitrophenol (NP) and 4â€™chloroacetanilide (CA) were obtained in four different alcohols and their effect as impurities on the crystallisation and solubility of paracetamol (PA) will be presented. HPLC measurements were used to determine the solution stability of these compounds. The solubility of NP appeared to be significantly higher than PA whereas the solubility of CA was lower than PA. The solubility difference between the impurities could be rationalised based on their molecular structure and hydrogen bonding interactions. Despite these significant differences between the impurities, their effect on the solubility of PA is the same, as each impurity slightly increases the solubility of PA. The solubility data was modelled using the empirical modified apelblat equation and thermodynamic models, which include the Margules, van-Laar, Wilson and NRTL models. The quality of fit of the tested thermodynamic models to the experimental data increases in the order Margules < van-Laar < Wilson < NRTL < Apelblat.

XRPD measurements were conducted to establish the polymorhpism of the compounds used in this study. Recrystallisation of PA from solution containing the highly soluble NP impurity resulted in small uniformly sized high purity PA form I crystals whereas the presence of the poorly soluble CA impurity induced the formation of large needle shaped crystals of PA form I. These differences in crystallisation are a consequence of the solubility difference and the different functional groups of PA and its impurities. The effect of the impurities on the nucleation kinetics of paracetamol was investigated using induction time measurements and showed that the nucleation process is inhibited for both impurities.

The presentation will cover experimental and modelling data and describes the effect of impurities on the solubility, crystallization and nucleation kinetics of paracetamol. Overall this work presents fundamental information for the development of crystallisation approaches for the purification of paracetamol from its main impurities.

525983 A Novel Approach into Secondary Nucleation and Crystal Growth, Controlled Particle Size Distribution, Using a Large Single Seed Crystal in Solution Crystallization

Abstract:

Secondary nucleation by means of seed crystals is extensively employed in crystallization process and remained a noteworthy consideration of research in recent years, but its in-depth understanding particularly at molecular level remained elusive until recently, molecular dynamic simulations provided an insight to it, i.e. crystal nuclei breeding. However, seed crystals are typically dispersed in solution, it becomes difficult to decouple nuclei breeding from attrition-enhanced secondary nucleation. In this work, a novel approach was used to experimentally investigate the mechanism of crystal nuclei breeding as an origin of secondary nucleation and its link with fluid shear, using a large single crystal of paracetamol (size = 15mm x 12mm) in propan-2-ol. The crystal was held stationary in the stirring vessel using a specially design crystal holder, to eliminate the possibility of enhanced secondary nucleation due to crystal attrition or breakage during agitation. The experiments were done in a low supersaturation (S_o = 1.0) at stirring speed of 300 and 370 rpm. Particle imaging velocimetry (PIV) was used to quantify the fluid shear rate at each stirring speed using a high-speed camera and laser illuminated sheet. Laser diffraction was used to perceive the onset of secondary nucleation using a focused beam of red laser passing through the vessel which was calibrated to detect the particle size as small as 20 Æžm. Focused beam reflectance measurement (FBRM) was used for real time tracking of particles which displayed a significant increase in particle counts with an increase in shear rate at higher rpm. At a high fluid shear, the MSZW exhibited a decrease in secondary nucleation threshold with a relative supersaturation ratio of 1.17 at 300 rpm and 1.08 at 370 rpm. Below this threshold, a growth only zone was apparent which was validated using the scanning electron microscopy (SEM). PSD data was measured on a volume basis with increasing number of fines at a higher share rate as expected. Since, no crystal attrition and breakage were involved in the onset of secondary nucleation, these results provided experimental evidence that support the hypothesized nuclei breeding mechanism. Further evidence from the SEM images of the seed crystal, showed the formation of alternate layered pattern of the surface growth and the existence of very fine crystallite layers attached to it. It is proposed that the surface of seed crystal served as a catalyst for the pre-existing clusters in the solution to attach and nucleate to form crystallites, which then sheared off by the fluid flow. The experiments presented an excellent repeatability over MSZW and growth only zone at specific shear rate, particularly the revelation of controlled particle size distribution elucidates the significance of fluid shear on the secondary nucleation using a single seed crystal. Controlled particle size distribution is an imperative quality factor in crystallization and the quality of crystalline product. This approach can offer a more precise modelling in process scale-up and manufacturing in industry.

533602 Crystal Growth Kinetics of Salicylamide Investigated Under Different Crystallisation Processes and Also Environmental Conditions

Abstract:

Crystal Growth Kinetics of Salicylamide Investigated under Different Crystallisation Processes and also Environmental Conditions

Crystallisation is a very important technique, especially in the Pharmaceutical Industry, as it is used in purification and recovery of the crystalline active pharmaceutical ingredients (API) [1]. Despite this, crystallisation fundamentals are still insufficiently understood. Efforts have been made by researchers to understand the kinetics' of salicylamide nucleation [2]. This research focuses on improving knowledge of crystal growth process and kinetics of salicylamide by investigating under rotating disk and also seeded desupersaturation agitated tank experiments. Observed the growth of individual crystals as well as populations of crystals. Crystal growth is also studied under different environmental conditions to detect any influence of the solvent, temperature and supersaturation.

The rotating disc experiments uses inverted optical microscope integrated with a camera, combined with a PC with image/ video capture and analysis software to capture micrographs of each crystal before and after growth. Multiple individual seed salicylamide crystals are fixed onto a rotating glass disc, and placed into a supersaturated solution for a growth period of 1 hour and rotated at 200rpm. Crystals are observed in-situ, so allows for the growth of specific faces to be measured and also view overall habit. The agitated tank experiments use a reactor vessel to perform seeded isothermal desupersaturation experiments. The concentration decay is recorded with a FTIR probe and first principles data treatment is applied to extract the concentration information from IR spectra; besides, FBRM is used for detection of nucleation during crystal growth. In the agitated tank experiments the average growth rate of a certain mass of seeds (better statistics) can be determined at hydrodynamic conditions closer to an industrial process.

The rotating disk crystal growth of salicylamide crystals at 15°C was measured as a function of relative supersaturation (S-1) in the range from 0.01 – 0.04, and in four different solvents (acetone AC, acetonitrile MeCN, ethyl acetate EA and methanol MeOH), rotating the disc at 200rpm. Different rotation speeds where tested in the range of 50 - 350rpm, and 200 rpm was found to be sufficient to essentially eliminate the film mass transfer resistance. The growth of a minimum of 25 crystals was determined for each system. The growth rates of the length and width of each individual crystal was measured using the maximum and minimum feret diameter respectively, also growth of the (200) face was measured. The studies revealed that salicylamide crystals grown from seed in different solvents, at different supersaturations and also within each solvent receive different shapes. There is a significant spread in the crystal growth rates obtained within each system, e.g. 0.05 µm/s up to 0.55 µm/s in one system; this is due to the crystal quality, where lower quality results in increased growth. The average growth rates increase with increasing supersaturation in each solvent. The average individual crystal growth rate behaviour, as a function of solvent and supersaturation, is similar for length and width. The average crystal growth rate of the (200) face of individual seed crystals decreases in the order MeOH>MeCN>AC>EA at supersaturation of 0.03. Salicylamide crystal surface was examined before and after the growth process using scanning electron microscopy; noticed growth steps and layers clearly present after growth indicating rough interfacial growth mechanism. The agitated tank experiments were also performed in all four solvents (AC, MeCN, EA and MeOH) and at 4 temperatures (10, 15, 20, 25°C). The empirical power law equation, spiral growth mechanism (BCF) equation and the birth & spread mechanism (B+S) equation were fitted to the data. In all four solvents, the growth rate increases with temperature as expected. The average crystal growth rate decreases in the order AC>MeOH>MeCN>EA at supersaturation of 0.03. The interfacial energy values for crystal growth depend on solvent, and where obtained by fitting the B+S model; the order is EA>AC>MeCN>MeOH. Due to the fact that low interfacial energy tends to lead to high crystal growth rate, which is in agreement with the result from the rotating disc experiments. The growth rate is faster in the agitated tank experiments as compared to rotating disc which possibly reflects an influence of mass transfer resistance in the latter.

In all growth processes, salicylamide was found to have the slowest growth kinetics in ethyl acetate. Molecular modelling was performed which high lightened the strong hydrogen bond propensity of ethyl acetate and due to its high molecular weight results in slow growth. These crystallisation studies with salicylamide in different solvents, supersaturations and temperatures have shown that crystal growth rates are affected significantly by the use of different environmental conditions.

Reference:

1. Lacmann, R., *Crystallization, Third Edition. J. W. MULLIN, Butterworth-Heinemann, Oxford 1997, 527 Seiten, zahlr. Abb. und ISBN 0-7506-3759-5.* Chemie Ingenieur Technik, 1998. **70**(11): p. 1468-1468.
2. Nordstrom, F.L., M. Svard, and A.C. Rasmuson, *Primary nucleation of salicylamide: the influence of process conditions and solvent on the metastable zone width.* CrystEngComm, 2013. **15**(36): p. 7285-7297.

539930 Fast Temperature Cycling via Microwave Heating Enables Enhanced Deracemization

Abstract:

Fast Temperature Cycling via Microwave Heating Enables Enhanced Deracemization

Christos Xiouras[1], Fabio Cameli[1], and Georgios D. Stefanidis[1]

[1] Process Engineering for Sustainable Systems, Department of Chemical Engineering, KU Leuven, Belgium;

Keywords: Deracemization, Crystallization, Temperature Cycling, Viedma ripening; Microwave Heating;

In 2005, Viedma [1] demonstrated a remarkable non-classical crystallization phenomenon according to which an initially racemic crystal population converts to a single chirality upon grinding the crystals in their saturated solution. The process was initially demonstrated for the achiral $NaClO_3$ which crystallizes as chiral crystals, but was later extended to intrinsically chiral compounds that crystallize as conglomerates (*i.e.* each individual crystal contains a single enantiomer) [2]. The underlying mechanisms of the phenomenon are still under discussion, but the process is considered to have high potential for application in the pharmaceutical industry, where separation of enantiomers is often required. However, the application of grinding at a larger industrial scale is rather complicated and could lead to problems in downstream processing. Recently, it was discovered that grinding could be replaced by periodic temperature fluctuations (temperature cycling), which is more convenient at a larger scale [3].

In contrast to conventional heating, which is slow and requires a heat transfer surface, microwave heating is rapid and volumetric in nature. Therefore, if combined with fast cooling, microwave activation can enable novel process operating windows for temperature cycling as compared to conventional heating. Figure 1 (left) shows that the application of such operation for the model system glutamic acid leads to an order of magnitude faster heating and cooling rates compared to conventional operation, which in turn leads to significant reduction in the total deracemization time required (Figure 2, right). Besides intensification of the deracemization process, the extreme operating windows applied here also lead to new insights into the obscure driving mechanisms of solid-state deracemization phenomena.

Figure 1. Left: Temperature profile using microwave heating (red dashed line) and conventional heating (black line). Right: Evolution of solid phase enantiomeric excess for in time in: microwave reactor (black points) and conventional heating configuration (red points). Model system: glutamic acid conglomerate crystals in acetic acid and catalytic salicylaldehyde.

References

[1] C. Viedma; Chiral Symmetry Breaking During Crystallization: Complete Chiral Purity Induced by Nonlinear Autocatalysis and Recycling *Phys. Rev. Lett.* 94 (2005), 3−6

[2] W.L. Noorduin, T. Izumi, A. Millemaggi, M. Leeman, H. Meekes, W.J.P. Van Enckevort, R.M. Kellogg, B. Kaptein, E. Vlieg, D.G. Blackmond, Emergence of a single solid chiral state from a nearly racemic amino acid derivative, *J. Am. Chem. Soc.*, 130 (2008), 1158-1159.

[3] K. Suwannasang, A. E. Flood., C. Rougeot, & G. Coquerel. Using programmed heating–cooling cycles with racemization in solution for complete symmetry breaking of a conglomerate forming system. *Cryst. Growth Des.*, 13 (2013), 3498-3504.

530887 Resolution of Conglomerates Using Preferential Crystallization and Enzymatic Racemization

Abstract:

Producing pure enantiomers is of great interest to pharmaceutical and fine chemicals industries, since these molecules interact differently with living organisms. Preferential crystallization (PC) is an important resolution technique to purify enantiomers. Nonetheless the separation can only reach maximum 50% yield. The addition of racemization of the non-desired enantiomer increases the theoretical process yield to 100%. Free and immobilized enzymes are attractive catalysts for this recycling step coupled with chiral separation processes. In this contribution, the separation of a model conglomerate forming system was investigated experimentally and the results were evaluated using a shortcut mathematical model for the coupling of enzymatic racemization with PC.

The amino acid racemase used was overproduced in bacteria and purified using affinity chromatography. The kinetic properties of the enzyme were investigated in preliminary studies. Low enantiomeric excesses produced by PC generate low driving forces for the racemization. This causes severe requirements related to the enzyme dosage. The optimal immobilization conditions were established to enable reuse of the catalyst and its easy separation from the products. Besides performing systematic experiments, a simple consistent mathematical model was developed assuming a lumped crystallization mechanism. The effective growth rate and the racemase kinetics were used to simulate the process combination. The performance of the biocatalyst was evaluated for the proposed application based on the experimental and simulated results.

Paper Number: 528336

ALKALINE CRYSTALLIZATION OF $CaCO_3$ IN A DIRECT AIR CAPTURE PROCESS

Caroline E. Giacomin[1], Thomas Holm[2], Luisa Burhenne[2], and Walter Mérida[2]

1) Chemical and Biological Engineering, University of British Columbia, Vancouver, BC, Canada
2) Clean Energy Research Centre, University of British Columbia, Vancouver, BC, Canada

Abstract

Crystallization of $CaCO_3$(s) from solution in an alkaline environment was studied as a function of temperature, $CaCO_3$ pellet loading, and $CaCO_3$ pellet size. Occurrence of crystallization as spontaneous nucleation or growth on pellets are both possible where growth is the desired outcome. Growth is quantified as retention. The crystallization on pellets was found to be mass transport controlled, while spontaneous nucleation was found to be kinetically controlled. The highest reported retention, 73%, was found at the conditions where mass transport was favored by high pellet loading and small pellet diameter, and kinetics were hindered by low temperature.

Introduction

Crystallization of $CaCO_3$ from alkaline solution is relevant for a process capturing CO_2 directly from air. Recently, the economic feasibility of this process was assessed, and it was found to be possible at a cost of less than \$100/tonCO_2 [1]. A sketch of the overall industrial process is shown in Figure 1, and the pelletizer (central in Figure 1) is a fluidized bed reactor that precipitates dissolved carbonate as $CaCO_3$(s). Precipitation can occur either on the solid $CaCO_3$ pellets or spontaneously in solution, where spontaneously nucleated particles are significantly smaller than the pellets and are referred to as fines. Precipitation on pellets is the desired outcome. Both forms of precipitation have the same driving force, the supersaturation, F_d, and this is defined in Equation (1) [2]:

$$F_d = [\mathrm{Ca}^{2+}][\mathrm{CO}_3^{2-}] - \frac{K_{sp}}{f_D^2} \qquad (1)$$

where, $[\mathrm{Ca}^{2+}]$ and $[\mathrm{CO}_3^{2-}]$ are the concentrations of calcium and carbonate in solution, K_{sp} is the thermodynamic solubility product, and f_D is the activity factor for divalent ions [3].

Crystallization of $CaCO_3$ has been studied in the water softening industry, where addition of dissolved carbonate is used to reduce the calcium content of water by precipitation on $CaCO_3$ pellets [3][4][5][6][7]. Tai et al. found that increased temperature, supersaturation, and pH increase the spontaneous nucleation rate and reduce pellet growth [4]. In works where the effect of surface area was studied, seed pellets with diameters between 0.6 and 4 μm were used, and it was found that the total available surface area significantly impacts calcium removal performance whereas pellet size showed no impact [6]. Also, Ševčík et al. used beaker experiments to show that increasing the probability of contacting a solid particle, via higher stirring velocities, increases pellet growth [5]. Therefore, conditions expected to be favorable for pellet growth are large total surface area, low temperature, low supersaturation, high stirring rate, and low pH.

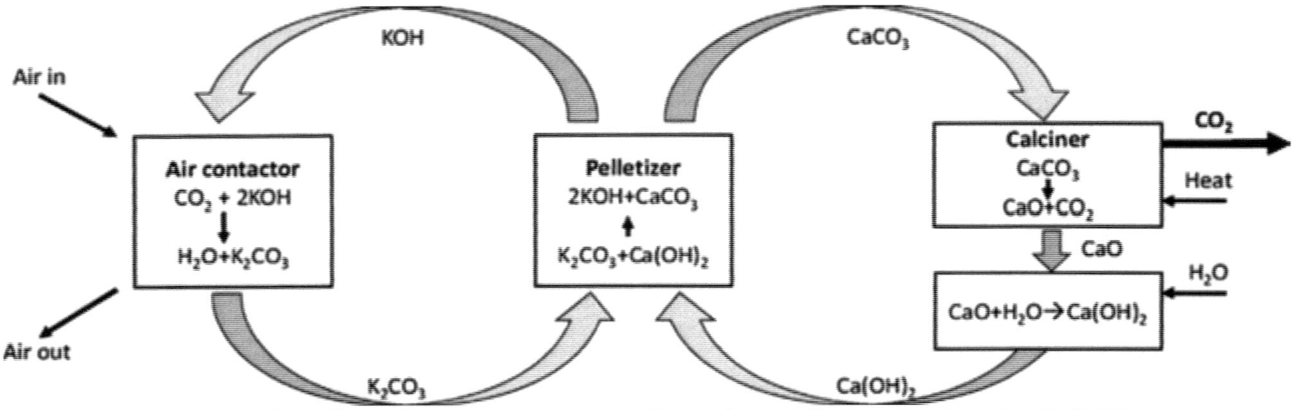

Figure 1: The direct air capture process and reactions of Carbon Engineering Ltd. [8].

The crystal growth on solid pellets was successfully modelled by van Schagen et al. [9] and the equation for the crystallization rate, C, is given in Equation (2):

$$C = \kappa A_s \left([\mathrm{Ca}^{2+}][\mathrm{CO}_3^{2-}] - \frac{K_{sp}}{f_D^2} \right) \tag{2}$$

where, κ is the rate constant, and A_s is the surface area. The reaction was described as a two-stage process where the first stage is the transportation of ions to the solid surface and the second stage is the crystallization on the pellets. Depending on the local conditions in the reaction vessel, the pellet growth reaction is limited by either mass transport, kinetics, or a combination of the two.

In contrast to the water softening process, the reaction studied in this work is operating at high pH (between 14-14.6), high carbonate concentration, and high supersaturation due to calcium being fed as solid $Ca(OH)_2$ particles. In addition, the pellets are large, more than 150 μm in diameter [8]. All these factors are expected to increase production of fines. However, high retention, of approximately 85%, has been demonstrated with high loading of $CaCO_3$ pellets even at high supersaturations [1][8]. Therefore, it is interesting to study pellet growth in a controlled environment to understand how the experimental parameters influence the retention.

In this work, $CaCO_3$ pellet growth was performed in beaker experiments controlled for pellet size, mass loading, total pellet surface area, and temperature in order to understand their effect on the overall retention. The most favorable conditions for high retention were for experiments with the smallest pellet size, highest mass loading and lowest temperature. Total pellet surface area had a limited effect on the retention. The experimental results suggest that crystallization on pellets is mass transport controlled, while spontaneous nucleation is kinetically controlled, and guides directions for further optimization of the direct air capture process.

Experimental Procedure

The reaction consists of three components: process solution, seed pellets, and lime slurry. The process solution used in these beaker experiments is a 1.0 M KOH (Sigma Aldrich 99% 45wt% solution) and 0.5 M K_2CO_3 (Sigma Aldrich 99% reagent grade) aqueous solution. The seed pellets consist of $CaCO_3$ and are sieved to yield a uniform size sample (see Table 1) and washed with deionized water to ensure all fine powder is removed (see Figure 2). Sieving is done using a Fritsch Analysette 3 PRO to shake approximately 150-200 g of pellets at an amplitude of 1.0 mm for 30 minutes. The lime slurry consists of a 2.0 M KOH solution that has $Ca(OH)_2$ powder (Sigma Aldrich, 95+%, A.C.S.) added to comprise 20wt% of the final slurry.

Table 1: These size groups represent the most common pellet sizes found within a bench scale reactor post-operation [8]. Seed is the average size fed; mid-size is about halfway grown and makes up the predominate size range in the reactor; and mature pellets are fully grown and ready for removal.

	Mesh	Diameter range [μm]	BET surface area [m^2/g]
Seed	60 - 50	250 – 300	39.7
Mid-size	45 - 40	355 - 425	27.6
Mature	25 - 14	700 - 1410	24.5

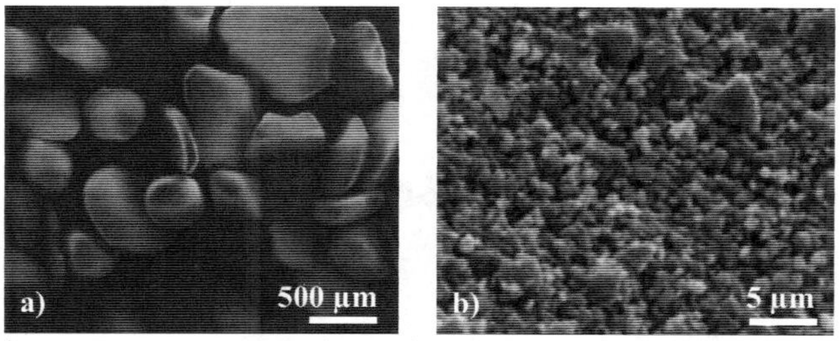

Figure 2: SEM images of mid-size pellets (a) and their surface (b) after being sieved to uniform diameters and washed of residual powder.

First, 250 mL of process solution is added to a 500 mL Pyrex beaker. The beaker was stirred using an overhead mixer (JJ-1 Precise Strength Power Mixer) with a two-bladed stainless steel impeller. A measured mass of calcium carbonate pellets is added from one size group listed in Table 1. Stirring is constant and sufficiently fluidizes the sample for all pellet sizes and loadings.

An aliquot of approximately 6.0 g of lime slurry was added. The precise mass and volume of lime slurry is recorded. After 30 minutes the beaker contents were poured through a wet sieving device consisting of a size 70 mesh (210μm) encased in PVC pipe. Effluent from the mesh was vacuum filtrated in a 9.0 cm Buchner funnel with a type 1 filter paper from Whatman. Solids found in the filter paper are the fines. By looking at the mid-sized pellets (see Figure 2) using a Nova NanoSEM their irregular shape and porous surface was observed. Surface area measurements and data for each size group were reported in Table 1.

Pellets and fines recovered are placed in separate glass petri dishes. These samples are dried overnight (at least 12 hours) at 105-110°C. Samples are allowed to cool to room temperature before mass gain is recorded. To account for relative humidity fluctuation, a control 50 g mid-size trial is run on all experiment days.

For temperature dependent trials, a type-K thermocouple attached to National Instruments LabView software was added to the reaction beaker and a Corning PC 420D heat plate was used as a heat source and temperature controller. The process solution was brought to the reaction temperature prior to pellet and slurry addition. Three trials were run for each set of conditions.

Results are reported as retention, R, calculated as in Equation (3):

$$R = \frac{m_{\text{pellets,after}} - m_{\text{pellets,before}}}{m_{\text{pellets,after}} + m_{\text{fines}}} \tag{3}$$

where $m_{\text{pellets,after}}$ is mass of pellets after reaction, $m_{\text{pellets,before}}$ is the original mass of $CaCO_3$ pellets added, and m_{fines} is mass of fines collected after reaction.

After reactions process solution concentrations are analyzed for their ion ratio, $[OH^-]:[CO_3^{2-}]$. SEM images were taken of both the pellets and fines produced. Powder X-ray diffraction (XRD) was also performed on both using a Bruker D8 Advance with a 2θ range of 10 to 90, step size 0.1, and scan time of 0.3s per step.

Results and Discussion

Ambient temperature

For all ambient temperature conditions reported, a positive retention rate was observed and $CaCO_3$ was formed both on pellets and as fines. In these experiments process solution concentration, amount of fed $Ca(OH)_2$ particles, and stirring rate were constant meaning that the supersaturation, F_d, was constant. Therefore, this set of experiments evaluates the influence of mass transport.

For varied mass loading with a constant pellet size (mid-size), Figure 3a, the retention varies from 52% at 25 g loading to 64% at 100 g loading. Higher loadings are expected to increase the surface area of $CaCO_3$ in the beaker and thus lead to higher retention. This trend was confirmed in our experimental data, but the effect is limited as lower loading did not seem to affect the retention. This suggests that other factors, such as pellet size, are more important.

When the pellet size was varied at constant mass loading, Figure 3b, the retention changed from 67% at seed size to 22% at mature size. The change of pellet from seed to mature increases the particle size and decreases the total surface area of $CaCO_3$ resulting in fewer particles present in the system. Both of these effects are expected to reduce the retention rate and therefore explains the trend observed in Figure 3b. In an attempt to deconvolute the two effects, a third set of experiments was done where the total surface area confirmed through BET analysis was kept constant (at 1,215 m^2). The results of these experiments are shown in Figure 3c. Here, it was evident that the pellet size has a large effect, and the retention changes from 58% at seed size to 22% at mature size. Notably, the seed size pellets and mid-size pellets give retentions within one standard deviation of each other.

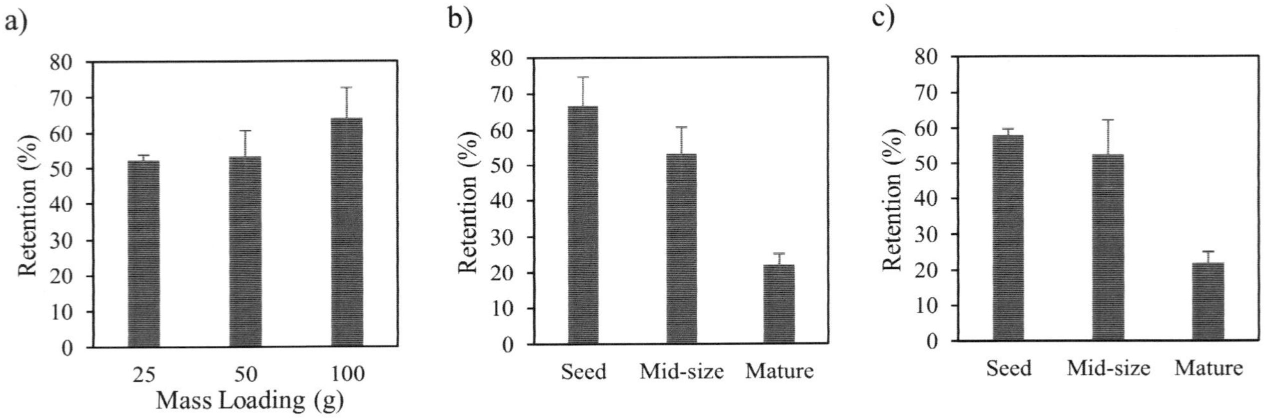

Figure 3: Experiments run at room temperature 20°C with the following variables being varied: a) mid-size pellets at different mass loadings of 25, 50, and 100 grams, b) constant mass loading with varied pellet size, and c) constant surface area with varied pellet size.

In order to understand this result and to separate the effect of pellet size and surface area, the retention rate was plotted as a function of surface area for all results reported in Figure 3. This is shown in Figure 4. Here, the trend in Figure 3c is confirmed where mature pellets give a low retention rate, while the seed and mid-size pellets get comparable retention rates. This can be expected if the actual pellet size is considered (Table 1), where the seed and mid-size pellets are close in diameter while the mature pellets have a diameter approximately 2.5-6 times the seed size and 1.75-4 times the mid-size pellets. From these results, it is evident that while Equation (2) can explain the overall reaction in some cases, care should be taken when interpreting the results solely with this equation. In particular, when $Ca(OH)_2$ is fed as solid particles, as is the case here, the actual supersaturation, F_d, is largely confined to the volume around the solid $Ca(OH)_2$ particles. Thus, to achieve high retention, one must ensure that solid $CaCO_3$ is present near the $Ca(OH)_2$ particles. From the results reported in Figure 3 and Figure 4, high mass loading and small pellet sizes is the best approach. To confirm this, an additional experiment was performed using 100 g of seed pellets resulting in 73% retention. This data is shown in Figure 4 as the rightmost triangle. Overall, this suggests that collisions, or near collisions, between the $Ca(OH)_2$ particles and $CaCO_3$ pellets are key to high retention and this gives a guideline for how to achieve an even higher retention rate at the pH and concentration used in this work.

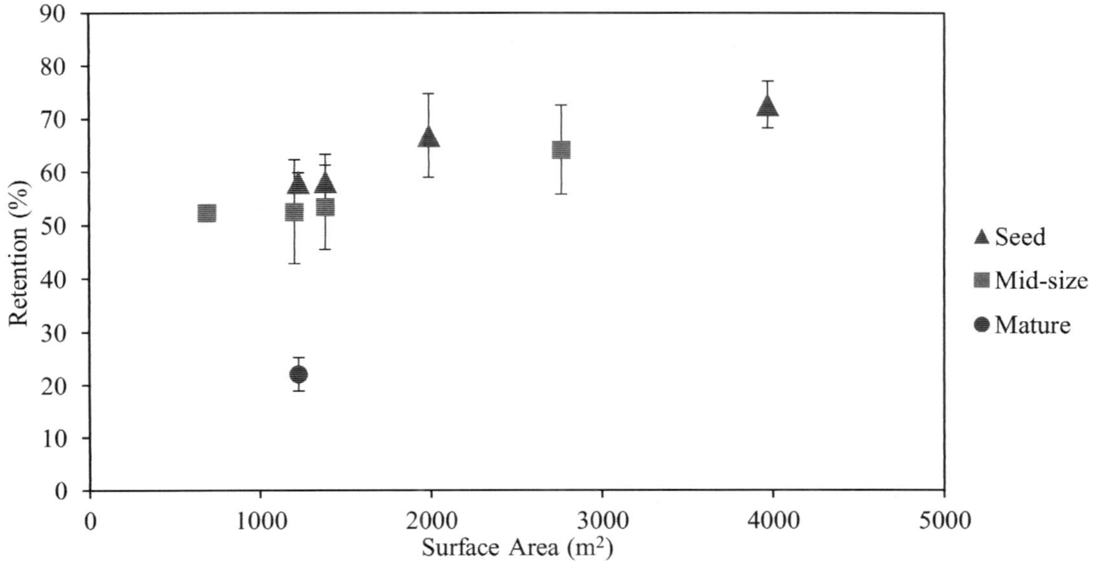

Figure 4: Retention for all trials shown in Figure 3 as well as best conditions experimental set. All shown with respect to surface area as calculated by BET isotherm measurements from the pellet size groups.

Elevated temperatures

Elevated temperature is expected to modify the kinetics of crystallization [3]. With mass transport variables held constant, increased crystallization rate is expected to increase spontaneous nucleation. In earlier work [5][10], at high temperatures, different $CaCO_3$ morphologies were reported and thus temperature can change the precipitation mechanism both for spontaneous nucleation and pellet growth. For example, the aragonite morphology has a needle-like structure that can help reduce the effect of increased temperature on the retention by assuring that a large surface area of $CaCO_3$ is present in the beaker. This morphology was reported at temperatures higher than 50°C [10].

In Figure 5, the retention rate at four temperatures is reported for mid-size pellets. Here, it is evident that the higher temperature gives significantly lower retention going from 53% at 20°C to -1.8% at 80°C. The negative retention at 80°C was within one standard deviation of no retention, and was assumed to be due to experimental uncertainty. If the limiting mechanism for pellet growth is physical proximity of $CaCO_3$ pellets to $Ca(OH)_2$ particles, an increase in temperature is expected to mainly influence the spontaneous nucleation rate. An increased spontaneous nucleation rate is expected to produce fines that are both more numerous and smaller in diameter than a slower spontaneous nucleation rate. In order to investigate this, SEM imaging was done on the fines at the various temperatures and the results are shown in Figure 6. Here, it is evident that the size of the fines decreases as the temperature is increased.

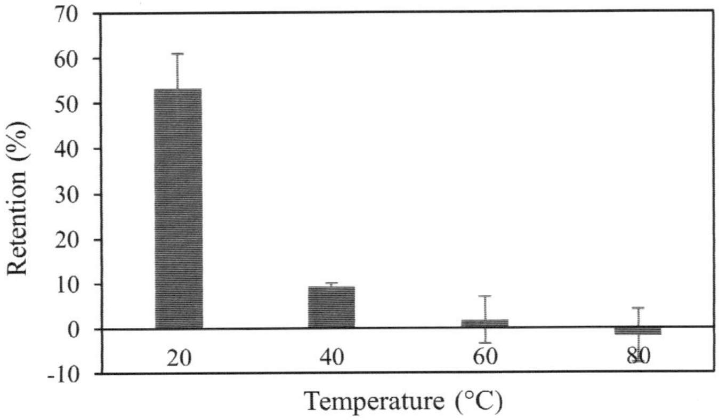

Figure 5: Percent of mass found on pellets over total mass collected after reaction for trials at different temperatures for mid-sized pellets.

Figure 6: $CaCO_3$ fines grown at a) 20°C, b) 40°C, c) 60°C, and d) 80°C.

SEM imaging was also done on the pellets after growth and is shown in Figure 7. No clear trend is visible in these images, and this can be expected as the amount of $CaCO_3$ deposited on a single pellet is relatively small compared to the overall pellet size (mass gain at 50% retention is 0.80 g or only 1.6% of total mass for a loading of 50 g). A long-term experiment might give different results and possibly different $CaCO_3$ morphologies, however that is outside the scope of this work but will be investigated in the future.

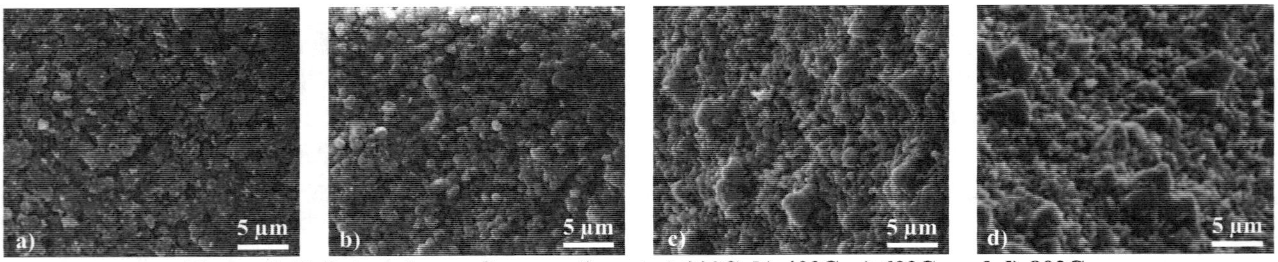

Figure 7: Pellet surfaces after reaction at a) 20°C, b) 40°C, c) 60°C, and d) 80°C.

The different $CaCO_3$ morphologies result in different shape of the particles so it may be visible in SEM images. However, if only small amounts of a morphology are present, it is hard to distinguish. Therefore, XRD was run to confirm the results of the SEM images. The XRD results are shown in Figure 8 for the a) pellets and b) fines and compared with reference spectra for three different $CaCO_3$ morphologies (calcite, vaterite, and aragonite) and $Ca(OH)_2$. For the pellets, this could be compared to the unreacted pellets. Here, it is evident that at all temperatures, the pellets are present mainly as calcite. Only a small peak appears in the 80°C results at a 2θ of 61. The peak position is not one of the main peaks for the vaterite or aragonite phases. Therefore, it is assumed that this is an impurity peak and that $CaCO_3$ is only present as calcite on the pellets.

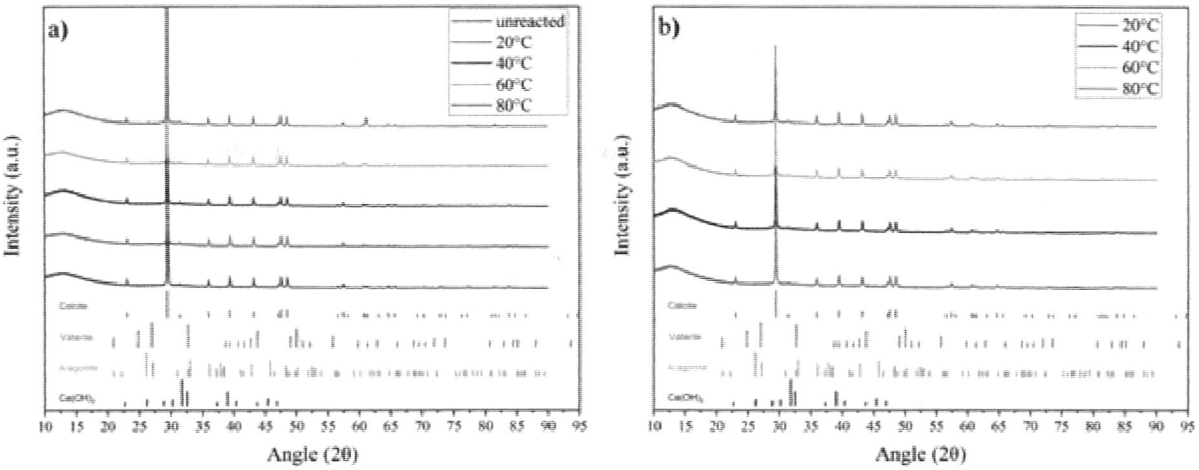

Figure 8: XRD results for (a) pellets, and (b) fines at different temperatures. Reference spectra for calcite (pink), vaterite (green), aragonite (orange), and $Ca(OH)_2$ (blue) are shown at the bottom.

For the fines, in Figure 8b, the XRD results were consistent with calcite at all temperatures. The fines are not seeded and consequently will form the morphology that is most kinetically and thermodynamically favorable. In our results, only calcite is observed, and this suggests that at the high pH used in this work, only calcite will be formed both through spontaneous nucleation and as precipitation on $CaCO_3$ pellets. Therefore, in contrast to work at lower pH values the temperature does not change the morphology of $CaCO_3$ produce, and only modifies the nucleation rate resulting in lower retention.

When results at ambient and elevated temperatures are compared, a framework for achieving high retention can be found. This builds on the understanding that high retention is achieved by improving the conditions for precipitation on $CaCO_3$ particles while hindering the spontaneous nucleation process. These two processes have equal driving force, the supersaturation, F_d. In the conditions used in this work, the supersaturation is high but locally confined around the $Ca(OH)_2$ particles. Thus, increased precipitation on the $CaCO_3$ pellets and resulting higher retention is improved by increasing the chance of a $CaCO_3$ pellet being present close to the $Ca(OH)_2$ particle. This is a pure mass transport constraint. On the other hand, the rate of spontaneous nucleation is increased by increasing the kinetic driving force, demonstrated here by the temperature. High retention is achieved when the mass transport conditions are favorable and the kinetic driving force is low. The highest retention reported in this work is 73% and is achieved for seed size pellets at 100 g loading and at 20°C. A further improvement of the retention can be expected for smaller particle sizes, higher mass loadings of $CaCO_3$ pellets, and lower temperatures.

Conclusion

Crystallization of $CaCO_3(s)$ from solution in an alkaline environment relevant for a direct CO_2 capture from air reactor was studied as a function of temperature, $CaCO_3$ pellet loading, and $CaCO_3$ pellet size. The results were evaluated in terms of retention: the fraction of $CaCO_3$ formed that grows on pellets. Higher mass loading and smaller pellet size resulted in higher retention. Higher temperatures resulted in lower retention. Higher temperatures also were shown to reduce the diameter of fines produced. The crystallization on pellets was found to be mass transport controlled, while spontaneous nucleation was found to be kinetically controlled. The highest retention, 73%, was found at high pellet loading, small pellet diameter, and low temperature.

The fluidized crystallization will next be looked at in a fluidized bed constructed at the lab scale to model the version used by Carbon Engineering Ltd. The work within fluidized bed reactor will provide longer reaction times, allowing larger net mass growth and conditions closer to the industrial process.

Acknowledgement

This work was supported by Carbon Engineering Ltd., the National Sciences and Engineering Research Council of Canada (NSERC), and Mitacs through the Mitacs Accelerate Program. The support of all colleagues and students at the University of British Columbia and Carbon Engineering Ltd. is greatly acknowledged, in particular we thank Jenny McCahill for many discussions pertaining to the direct air capture process of Carbon Engineering Ltd.

References

[1] D. W. Keith, G. Holmes, D. St. Angelo, and K. Heidel, "A Process for Capturing CO2 from the Atmosphere," *Joule*, 2018.

[2] N. Spanos and P. G. Koutsoukos, "Kinetics of Precipitation of Calcium Carbonate in Alkaline pH at Constant Supersaturation. Spontaneous and Seeded Growth," *J. Phys. Chem. B*, vol. 102, no. 34, pp. 6679–6684, 1998.

[3] H. N. S. Wiechers, P. Sturrock, and G. v R. Marais, "Calcium carbonate crystallization kinetics," *Water Res.*, vol. 9, no. 9, pp. 835–845, 1975.

[4] C. Y. Tai and H. P. Hsu, "Crystal growth kinetics of calcite and its comparison with readily soluble salts," *Powder Technol.*, vol. 121, no. 1, pp. 60–67, 2001.

[5] R. Ševčík, M. Pérez-Estébanez, A. Viani, P. Šašek, and P. Mácová, "Characterization of vaterite synthesized at various temperatures and stirring velocities without use of additives," *Powder Technol.*, vol. 284, pp. 265–271, 2015.

[6] K. L. Mercer, Y. P. Lin, and P. C. Singer, "Enhancing calcium carbonate precipitation by heterogeneous nucleation during chemical softening," *Journal / American Water Works Association*, vol. 97, no. 12. 2005.

[7] G. H. Nancollas and M. M. Reddy, "The crystallization of calcium carbonate. II. Calcite growth mechanism," *J. Colloid Interface Sci.*, vol. 37, no. 4, pp. 824–830, 1971.

[8] L. Burhenne, C. Giacomin, T. Follett, J. Ritchie, J. S. J. McCahill, and W. Mérida, "Characterization of reactive CaCO3 crystallization in a fluidized bed reactor as a central process of direct air capture," *J. Environ. Chem. Eng.*, vol. 5, no. 6, pp. 5968–5977, 2017.

[9] K. M. Van Schagen, L. C. Rietveld, and R. Babuška, "Dynamic modelling for optimisation of pellet softening," *J. Water Supply Res. Technol. - AQUA*, vol. 57, no. 1, pp. 45–56, 2008.

[10] Y. F. Ma, Y. H. Gao, and Q. L. Feng, "Effects of pH and temperature on CaCO3 crystallization in aqueous solution with water soluble matrix of pearls," *J. Cryst. Growth*, vol. 312, no. 21, pp. 3165–3170, 2010.

AUTHOR INDEX

AUTHOR INDEX

AUTHOR INDEX

AUTHOR INDEX

AUTHOR INDEX

AUTHOR INDEX

AUTHOR INDEX

AUTHOR INDEX

AUTHOR INDEX

AUTHOR INDEX